Springer Finance

Springer Finance

Springer Finance is a programme of books addressing students, academics and practitioners working on increasingly technical approaches to the analysis of financial markets. It aims to cover a variety of topics, not only mathematical finance but foreign exchanges, term structure, risk management, portfolio theory, equity derivatives, and financial economics.

Ammann M., Credit Risk Valuation: Methods, Models, and Application (2001)
Back K., A Course in Derivative Securities: Introduction to Theory and Computation (2005)
Barucci E., Financial Markets Theory. Equilibrium, Efficiency and Information (2003)
Bielecki T.R. and Rutkowski M., Credit Risk: Modeling, Valuation and Hedging (2002)
Bingham N.H. and Kiesel R., Risk-Neutral Valuation: Pricing and Hedging of Financial Derivatives (1998, 2nd ed. 2004)
Brigo D. and Mercurio F., Interest Rate Models: Theory and Practice (2001, 2nd ed. 2006)
Buff R., Uncertain Volatility Models – Theory and Application (2002)
Carmona R.A. and Tehranchi M.R., Interest Rate Models: An Infinite Dimensional Stochastic Analysis Perspective (2006)
Dana R.-A. and Jeanblanc M., Financial Markets in Continuous Time (2003)
Deboeck G. and Kohonen T. (Editors), Visual Explorations in Finance with Self-Organizing Maps (1998)
Delbaen F. and Schachermayer W., The Mathematics of Arbitrage (2005)
Elliott R.J. and Kopp P.E., Mathematics of Financial Markets (1999, 2nd ed. 2005)
Fengler M.R., Semiparametric Modeling of Implied Volatility (2005)
Filipović D., Term-Structure Models (2009)
Fusai G. and Roncoroni A., Implementing Models in Quantitative Finance (2008)
Geman H., Madan D., Pliska S.R. and Vorst T. (Editors), Mathematical Finance – Bachelier Congress 2000 (2001)
Gundlach M. and Lehrbass F. (Editors), CreditRisk$^+$ in the Banking Industry (2004)
Jeanblanc M., Yor M., Chesney M., Mathematical Methods for Financial Markets (2009)
Jondeau E., Financial Modeling Under Non-Gaussian Distributions (2007)
Kabanov Y.A. and Safarian M., Markets with Transaction Costs (2009)
Kellerhals B.P., Asset Pricing (2004)
Külpmann M., Irrational Exuberance Reconsidered (2004)
Kwok Y.-K., Mathematical Models of Financial Derivatives (1998, 2nd ed. 2008)
Malliavin P. and Thalmaier A., Stochastic Calculus of Variations in Mathematical Finance (2005)
Meucci A., Risk and Asset Allocation (2005, corr. 2nd printing 2007, softcover 2009)
Pelsser A., Efficient Methods for Valuing Interest Rate Derivatives (2000)
Platen E. and Heath D., A Benchmark Approach to Quantitative Finance (2006, corr. printing 2010)
Prigent J.-L., Weak Convergence of Financial Markets (2003)
Schmid B., Credit Risk Pricing Models (2004)
Shreve S.E., Stochastic Calculus for Finance I (2004)
Shreve S.E., Stochastic Calculus for Finance II (2004)
Yor M., Exponential Functionals of Brownian Motion and Related Processes (2001)
Zagst R., Interest-Rate Management (2002)
Zhu Y.-L., Wu X., Chern I.-L., Derivative Securities and Difference Methods (2004)
Ziegler A., Incomplete Information and Heterogeneous Beliefs in Continuous-time Finance (2003)
Ziegler A., A Game Theory Analysis of Options (2004)

Yuri Kabanov • Mher Safarian

Markets with Transaction Costs

Mathematical Theory

Yuri Kabanov
Laboratoire de Mathématiques
Univerisité de Franche-Comté
16 Route de Gray
25030 Besançon Cedex
France
youri.kabanov@univ-fcomte.fr

and

Central Economics and
Mathematics Institute
Russian Academy of Sciences
Moscow
Russia

Mher Safarian
Institut für Statistik und
Mathematische Wirtschaftstheorie
Fakultät für Wirtschaftswissenschaften
Universität Karlsruhe
76128 Karlsruhe
Germany
safarian@dr-safarian.eu

and

Landesbank Baden-Württemberg
(LBBW)
Fritz-Elsas-Straße 31
70144 Stuttgart
Germany

ISBN 978-3-540-68120-5 e-ISBN 978-3-540-68121-2
DOI 10.1007/978-3-540-68121-2
Springer Heidelberg Dordrecht London New York

Library of Congress Control Number: 2009940793

Mathematics Subject Classification (2000): 60G42, 60G44, 91B28
JEL Classification: G10, G11, G13

Cover illustration: © Margarita Kabanova, http://kabanova.neuf.fr
Cover design: WMXDesign GmbH

Printed on acid-free paper

Springer is part of Springer Science+Business Media (www.springer.com)

Preface

This book contains an introduction to the mathematical theory of financial markets with proportional transaction costs. Traditionally, a theoretical analysis of models with market imperfections was considered as the most challenging and difficult chapter of mathematical finance. Nowadays there are hundreds of papers on this subject, but it still is not covered by monographic literature fixing the main achievements. We propose here a highly subjective selection of results, which, as we hope, give an idea what is going on and which may serve as a platform for further studies. The main topics are: approximative hedging, arbitrage theory, and consumption–investment problems.

Our interest in the subject originates from the famous paper of Heyne Leland, who suggested a method how to price contingent claims on a market with constant proportional transaction costs and gave to the traders a practically important benchmark. From the economical perspectives his idea is to replace the classical Black–Scholes principle of "pricing by replication" (coinciding, in the case of complete markets, with the principle of "pricing by arbitrage") by the principle of "pricing by an approximative hedging." This approximative hedging can be realized in various ways. Leland's suggestion was to apply the common algorithm of periodical portfolio revisions using the common Black–Scholes formulae but with an appropriately enlarged volatility. This method happened to be very efficient for practical values of the model parameters where the transaction costs are small. It is widely accepted in the financial industry. However, a more detailed analysis shows some interesting mathematical aspects of this approach: even for the standard call option, the terminal value of the replicating portfolio does not converge to the terminal pay-off if the transaction cost coefficients do not depend on the number of revisions (tending to infinity). The limiting discrepancy can be calculated explicitly. This is a rigorous mathematical result disclaiming but not discarding Leland's approach. In his paper Leland conjectured that the convergence takes place when the transaction costs are decreasing to zero as $n^{-1/2}$, where n is the number of revisions. This was confirmed in the thesis of Klaus Lott, who provided necessary mathematical arguments; we refer to this particular case where the enlarged volatility does not depend on n as to the Leland–Lott model. In fact, the approximation error always tends to zero if the transaction costs tend to zero (with any rate). The latter property explains the applicability of the Leland idea and its importance for the practical purposes: the trader may assume that the real-world market is described by a model with a particular but sufficiently large number n of revision intervals for which the transaction cost coefficients k_n are small enough.

The case investigated by Lott seems to be the most important. This is the reason why we concentrate our efforts on the analysis of the asymptotical behavior of the hedging error. We obtain the exact rate of asymptotics of the L^2-norm of this error for a large class of options with convex pay-off functions. We consider the setting with nonuniform revision intervals and establish an asymptotic expansion when the revision dates are $t_i^n = g(i/n)$, where the strictly increasing scale function $g : [0, 1] \to [0, 1]$ and its inverse f are continuous with their first and second derivatives on the whole interval or $g(t) = 1-(1-t)^\beta$, $\beta \geq 1$. We show that the sequence $n^{1/2}(V_T^n - V_T)$ converges in law to a random variable which is the terminal value of a component of a two-dimensional Markov diffusion process and calculate the limit.

It is worth noticing that the result that there is no convergence in the case where the transaction costs do not depend on the model number also has some practical implications: the revealed structure of the discrepancy explains the empirical fact that the precision of approximation is worse in the case where the stock at the maturity date evolves near the pay-off. We discuss this and other aspects of the Leland strategy in our first chapter. In particular, we provide a formulation of the Pergamenshchikov limit theorem which gives a limiting distribution of the approximation error corrected by the discrepancy.

We present also the beautiful proof due to Skorokhod and Levental of the Davis–Norman conjecture that in the presence of market friction the hedging (super-replication) of a call option on the stock evolving as a geometric Brownian motion can be achieved by a single buy-and-hold transaction, at the beginning of the trading period, without further trading.

In the second chapter we develop an arbitrage theory for financial markets without friction.

First, we recall the classical arbitrage theory for frictionless market models in discrete time, providing a self-contained and rather exhaustive synthesis of the known results. We give a detailed analysis of the Dalang–Morton–Willinger theorem in its modern formulation, which is a list of equivalent conditions, and show that for the model with restricted information, the list is necessarily shorter. Our presentation is adapted for the treatment of more delicate problems for transaction cost model. Note that the mentioned classical topic is rarely discussed in textbooks on mathematical finance being considered as too complicated. By this reason we present a "fast" and "elementary" proof of the major equivalence suitable for lecture courses. It is based on a combination of the original approach due to Chris Rogers with a lemma on convergent measurable subsequences.

We also discuss the structure of equivalent martingale measures and prove the theorem that in the case where the reference measure is a martingale one, the martingale measures with bounded densities are norm-dense in the set of all martingale measures. This implies, in particular, that the set of martingale measures with finite entropy, if nonempty, is dense in the set of all martingale measures. This section also contains a simple proof of the optional decomposition theorem, which is, in the discrete-time setting, a very simple result. We also prove a hedging theorem for European options, which asserts that the set of initial endowments for (self-financing) portfolios super-replicating a given contingent claim is a closed interval. Its left extremity is the supremum of expectations of the contingent claim with respect to the set of all martingale measures. We go beyond finite-horizon setting and prove some no-arbitrage criteria for infinite-horizon models. We conclude the section by an example of application of the duality theory to a utility maximization problem and a brief comment on continuous-time models.

With the above preliminaries, in the third chapter we attack the problem of no-arbitrage conditions for markets with proportional transaction costs.

As a mathematical description for the latter, we use a general scheme of two adapted cone-valued processes in convex duality, giving, at least for mathematicians, a comprehensive "parameter-free" description of the main objects of the theory. In the financial context the values of the primary processes are polyhedral cones, describing solvency regions (evolving in time and depending on the state of the nature). The portfolio processes are vector-valued; they can be viewed either in terms of quotes (i.e., units of a certain numéraire) or in terms of "physical units" (e.g., for models of currency markets, positions in euros, dollars, yens, etc.); both descriptions are related in the obvious way. It is convenient to treat the no-arbitrage conditions in "physical units domain". The crucial observation is that the natural analog of the density processes of equivalent martingale measures in this general setting are strictly positive martingales evolving in the dual cone-valued process. The considered framework covers the majority of models considered in the literature, including the pioneering paper by Jouini and Kallal.

We discuss three types of no-arbitrage properties: weak (NA^w), strict (NA^s), and robust (NA^r), all coinciding with the classical one when the transaction cost coefficients are zero. The most natural generalization is the weak NA-property, claiming the absence of strict arbitrage opportunities, i.e., portfolio processes starting from zero and having as the terminal values a nontrivial random vector with positive coordinates. For finite probability space, the NA^w-criterion can be established in a very easy way, by the same arguments as the Harrison–Pliska theorem. Its formulation is simple: NA^w holds if and only if there exists a strictly positive martingale with values in the dual of solvency cones. Surprisingly, an extension to an arbitrary probability space, i.e., an analog of the Dalang–Morton–Willinger theorem, as it was shown by Schachermayer, fails to be true in general. By this reason other definitions of no-arbitrage were investigated by a number of authors. A particularly fruitful idea, due to Schachermayer, is to consider as arbitrage-free the models for which the NA^w-property still holds even under better investment opportunities, i.e., with "larger" solvency cones. Such a "robust" no-arbitrage property, referred to as NA^r, allows for an equivalent ("dual") description without any assumptions on the underlying probability space. We present another surprising result, due to Grigoriev: in the two-asset model the NA^w-criterion in the above formulation still holds.

Another interesting feature of models with transaction costs is the presence of arbitrage of the second kind. The latter notion serves to describe the situation that the initial endowments of the investor lie outside the solvency cone. Nevertheless, there is a self-financing portfolio which ends up in the solvency cone. It happens that the absence of arbitrage of the second kind is equivalent to the existence of martingales evolving in the interiors of dual of the solvency cones with arbitrary starting points.

The hedging problem for European contingent claims for markets with transaction costs can be formulated as follows. The contingent claim ξ is a vector of liabilities expressed in the units of corresponding assets. The investor wants to know whether he can super-replicate the contingent claim (in the sense of partial ordering generated by the solvency cone at the terminal date) by a self-financing portfolio starting from the initial endowment x. The answer is: the initial endowment x allows this if and only if its value Z_0x is not less than the expected value of the contingent claim $EZ_T\xi$ whatever is the process Z evolving in the dual to the solvency cone (a suggestive name for such a process Z: consistent price system).

The American contingent claim is a process. The option seller is interested to determine whether his initial endowment x suits to start a portfolio super-replication the American contingent claim at all dates. We present in Chap. 3 a hedging theorem which involves a class of coherent price systems which is larger than the class of consistent price systems.

We explain, following Bouchard, that models where the investor's information is delayed or restricted can be treated in the space of orders which is of higher dimension and give no-arbitrage criteria for such a situation.

We provide some results on the other important theoretical problem, hedging theorem in continuous-time framework. The latter gives a description of the set of initial vector-valued endowments ensuring the existence of a hedging portfolio for a given, also vector-valued, contingent claim. The situation here is more complicated than in the discrete-time setting: one needs an appropriate definition of admissibility and even that of portfolio processes. We present here the recent result due to Campi and Schachermayer explaining that the requirement that the portfolio process is càdlàg is too restrictive to get a "good" hedging theorem and should be replaced. We complete the chapter by a hedging theorem for American options.

The concluding chapter is devoted to Davis–Norman consumption–investment problem in a multi-asset framework. We start by recalling the classical Merton problem for a power utility function. From the point of view of experts in stochastic control the latter is trivial. Indeed, the verification theorem (which is itself a very simple result) requires to find a solution of the Hamilton–Jacobi–Bellman (HJB) equation. An easy argument shows that the Bellman function inherits the homogeneity property of the utility function and, thus, if finite, it is the same utility function up to a multiplicative constant, and the HJB equation is reduced immediately to an algebraic one to determine the latter.

The situation becomes quite different in the transaction cost setting. The problem is far from being trivial even in the case of powerful utility function. There is no smooth solution, and, therefore, the usual verification theorem does not work. The remedy comes from the theory of viscosity solutions. However, one should first check that the Bellman function is a viscosity solution of the HJB equation. Though the general lines of the arguments are well known, they are quite lengthy and rarely presented in detail. Moreover, there are many definitions of the viscosity solutions. We used the simplest one adapted for positive utility functions. In this chapter we give a rigorous proof of the Dynamic Programming Principle and derive that the Bellman function is a viscosity solution of the HJB equation. We proof that the latter has a unique solution in a class determined by a suitably defined Lyapunov function. Following Soner–Shreve, we analyze the structure of the Bellman function in the case of the two-asset model and conclude by presenting the results of Shreve and Janeček on the asymptotics of the solution when the transaction costs tend to zero.

In the Appendix we collect various auxiliary results from convex geometry, functional analysis, probability theory, measurable selection, and stochastic differential equations with reflection. Of course, our bibliographical comments are not exhaustive, and we apologize in advance for missing references.

This book could not be written without fruitful collaboration with our colleagues: Christophe Stricker, Freddy Delbaen, Esko Valkeila, Günter Last, Nizar Touzi, Bruno Bouchard, Claudia Klüppelberg. We are grateful to Thomas Keller and Iris Reinelt for their constructive comments and suggestions. We express our thanks to Emmanuel Denis and Dimitri De Vallière for their contribution and attentive reading of the manuscript.

A large portion of manuscript was completed during the stay of the first author at the Kyoto University and Tokyo Metropolitan University.

Besançon Cedex, France — Yuri Kabanov
Moscow, Russia

Karlsruhe, Germany — Mher Safarian
Stuttgart, Germany

Contents

1

Approximative Hedging

1.1 Black–Scholes Formula Revisited

1.1.1 Pricing by Replication

Certainly, the reader of this book is well acquainted with foundations of the option pricing. Nevertheless, having in mind that the theory we develop in this chapter will deviate from the standard approach, we start our presentation with a short discussion of the Black–Scholes model and principle of option pricing by replication.

In the classical Black–Scholes model it is assumed that there are only two securities in the market, the bank account (or bond) and the stock. Their price increments are respectively:

$$\begin{aligned} dB_t &= rB_t\,dt, \\ dS_t &= aS_t\,dt + \sigma S_t\,dw_t, \end{aligned}$$

where w is a Wiener process. The first relation means that the deposit on the bank account (in the other interpretation, the amount invested in bond) is increasing exponentially if the interest rate $r > 0$ and remains constant if $r = 0$:

$$B_t := B_0 e^{rt}.$$

The second equation expresses the idea that the relative increments dS_t/S_t of the price of risky security are due to a deterministic instantaneous rate of return a and Gaussian fluctuations with an amplitude characterized by the volatility σ. This equation also can be solved, and the resulting explicit formula shows that the price of the risky asset evolves as the geometric Brownian motion:

$$S_t = S_0 e^{(a-\frac{1}{2}\sigma^2)t+\sigma w_t}.$$

In our study, to simplify the notation, we suppose immediately that $r = 0$. This is not a restriction and simply means that the bank account is chosen as

Yu. Kabanov, M. Safarian, *Markets with Transaction Costs*,
Springer Finance, DOI 10.1007/978-3-540-68121-2_1,

the *numéraire*: the price of the stock is measured in its units. The translation into other units (of the "cash") is so simple that there is absolutely no necessity to keep r in the presentation.[1]

In the Black–Scholes model it is assumed that the price process carries all possible information. Mathematically speaking, this means that the filtration $\mathbf{F} = (\mathcal{F}_t)$ is generated by the process S (or, equivalently, by w).

The self-financing portfolio with the initial capital p is given by its value process

$$X_t := p + H \cdot S_t = p + \int_0^t H_u \, dS_u,$$

where H is a predictable (or adapted) process taken from a certain class of "admissible" integrands; $H \cdot S$ is a common (and convenient) notation for the stochastic integral.

We shall discuss a bit later the important (and nontrivial) question how to specify these "admissible" integrands. We also postpone a short discussion on the interpretation of the introduced terminology.

Let us examine the problem of option pricing, considering as an example the European call option. This is a contract between two agents. At the end of the contract, at the maturity (exercise) date T, the option seller will be faced the contingent claim (random pay-off) $\zeta = (S_T - K)^+$. In a market model without friction this is equivalent to sell one unit of stock at the price $K > 0$. Note that in the presence of transaction costs there is a difference whether the stock will be delivered or not.

The maturity date T and the parameter K ("strike") are stipulated by the contract.

How to price such an option? The classical answer in the Black–Scholes theory is the following: price it by replication. This means that for the option writer, it would be desirable to sell the option for an amount p with which he could start a self-financing portfolio the terminal value of which is the pay-off ξ. Mathematically, the problem is: to find a real number x and an admissible strategy H such that

$$p + H \cdot S_T = (S_T - K)^+.$$

Being "well formulated," this problem can be solved by a straightforward reference. A short search through a textbook on stochastic calculus reveals that the predictable representation theorem is a result exactly in this spirit. We formulate it in the following way, denoting by $\mathcal{P}$ the predictable σ-algebra:

Predictable representation theorem. *Any random variable* $\zeta \in L^2(\mathcal{F}_T)$ *admits a representation*

$$\zeta = E\zeta + G \cdot w_T$$

with (uniquely defined) $G \in L^2(\Omega \times [0,T], \mathcal{P}, dP \times dt)$.

[1] Unfortunately, this tradition is persistent. This phenomena might be explained by a desire of some authors to keep the presentation closer to "practically looking" formulae.

Remarkably, for the case where the instantaneous rate of return a is zero, the direct application of this theorem already gives the answer: the replication price is $p = E\zeta$, and the replication strategy is $H = G/(\sigma S)$. For $\zeta = (S_T - K)^+$, the calculation of the above expectation is easy, and only minor efforts are needed to obtain an explicit expression for G.

For the general case, one needs to apply at the first step the Girsanov measure transformation removing the drift. We need it in the simplest form[2] claiming that the process $\tilde{w}_t := w_t - \vartheta t$ on the interval $[0, T]$ is a standard Wiener process under the probability measure $\tilde{P}$ (sometimes called the risk-neutral measure) given by the formula

$$\tilde{P} = e^{\vartheta w_T - \frac{1}{2}\vartheta^2 T} P. \tag{1.1.1}$$

Thus, with $\vartheta = -a/\sigma$, we obtain that $dS_t = \sigma S_t \, d\tilde{w}_t$ and

$$S_t = S_0 e^{\sigma \tilde{w}_t - \frac{1}{2}\sigma^2 t}, \tag{1.1.2}$$

where $\tilde{w}$ is a standard Wiener process under the measure $\tilde{P}$. The general case is reduced in this way to the particular one considered above. We conclude that the replication price $p = \tilde{E}(S_T - K)^+$.

1.1.2 Explicit Formulae

Let $\eta \sim \mathcal{N}(0,1)$, and let b and A be positive constants. The difference $\eta - b$ under the measure $P' = e^{b\eta - \frac{1}{2}b^2} P$ has the same distribution as η under P; this simple observation is an ancestor of the Girsanov theorem. Therefore, we have

$$\begin{aligned} E\big(e^{b\eta - \frac{1}{2}b^2} - A\big)^+ &= E e^{b\eta - \frac{1}{2}b^2} I_{\{\eta \geq \frac{1}{b}\ln A + \frac{1}{2}b\}} - AEI_{\{\eta \geq \frac{1}{b}\ln A + \frac{1}{2}b\}} \\ &= EI_{\{\eta \geq \frac{1}{b}\ln A - \frac{1}{2}b\}} - AEI_{\{\eta \geq \frac{1}{b}\ln A + \frac{1}{2}b\}} \\ &= \Phi\Big(-\frac{1}{b}\ln A + \frac{1}{2}b\Big) - A\Phi\Big(-\frac{1}{b}\ln A - \frac{1}{2}b\Big). \end{aligned}$$

Noticing that $\tilde{w}_T/\sqrt{T} \sim \mathcal{N}(0,1)$ under $\tilde{P}$, from this and from (1.1.2) we obtain that the replication price $p = \tilde{E}(S_T - K)^+$ when $S_0 = x$ is given by the first Black–Scholes formula

$$C(0, x, \sigma) = x\Phi\Big(\frac{1}{\sigma\sqrt{T}}\ln\frac{x}{K} + \frac{1}{2}\sigma\sqrt{T}\Big) - K\Phi\Big(\frac{1}{\sigma\sqrt{T}}\ln\frac{x}{K} - \frac{1}{2}\sigma\sqrt{T}\Big). \tag{1.1.3}$$

---

[2] It was known a long time before Girsanov. The fact that, under $\tilde{P}$, the increments are centered independent Gaussian r.v.'s follows immediately from the expression for the characteristic functions. This was already known to Escher in 1930s.

Respectively, if the contract starts at time $t < T$ where T is the maturity time and $T-t$ is the time to maturity, the pricing formula can be written as follows:

$$C(t,x) = C(t,x,\sigma) = x\Phi(d) - K\Phi(d - \sigma\sqrt{T-t}), \tag{1.1.4}$$

where

$$d = d(t,x) = d(t,x,\sigma) = \frac{1}{\sigma\sqrt{T-t}} \ln\frac{x}{K} + \frac{1}{2}\sigma\sqrt{T-t}. \tag{1.1.5}$$

We also put $C(T,x) = (x-K)^+$. With this definition, the function C is continuous at every point $(t,x) \in [0,T] \times]0,\infty[$ except (T,K). The singularity at the point (T,K) is of no importance in the classical theory but, as we shall see further, has dramatic consequences for models with transaction costs.

The option price process $C(t,S_t)$ is a $\tilde{P}$-martingale. To see this, we calculate the derivatives

$$C_x(t,x) = \Phi\big(d(t,x)\big), \tag{1.1.6}$$

$$C_t(t,x) = -\frac{\sigma x}{2\sqrt{T-t}}\varphi\big(d(t,x)\big), \tag{1.1.7}$$

$$C_{xx}(t,x) = \frac{1}{x\sigma\sqrt{T-t}}\varphi\big(d(t,x)\big). \tag{1.1.8}$$

Notice that

$$C_t(t,x) + (1/2)\sigma^2 x^2 C_{xx}(t,x) = 0. \tag{1.1.9}$$

Applying the Itô formula, we have that, on $[0,T[$,

$$\begin{aligned} C(t,S_t) = C(0,S_0) &+ \int_0^t C_x(u,S_u)\,dS_u \\ &+ \int_0^t \left[C_t(u,S_u) + (1/2)S_u^2 C_{xx}(u,S_u)\right] du, \end{aligned}$$

and, therefore, due to the above equation,

$$C(t,S_t) = C(0,S_0) + \int_0^t C_x(u,S_u)\,dS_u.$$

Since the process S is a square-integrable martingale with respect to $\tilde{P}$ on $[0,T]$, so is the integral in the right-hand side of the resulting identity (the integrand is bounded). As is easy to see, the limit of the left-hand side exists a.s. and is equal to $(S_T - K)^+$. We arrive at the second Black–Scholes formula

$$(S_T - K)^+ = C(0,S_0) + \int_0^T C_x(u,S_u)\,dS_u. \tag{1.1.10}$$

The formula (1.1.6) is easy to memorize: it looks like if we would forgotten, differentiating (1.1.4), that d depends on x. It is worth emphasizing that for the call option, the instantaneous holding of stock in stock units, namely,

$$H_t = C_x(t, S_t) = \Phi\big(d(t, S_t)\big),$$

is a random process evolving in the interval $[0, 1]$, i.e., for the replication purposes, the short selling is not needed. The holding of the stock in units of the numéraire is $S_t\Phi(d(t, S_t))$. Hence, the holding in the numéraire is $V_t - S_t\Phi(d(t, S_t))$.

1.1.3 Discussion

As the reader may observe, our derivation of the Black–Scholes formulae (we prefer to use the plural) is based on a rather simple mathematics and... a murky economical background.

In fact, we choose this presentation as the quickest one to create a platform for further analysis and discussion of mathematical and financial aspects. The following considerations do not pretend to be rigorous, but we provide them to sketch some ideas exploited in the theory of option pricing in frictionless financial markets.

1. The first point is that the choice of a geometric Brownian motion as the model of price process is ad hoc. Numerous statistical tests definitely reject this model: the logarithms of price increments are not Gaussian random variables. Stable processes give much better fit, but their use is still limited since models based on them fail to be complete.

2. The pricing principle (that the option price is the expectation with respect to the martingale measure $\tilde{P}$) is universal in the Black–Scholes model. It works also for the path-dependent option and allows one to calculate easily the price for options with complicated structure using Monte Carlo methods. The problem of finding the replicating strategies is more difficult. Fortunately, for pay-offs which are contingent with the stock price value at the exercise date, i.e., of the form $\zeta = g(S_T)$ where g is a "reasonable" function, this can be done in a regular way, namely, by finding the solution of the Cauchy problem (1.1.9) with the terminal condition $C(T, x) = g(x)$. This idea works well also for more general diffusion price processes like that given by the formula (1.1.11) below. Methods of partial differential equations seems to be helpful also for models with transaction costs.

3. The representation theorem, playing such an important role in the option pricing, holds true only for a very restrictive class of processes. For example, in the realm of stochastically continuous scalar processes with independent increments, only the Wiener process and the Poisson process generate filtrations satisfying the predictable representation property. However, the price process can be modeled as the solution of the stochastic equation

$$dS_t = a(t, S_t)S_t\,dt + \sigma(t, S_t)S_t\,dw_t \tag{1.1.11}$$

with "reasonable" coefficients. This ensures an additional flexibility of the Brownian framework.

4. It is clear that if the buyer agrees for a larger price than p, then the seller has a free lunch. A moment reflection shows that in the market where the short selling is permitted, if the contract price is lower than p, the buyer of the option will have a free lunch. One can try to define a "fair" price of the derivative as such that it does not give arbitrage opportunities to any of two counterparts. Unfortunately, in more complicated models such a "fair" price is not unique.

5. To be consistent with statistical tests, financial theory needs models where the predictable representation property for the price processes may fail. This means that certain contingent claims cannot be replicated, i.e., represented as the terminal values of self-financing portfolios. In such models (usually referred to as models of incomplete markets) there are many equivalent martingale measures. One can find in the literature a lot of ideas how to price contingent claims in incomplete markets; there exists a number of "prices." We mentioned here only the super-replication price: the minimal initial capital for a self-financing portfolio the terminal value of which in all cases dominates the terminal pay-off. Under very general assumptions, it is the infimum over the set of expectations of the pay-off with respect to the equivalent martingale measures. The idea of super-replication can be extended to markets with transaction costs. It will be extensively discussed in this book.

6. How to choose a convenient class of admissible strategies? The predictable representation theorem gives a hint: the process HS should be in the space L^2 with respect to the product measure $d\tilde{P}\,dt$. When the martingale measure is not unique, the situation becomes much more complicated. However, the definition of admissibility requiring the boundedness from below of the value processes works well in a number of problems.

7. How to use the Black–Scholes formulae? The first one, (1.1.4), gives a "theoretical price," i.e., a reference point for the trader. The second prescribes the fraction of the stock to be hold. Since the continuous trading is not possible—it is nothing but a mathematical idealization—the trader, in practice, revises the portfolio time to time. The simplest method is to do these revisions periodically, by dividing the interval $[0,T]$ into n intervals and making the revisions in accordance with the Black–Scholes prescription. In this case the portfolio process can be described as follows:

$$V_t^n = C(0,S_0) + \int_0^t H_r^n\,dS_r, \tag{1.1.12}$$

where

$$H^n = \sum C_x(t_{i-1}, S_{t_{i-1}}, \sigma) I_{]t_{i-1},t_i]}, \tag{1.1.13}$$

$t_i = (i/n)T$, $i = 0,\dots,n-1$. When the number of the revision intervals converges to infinity, $V_T^n \to V_T = (S_T - K)^+$ in $L^2(\tilde{P})$ (just by the construction of the stochastic integral). Thus, $V_T^n \to V_T = (S_T - K)^+$ also in P-probability.

The parameters T and K are stipulated in the contract. The only remaining parameter, the volatility σ, should be extracted somehow from the market data. In the Black–Scholes world of continuous trading, σ can be recovered from any, arbitrary small, part of the trajectory of S. In reality, by various reasons (discrete observations, model misspecification, round-off errors, and many others), this is not possible. There are many ideas how to estimate σ, but this discussion (involving the problem of historical and implied volatilities) is beyond the scope of our book. We shall assume that the model parameter σ is available.

8. We leave aside the question why the derivatives do exist and traded actively in many markets. In the Black–Scholes world they are redundant securities. As we shall see further, in the natural extension of classical model by including transaction costs for the super-replication of most common options, even the continuous trading can be avoided.

1.2 Leland–Lott Theorem

1.2.1 Formulation and Comments

As early as in their famous paper of 1973 Black and Scholes noticed a discrepancy between the really observed option price and the "theoretical" one given by the formula. Between other suggestions, they indicated that this may be due to the transaction costs. Indeed, though the percentage of the transaction volume payed as the brokerage fee individually can be considered as negligible, the total sum after hundreds and thousands portfolio revisions is far from being such: in the continuous trading (i.e., in the limit) the Black–Scholes prescription leads to the explosion of the accumulated transaction cost payments.

In 1985 Heyne Leland suggested a new prescription for the standard call option based on the idea of approximate replication. He observed that for small proportional transaction costs, the terminal pay-off $\zeta = (S_T - K)^+$ is close to the terminal value of the portfolio using the strategy to keep on the interval $]t_{i-1}, t_i]$ between two consecutive revisions, instead of $C_x(t_{i-1}, S_{t_{i-1}}, \sigma)$ units of stock as in (1.1.12) and (1.1.13), $C_x(t_{i-1}, S_{t_{i-1}}, \widehat{\sigma}_n)$ units. The input parameter $\widehat{\sigma}_n$ here is a multiple of the true volatility σ by a certain magnifying factor depending of the transaction constant coefficient, the volatility itself, and the number n of the revision intervals.

Leland's conclusion was very important for practitioners and it remains such because it provides a reference point for pricing contingent claims in real-world markets. Its great advantage is an easy implementation. Numerical simulations reveal good correspondence of the calculated prices with the market option prices.

Unfortunately, Leland could not provide a mathematically correct confirmation for his prescription. In his basic setting he considered a kind of scheme of series, in number n of the revision intervals, where the proportional transaction cost coefficient is constant in n. His only theorem claimed that the terminal values of portfolios converge to the pay-off. This assertion is false: the convergence holds but not to the terminal pay-off indicated in the contract. There is a nontrivial discrepancy, which we shall study in the next section.[3] Leland also made a remark, without providing arguments, that the convergence holds also in the model where the transaction cost coefficient is a function of the number of revisions decreasing as $n^{-1/2}$. This conjecture is correct. It was proven in the thesis of Klaus Lott. The Lott result is the first rigorous explanation why the Leland strategy does work in practical situations of small transaction costs and not very high frequencies of portfolio revisions.

We present here a slightly more general convergence result showing that the convergence holds when the coefficient decreases as $n^{-\alpha}$ for any $\alpha > 0$. An inspection of the proof below shows that the same conclusion holds whenever $k_n \to 0$ as $n \to \infty$.

In the two-asset model of the previous section (extended to take into account proportional transaction costs), the current value of the portfolio process at time t, corresponding to the regular revisions at $t_i = (i/n)T$, can be described (for $t < T$) as follows:

$$V_t^n = V_0^n + \int_0^t H_u^n \, dS_u - \sum_{t_i \le t} S_{t_i} f(H_{t_i}^n - H_{t_{i-1}}^n), \tag{1.2.1}$$

where H^n is a predictable piecewise constant process of the form

$$H^n = \sum_{i=1}^n H_{t_{i-1}}^n I_{]t_{i-1},t_i]}.$$

$H_{t_i}^n$ are $\mathcal{F}_{t_i}$-measurable random variables, and

$$f(x) = k_b x I_{\{x>0\}} - k_s x I_{\{x<0\}}$$

for some constants $k_s, k_b > 0$. We shall consider the case where the buying and selling are equally charged, i.e., $k_b = k_s = k$ and $f(x) = k|x|$. The coefficient may depend on n. Namely, we shall assume that, for $n \ge 1$,

$$k = k_n = k_0 n^{-\alpha}, \quad \alpha \in [0, 1/2]. \tag{1.2.2}$$

The Leland strategy is given by the process H^n with

$$H_{t_i}^n = C_x(t_i, S_{t_i}, \widehat{\sigma}_n),$$

where

$$\widehat{\sigma}_n^2 := \sigma^2 \left(1 + \frac{\gamma_n}{\sigma}\right), \quad \gamma_n := \sqrt{\frac{8}{\pi}} k_n n^{1/2} = \sqrt{\frac{8}{\pi}} k_0 n^{1/2-\alpha}. \tag{1.2.3}$$

---

[3] Warning: the reader should take care that there are still papers where authors believe that the discrepancy is zero.

Remarks on notation. In the case $\alpha = 1/2$ (and only in this case) the parameters γ and $\widehat{\sigma}$ do not depend of n. However, to alleviate formula, we shall often omit the index n also in other cases, hoping that this will not lead to ambiguities. We assume that the time unit is such that $T = 1$. We use also the abbreviations

$$\widehat{C}(t,x) := C(t,x,\widehat{\sigma}), \qquad \widehat{H}_t := \widehat{C}_x(t,S_t), \qquad \widehat{h}_t := \widehat{C}_{xx}(t,S_t).$$

Note that, in contrast to the piecewise constant process H^n, the process $\widehat{H} = \widehat{H}^n$ is continuous, but their values at the revision dates t_i coincide.

Prerequisites. In this chapter we use rather elementary mathematical tools. Our arguments require only one result from stochastic calculus beyond standard facts used in the derivation of the Black–Scholes formula. Namely, we use the following version of the Lenglart inequality, which can be found in any textbook on martingales: if M is a square-integrable martingale and $\langle M\rangle$ is its quadratic characteristics, then for all $a, b > 0$,

$$P\big(|M_T| > a\big) \le a^{-2}E\big(b \wedge \langle M\rangle_T\big) + P\big(\langle M\rangle_T \ge b\big).$$

This inequality implies immediately that if M^n is a sequence of square-integrable martingales (not necessarily with respect to the same filtration) such that $\langle M^n\rangle_T \to 0$ in probability as $n \to 0$, then M^n_T also converges to zero in probability.

The assertion of the following theorem for $\alpha = 1/2$ was conjectured by Leland and proved by Lott. The extension to the case $\alpha \in]0,1/2[$ was established in [130].

Theorem 1.2.1 *Let $k_n = k_0 n^{-\alpha}$, where $k_0 > 0$ and $\alpha \in]0,1/2]$. Then*

$$P\text{-}\lim_n V_1^n = (S_1 - K)^+. \tag{1.2.4}$$

1.2.2 Proof

The convergence in probability is invariant under the equivalent changes of probability measure. By this reason we assume from the very beginning that P is the *martingale measure* (i.e., the drift coefficient of the price process is zero).

By the Itô formula we get that

$$\widehat{C}_x(t,S_t) = \widehat{C}_x(0,S_0) + M_t^n + A_t^n, \tag{1.2.5}$$

where

$$M_t^n := \int_0^t \widehat{C}_{xx}(u,S_u)\,dS_u = \int_0^t \sigma S_u \widehat{C}_{xx}(u,S_u)\,dw_u = \int_0^t \sigma S_u \widehat{h}_u\,dw_u,$$

$$A_t^n := \int_0^t \left[\widehat{C}_{xt}(u,S_u) + \frac{1}{2}\sigma^2 S_u^2 \widehat{C}_{xxx}(u,S_u)\right] du.$$

The process M^n is a square-integrable martingale on $[0,1]$. In virtue of the expression for C_{xx} given by (1.1.8),

$$\langle M^n \rangle_t = \frac{1}{2\pi} \int_0^t \frac{\sigma^2}{\widehat{\sigma}^2(1-u)} \exp\{-d^2(u, S_u, \widehat{\sigma})\}\, du.$$

Now we have to represent the difference $V_1^n - \zeta$ in a form convenient for the asymptotic analysis.

Lemma 1.2.2 *We have* $V_1^n - \zeta = F_1^n + F_2^n$, *where*

$$F_1^n := \int_0^1 (H_t^n - \widehat{H}_t)\, dS_t = \sum_{i=1}^n \int_{t_{i-1}}^{t_i} (\widehat{C}_x(t_{i-1}, S_{t_{i-1}}) - \widehat{C}_x(t, S_t))\, dS_t, \quad (1.2.6)$$

$$F_2^n := \frac{1}{2}\gamma\sigma \int_0^1 S_t^2 \widehat{C}_{xx}(t, S_t)\, dt - k \sum_{i=1}^n |H_{t_i}^n - H_{t_{i-1}}^n| S_{t_i}. \quad (1.2.7)$$

Proof. According to the Black–Scholes formulae, the contingent claim ζ admits the representation

$$\zeta = C(0, S_0) + \int_0^1 H_t\, dS_t, \quad (1.2.8)$$

where, we recall, $H_t = C_x(t, S_t, \sigma)$.

Taking the difference of (1.2.1) and (1.2.8), we obtain that

$$V_1^n - \zeta = \int_0^1 (H_t^n - H_t)\, dS_t + \widehat{C}(0, S_0) - C(0, S_0) - k \sum_{i=1}^n |H_{t_i}^n - H_{t_{i-1}}^n| S_{t_i}.$$

It remains to check that

$$\widehat{C}(0, S_0) - C(0, S_0) = \int_0^1 (H_t - \widehat{H}_t)\, dS_t + \frac{1}{2}\gamma\sigma \int_0^1 S_t^2 \widehat{C}_{xx}(t, S_t)\, dt.$$

This identity holds because by the Itô formula

$$\begin{aligned} \widehat{C}(0, S_0) - C(0, S_0) &= C(1, S_1) - C(0, S_0) - \left[\widehat{C}(1, S_1) - \widehat{C}(0, S_0)\right] \\ &= \int_0^1 C_x(t, S_t)\, dS_t - \int_0^1 \widehat{C}_x(t, S_t)\, dS_t \\ &\quad + \frac{1}{2} \int_0^1 (\widehat{\sigma}_t^2 - \sigma^2) S_t^2 \widehat{C}_{xx}(t, S_t)\, dt, \end{aligned}$$

where the simplification in the right-hand side is due to the fact that $C(t, x)$ is the solution of the parabolic equation $(1/2)\sigma^2 x^2 C_{xx} + C_t = 0$ with the boundary condition $C(1, x) = (x - K)^+$, while $\widehat{C}(t, x)$ is the solution of $(1/2)\widehat{\sigma}^2 x^2 \widehat{C}_{xx} + \widehat{C}_t = 0$ with the same boundary condition. □

Note that there is no need in portfolio rebalancing at the maturity date, and that is why the summation in (1.2.7) is taken only up to $n-1$.

Lemma 1.2.3 *For any* $\alpha \in [0,1/2]$,

$$P\text{-}\lim_n F_1^n = 0. \tag{1.2.9}$$

Proof. In virtue of the basic (isometry) property of the stochastic integral with respect to a Wiener process, it is sufficient to verify that

$$\lim_n E\int_0^1 \sigma^2 S_t^2 \big(H_t^n - \widehat{H}_t\big)^2\, dt = 0.$$

But this is almost obvious. In the case $\alpha \in [0,1/2[$, the parameter $\widehat{\sigma}_n$ increases to infinity, and both H_t^n and $\widehat{H}_t$ tend to unit for $t < 1$. If $\alpha = 1/2$, the parameter $\widehat{\sigma}_n$ remains constant, but still $H_t^n - \widehat{H}_t \to 0$ for $t < 1$. □

Comment on the intuition. It seems to be rather clear that the integral in the definition of F_2^n can be well approximated by the Riemann sums with terms $S_{t_{i-1}}^2 C_{xx}(t_{i-1}, S_{t_{i-1}})$. On the other hand, $E|w_{t_i} - w_{t_{i-1}}| = \sqrt{2/\pi}\sqrt{\Delta t}$ and

$$\begin{aligned}|H_{t_i}^n - H_{t_{i-1}}^n| S_{t_i} &\approx \sigma S_{t_{i-1}}^2 C_{xx}(t_{i-1}, S_{t_{i-1}})|w_{t_i} - w_{t_{i-1}}| \\ &\approx \sigma S_{t_{i-1}}^2 C_{xx}(t_{i-1}, S_{t_{i-1}})\sqrt{2/\pi}\sqrt{\Delta t}.\end{aligned}$$

So, it is reasonable to expect that the choice $\gamma = \gamma_n = 2\sqrt{2/\pi} n^{1/2}$ will lead to an appropriate compensation and F_2^n can be represented as $\sum_i \leq n\xi_i^n$, where for each i, the sequence $(\xi_i^n)_{n\geq 1}$ converges to zero sufficiently fast to ensure that the whole sum converges to zero. Indeed, this is the case for $\alpha \in]0,1/2]$. A rigorous study of F_2^n is rather delicate and requires some patience.

Put $\Delta t := 1/n$. It is easily seen that $F_2^n = \sum_{i=1}^5 L_i^n$, where

$$L_1^n := \frac{1}{2}\gamma\sigma \int_0^1 S_t^2 \widehat{h}_t\, dt - \frac{1}{2}\gamma\sigma \int_0^1 \sum_{i=1}^n S_{t_{i-1}}^2 \widehat{h}_{t_{i-1}} I_{]t_{i-1},t_i]}(t)\, dt,$$

$$L_2^n := \frac{1}{2}\gamma\sigma \sum_{i=1}^n S_{t_{i-1}}^2 \widehat{h}_{t_{i-1}} \Delta t - k\sigma \sum_{i=1}^n S_{t_{i-1}}^2 \widehat{h}_{t_{i-1}} |\Delta w_{t_i}|,$$

$$L_3^n := k\sigma \sum_{i=1}^n S_{t_{i-1}}^2 \widehat{h}_{t_{i-1}} |\Delta w_{t_i}| - k\sum_{i=1}^n S_{t_{i-1}} |\Delta M_{t_i}|,$$

$$L_4^n := k\sum_{i=1}^n S_{t_{i-1}} |\Delta M_{t_i}| - k\sum_{i=1}^n S_{t_{i-1}} |\Delta \widehat{H}_{t_i}|,$$

$$L_5^n := -k\sum_{i=1}^n \Delta S_{t_i} |\Delta \widehat{H}_{t_i}|,$$

$\Delta w_{t_i} := w_{t_i} - w_{t_{i-1}}$, $\Delta S_{t_i} := S_{t_i} - S_{t_{i-1}}$, and $\Delta \widehat{H}_{t_i} := \widehat{H}_{t_i} - \widehat{H}_{t_{i-1}}$.

Lemma 1.2.4 *For any $\alpha \in [0,1/2[$, both terms the difference of which defines L_1^n converge almost surely, as $n \to \infty$, to $J_1 := S_1 \wedge K$, and, therefore, $L_1^n \to 0$ a.s. The latter property also holds for $\alpha = 1/2$.*

Proof. We shall argue for ω outside the null set $\{S_1 = K\}$. Recalling the definition $\widehat{h}_t = C_{xx}(t, S_t, \widehat{\sigma})$ and using formula (1.1.8), we obtain, after the substitution $v = \widehat{\sigma}^2(1-t)$, that the first term in the representation of L_1^n can be written as

$$\frac{1}{2}\frac{\gamma\sigma}{\widehat{\sigma}^2}\int_0^{\widehat{\sigma}^2} \frac{S_{1-v/\widehat{\sigma}^2}}{\sqrt{v}}\varphi\Big(\frac{\ln(S_{1-v/\widehat{\sigma}^2}/K)}{\sqrt{v}} + \frac{1}{2}\sqrt{v}\Big)\,dv.$$

In a similar way we rewrite the second term:

$$\frac{1}{2}\frac{\gamma\sigma}{\widehat{\sigma}^2}\int_0^{\widehat{\sigma}^2} \sum_{i=1}^{n} \frac{S_{1-v_{i-1}/\widehat{\sigma}^2}}{\sqrt{v}_{i-1}}\varphi\Big(\frac{\ln(S_{1-v_{i-1}/\widehat{\sigma}^2}/K)}{\sqrt{v}_{i-1}} + \frac{1}{2}\sqrt{v}_{i-1}\Big)I_{]v_{i-1},v_i]}(v)\,dv,$$

where $v_i := \widehat{\sigma}^2(1-t_i)$. In the case $\alpha = 1/2$, the adjusted volatility $\widehat{\sigma}$ does not depend on n, and $L_1^n \to 0$ because of the convergence of the Riemann sum to the integral. If $\alpha \in [0,1/2[$, then $\widehat{\sigma}^2 \to \infty$ and $\gamma\sigma/\widehat{\sigma}^2 \to 1$.

The following simple observation is important (and will be used several times in future): for every ω outside the null-set $\{S_1 = K\}$, the integrands above are dominated by a continuous function dependent on ω which decreases to zero at zero and infinity exponentially fast and hence is integrable on $[0,\infty[$. By the dominated convergence both integrals converge to

$$\frac{1}{\sqrt{8\pi}}\int_0^{\infty} \frac{S_1}{\sqrt{v}}\exp\Big\{-\frac{v}{2}\Big(\frac{\ln(S_1/K)}{v} + \frac{1}{2}\Big)^2\Big\}\,dv = S_1 \wedge K. \tag{1.2.10}$$

The (unexpectedly simple) expression for the above integral can be easily deduced from the formula

$$f(a) := \int_0^{\infty} e^{-(u^2+a^2/u^2)}\,du = \frac{\sqrt{\pi}}{2}e^{-2|a|}. \tag{1.2.11}$$

To verify the latter, it is sufficient to consider the case $a \ge 0$ and notice that $f' = -2f$ and $f(0) = \sqrt{\pi}/2$. □

Our next step is to establish the following:

Lemma 1.2.5 *For any $\alpha \in [0,1/2]$, we have P-$\lim_n L_2^n = 0$.*

Proof. Let us consider the discrete-time process $M^n = (M_m^n)$ with

$$M_m^n := \sigma k \sum_{i=1}^{m} \widehat{h}_{t_{i-1}}\big(|\Delta w_{t_i}| - n^{-1/2}\sqrt{2/\pi}\big).$$

Taking into account the independence of increments of the Wiener process and the equalities

$$E|\Delta w_{t_i}| = n^{-1/2}\sqrt{2/\pi},$$
$$E\big(|\Delta w_{t_i}| - n^{-1/2}\sqrt{2/\pi}\big)^2 = (1-2/\pi)n^{-1} = (1-2/\pi)\Delta t,$$

we infer that $M^n = (M^n_m)$ is a square-integrable martingale with the characteristics

$$\langle M^n\rangle_m = \sigma^2(1-2/\pi)k^2\sum_{i=1}^{m} S^4_{t_{i-1}}\widehat{C}^2_{xx}(t_{i-1}, S_{t_{i-1}})\Delta t.$$

Notice that $\langle M^n\rangle_n \to 0$ as $n\to\infty$. Indeed, as in the proof above, we treat separately the case $\alpha\in[0,1/2[$ (in which we obtain, using the same change of variable, that the convergence is of order $(k^2/\widehat{\sigma}^2)\ln n = O(n^{-1/2-\alpha}\ln n)$) and the case $\alpha = 1/2$, where the convergence is of order $k^2 = O(n^{-1})$.

Now, since $\langle M^n\rangle_n \to 0$ in probability, we obtain from the Lenglart inequality that also $M^n_n \to 0$ in probability, and the result follows. □

Now we need again to exercise in calculations.

Lemma 1.2.6 *For $t\in[0,1[$, we have*

$$ES_t^2\widehat{h}_t^2 = ES_t^2\widehat{C}^2_{xx}(t,S_t) = \frac{1}{2\pi\widehat{\sigma}^2(1-t)}\frac{1}{\sqrt{2a^2+1}}\exp\left\{-\frac{b^2}{2a^2+1}\right\}, \qquad (1.2.12)$$

where

$$a := \frac{\sigma\sqrt{t}}{\widehat{\sigma}\sqrt{1-t}}, \qquad b := \frac{\ln(S_0/K) - \sigma^2 t/2}{\widehat{\sigma}\sqrt{1-t}} + \frac{1}{2}\widehat{\sigma}\sqrt{1-t}.$$

Proof. Let $\eta\in\mathcal{N}(0,1)$. Then, for any real numbers a and b,

$$E\exp\{-(a\eta+b)^2\} = \frac{1}{\sqrt{2a^2+1}}\exp\left\{-\frac{b^2}{2a^2+1}\right\}.$$

Since

$$2\pi\widehat{\sigma}^2(1-t)S_t^2\widehat{C}^2_{xx}(t,S_t) = \exp\{-(aw_t/\sqrt{t}+b)^2\},$$

the result follows. □

As a corollary, we get that, for $t\in[1/2,1[$,

$$ES_t^2\widehat{C}^2_{xx}(t,S_t) \le \frac{c}{\widehat{\sigma}\sqrt{1-t}}. \qquad (1.2.13)$$

We shall need some bounds for higher-order derivatives of the function $\widehat{C}_x(t,x) = \Phi(\widehat{d})$, where

$$\widehat{d} = d(t,x,\widehat{\sigma}) = \frac{\ln(x/K)}{\widehat{\sigma}\sqrt{1-t}} + \frac{1}{2}\widehat{\sigma}\sqrt{1-t}.$$

Using the abbreviation

$$\widehat{\delta} := \widehat{\delta}_t := \frac{1}{\widehat{\sigma}\sqrt{1-t}}$$

and noticing that

$$\frac{\partial \widehat{d}}{\partial x} = \frac{1}{x}\widehat{\delta}, \qquad \frac{\partial \varphi(\widehat{d})}{\partial x} = -\frac{1}{x}\widehat{\delta}\widehat{d}\varphi(\widehat{d}),$$

we get easily the needed derivatives:

$$\begin{aligned}
\widehat{C}_{xx}(t,x) &:= \frac{1}{x}\widehat{\delta}\varphi(\widehat{d}),\\
\widehat{C}_{xxx}(t,x) &:= -\frac{1}{x^2}\widehat{\delta}(1+\widehat{\delta}\widehat{d})\varphi(\widehat{d}),\\
\widehat{C}_{xxxx}(t,x) &:= \frac{1}{x^3}\left[2\widehat{\delta}(1+\widehat{\delta}\widehat{d}) + \widehat{\delta}^2\widehat{d}(1+\widehat{\delta}\widehat{d}) - \widehat{\delta}^3\right]\varphi(\widehat{d}),\\
\widehat{C}_{xt}(t,x) &:= \widehat{\delta}\left(\frac{1}{2}\frac{1}{1-t}\ln(x/K) - \frac{1}{4}\widehat{\sigma}^2\right)\varphi(\widehat{d}),\\
\widehat{C}_{xxt}(t,x) &:= \frac{1}{x}\widehat{\delta}\left[\frac{1}{2}\frac{1}{1-t} - \widehat{\delta}\widehat{d}\left(\frac{1}{2}\frac{1}{1-t}\ln(x/K) - \frac{1}{4}\widehat{\sigma}^2\right)\right]\varphi(\widehat{d}).
\end{aligned}$$

It follows that for $t < 1$, we have, with some constant c, the following bounds:

$$\begin{aligned}
\left|\widehat{C}_{xxx}(t,x)\right| &\le c\frac{1}{x^2}(\widehat{\delta} + \widehat{\delta}^2),\\
\left|\widehat{C}_{xxxx}(t,x)\right| &\le c\frac{1}{x^3}(\widehat{\delta} + \widehat{\delta}^2 + \widehat{\delta}^3),\\
\left|\widehat{C}_{xt}(t,x)\right| &\le c\left(\frac{1}{1-t} + \widehat{\sigma}^2\widehat{\delta}\right),\\
\left|\widehat{C}_{xxt}(t,x)\right| &\le c\frac{1}{x(1-t)}(\widehat{\delta} + 1)
\end{aligned}$$

(to estimate these derivatives, we use the boundedness of the function $\widehat{d}^p\varphi(\widehat{d})$).

Lemma 1.2.7 *For any* $\alpha \in [0, 1/2]$, *we have* $P\text{-}\lim_n L_3^n = 0$.

Proof. Using the elementary inequality $||a_1| - |a_2|| \le |a_1 - a_2|$, we get that

$$|L_n^3| \le k\sigma \sum_{i=1}^n S_{t_{i-1}} \left| \int_{t_{i-1}}^{t_i} (S_{t_{i-1}}\widehat{h}_{t_{i-1}} - S_u\widehat{h}_u)\, dw_u \right| \le k\sigma \max_{u\le 1} S_u \sum_{i=1}^n \xi_i,$$

where we use the abbreviation $\xi_i = \xi_i^n$ for the absolute value of the stochastic integral. With this remark, it is sufficient to check that $k\sum_{i=1}^n \xi_i \to 0$ in L^1. We verify a stronger property, namely, that $k\sum_{i=1}^n \|\xi_i\|_{L^2} \to 0$. Notice that the last terms involving a singularity can be omitted because

$$E\xi_n^2 \le 2n^{-1}ES_{t_{n-1}}^2\widehat{h}_{t_{n-1}}^2 + 2n^{-1}\int_{1-n^{-1}}^1 ES_u^2\widehat{h}_u^2\,du \to 0$$

in virtue of (1.2.13).

$$E\left|\int_{t_{i-1}}^{t_i}(S_{t_{i-1}}\widehat{h}_{t_{i-1}} - S_u\widehat{h}_u)\,dw_u\right| \le \left(\int_{t_{i-1}}^{t_i} E(S_{t_{i-1}}\widehat{h}_{t_{i-1}} - S_u\widehat{h}_u)^2\,du\right)^{1/2}.$$

By the Itô formula,

$$dS_t\widehat{h}_t = d\big[S_t\widehat{C}_{xx}(t,S_t)\big] = f_t\,dw_t + g_t\,dt,$$

where

$$f_t := \sigma S_t\widehat{C}_{xx}(t,S_t) + \sigma^2S_t^2\widehat{C}_{xxx}(t,S_t),$$
$$g_t := S_t\widehat{C}_{xxt}(t,S_t) + \frac{1}{2}\sigma^2S_t^3\widehat{C}_{xxxx}(t,S_t) + \sigma^2S_t^2\widehat{C}_{xxx}(t,S_t).$$

Thus,

$$E(S_{t_{i-1}}\widehat{h}_{t_{i-1}} - S_t\widehat{h}_t)^2 = E\left[\int_{t_{i-1}}^t f_u\,dw_u + \int_{t_{i-1}}^t g_u\,du\right]^2$$
$$\le 2\int_{t_{i-1}}^{t_i} Ef_u^2\,du + 2\Delta t\int_{t_{i-1}}^{t_i} Eg_u^2\,du.$$

Using the prepared bounds for the derivatives, we have, for $i < n$,

$$\|\xi_i\|_{L^2}^2 = \int_{t_{i-1}}^{t_i} E(S_{t_{i-1}}\widehat{h}_{t_{i-1}} - S_u\widehat{h}_u)^2\,du$$
$$\le c(\Delta t)^2(\widehat{\delta}_{t_i}^2 + \widehat{\delta}_{t_i}^4) + c(\Delta t)^3\left(\frac{(1+\widehat{\delta}_{t_i})^2}{(1-t_i)^2} + \widehat{\delta}_{t_i}^6\right).$$

It follows that

$$\|\xi_i\|_{L^2} \le c\Delta t(\widehat{\delta}_{t_i} + \widehat{\delta}_{t_i}^2) + c(\Delta t)^{3/2}\frac{1+\widehat{\delta}_{t_i}}{1-t_i}.$$

It is clear that

$$\sum_{i=1}^{n-1}\frac{\Delta t}{\widehat{\sigma}(1-t_i)^{1/2}} \sim \widehat{\sigma}^{-1}\int_0^1\frac{dt}{(1-t)^{1/2}},$$
$$\sum_{i=1}^{n-1}\frac{\Delta t}{\widehat{\sigma}^2(1-t_i)} \sim \widehat{\sigma}^{-2}\ln n,$$
$$(\Delta t)^{1/2}\sum_{i=1}^{n-1}\frac{\Delta t}{\widehat{\sigma}^{1/2}(1-t_i)^{3/2}} \sim 2n^{-1/2}\widehat{\sigma}^{-1/2}n^{1/2},$$
$$(\Delta t)^{1/2}\sum_{i=1}^{n-1}\frac{\Delta t}{1-t_i} \sim n^{-1/2}\ln n.$$

All these terms converge to zero if $\alpha \in [0,1/2[$, even without multiplying by $k = k_n$. The case $\alpha = 1/2$ is special, but then, luckily, $k_n = k_0 n^{-1/2}$, and we always have the convergence $k\sum_{i=1}^n \|\xi\|_{L^2} \to 0$. □

Lemma 1.2.8 *For any* $\alpha \in]0,1/2]$, *we have* $P\text{-}\lim_n L_4^n = 0$. *For* $\alpha = 0$, *the sequence* L_n^4 *is bounded in probability.*

Proof. Using again the inequality $||a_1| - |a_2|| \le |a_1 - a_2|$, we get that

$$\begin{aligned} |L_n^4| &\le k\sum_{i=1}^n S_{t_{i-1}}|A_{t_i} - A_{t_{i-1}}| \\ &\le k\xi\int_0^1 |\widehat{C}_{xt}(u,S_u)|\,du + k\xi\sum_{i=1}^n \int_{t_{i-1}}^{t_i} \sigma^2 S_u^2 |\widehat{C}_{xxx}(u,S_u)|\,du, \end{aligned}$$

where $\xi \ge 0$ is a finite random variable.

Both terms in the right-hand side of the above bound converge to zero in the case $\alpha = 1/2$ (where $k_n = k_0 n^{-1/2}$, and $\widehat{C}$ does not depend on n).

When $\alpha \in [0,1/2[$, by the same reasoning as in Lemma 1.2.4 we conclude that, outside of the null set $\{S_1 = K\}$, the first term in the right-hand side of the above bound converges to the finite limit

$$\xi k_\infty \frac{1}{\sqrt{2\pi}}\int_0^\infty \left|\frac{1}{2}\frac{\ln(S_1/K)}{v\sqrt{v}} - \frac{1}{4\sqrt{v}}\right| \exp\left\{-\frac{v}{2}\left(\frac{\ln(S_1/K)}{v} + \frac{1}{2}\right)\right\} dv,$$

where $k_\infty = \lim k_n$ is equal to zero for $\alpha > 0$ and $k_\infty = k_0$ for $\alpha = 0$.

To establish the convergence to zero of the second term when $\alpha \in [0,1/2[$, we use the estimate of $C_{xxx}(t, S_t, \widehat{\sigma}_n)$ prepared earlier. We have

$$\begin{aligned} \sum_{i=1}^{n-1}\int_{t_{i-1}}^{t_i} \sigma^2 S_u^2 |\widehat{C}_{xxx}(u,S_u)|\,du &\le c\sum_{i=1}^{n-1}(\widehat{\delta} + \widehat{\delta}^2) \\ &\le (\widehat{\sigma}^{-1} + \widehat{\sigma}^{-2}\ln n) \to 0. \end{aligned}$$

The remaining summand converges to zero a.s. because, for any ω for which $S_1(\omega) \ne K$, the sequence of functions $C_{xxx}(u, S_u(\omega))$ is bounded near the terminal date $t = 1$. □

Lemma 1.2.9 *For any* $\alpha \in [0,1/2]$, *we have* $P\text{-}\lim_n L_5^n = 0$.

Proof. Since $\max_i |\Delta S_{t_i}| \to 0$ as $n \to \infty$, it is sufficient to verify that the sequence $k\sum_{i=1}^n |\Delta\widehat{H}_{t_i}|$ is bounded in probability or, equivalently, that $k\sum_{i=1}^n S_{t_{i-1}}|\Delta\widehat{H}_{t_i}|$ has this property. But the latter fact follows from the preceding lemmas. □

Inspecting the formulations of above lemmas, we conclude that all terms L_j^n tend to zero in probability in the case where $\alpha \in]0,1/2]$, and the proof of Theorem 1.2.1 is finished.

1.3 Constant Coefficient: Discrepancy

1.3.1 Main Result

A detailed view of the above proof reveals that almost all steps in the arguments works well also when $\alpha = 0$, i.e., when the transaction cost coefficient does not depend on the number of portfolio revisions, but in Lemma 1.2.8 in this case we have nontrivial limits. This observation leads to the following result.

Theorem 1.3.1 *Let $k = k_0 \geq 0$ (i.e., $\alpha = 0$). Then*

$$P\text{-}\lim_n V_1^n = (S_1 - K)^+ + J_1 - J_2(k_0), \tag{1.3.1}$$

where $J_1 := S_1 \wedge K$ and $J_2(k_0) := S_1 F(\ln(S_1/K), k_0)$ with

$$F(y, k_0) := \frac{1}{4}\int_0^\infty \frac{1}{\sqrt{v}} G(y, v, k_0) \exp\left\{-\frac{v}{2}\left(\frac{y}{v} + \frac{1}{2}\right)^2\right\} dv, \tag{1.3.2}$$

$$G(y, v, k_0) := \frac{1}{\sqrt{2\pi}}\int_{-\infty}^\infty \left| x - \frac{2k_0 y}{\sqrt{2\pi v}} + \frac{k_0}{\sqrt{2\pi}} \right| e^{-x^2/2}\, dx. \tag{1.3.3}$$

Since $(x - K)^+ + x \wedge K = x$, the right-hand side of (1.3.1) is equal to $S_1 - J_2(k_0)$.

Notice that $G(y, v, 0) = 2/\sqrt{2\pi}$ and, hence, according to the formula (1.2.10),

$$J_2(0) = S_1 F\big(\ln(S_1/K), 0\big) = S_1 \wedge K = J_1.$$

We recover the known result that, in the absence of transaction costs, the considered strategy leads to the replication in the limit.

Proof. In virtue of Lemmas 1.2.4–1.2.7 for $\alpha = 0$, each of the "chained" terms L_1^n, L_2^n, and L_3^n is a difference of sequences of random variables converging to the common limit J_1. Thus, in our representation of L_4^n the first component also converges to J_1, and it remains to check the convergence property for the second component, i.e., that

$$k_0 \sum_{i=1}^n S_{t_{i-1}} |\widehat{H}_{t_i} - \widehat{H}_{t_{i-1}}| \to J_2(k_0). \tag{1.3.4}$$

Put

$$Z_i^n := \left| \sqrt{n}(w_{t_i} - w_{t_{i-1}}) - \frac{\ln(S_{t_{i-1}}/K)}{2\sigma(1 - t_{i-1})\sqrt{n}} + \frac{\widehat{\sigma}^2}{4\sigma\sqrt{n}} \right|.$$

We shall work with the identity

$$\sum_{i=1}^n S_{t_{i-1}} |\widehat{H}_{t_i} - \widehat{H}_{t_{i-1}}| - \frac{1}{k_0} J_2(k_0) = I_1^n + I_2^n + I_3^n - \frac{1}{k_0} J_2(k_0),$$

where

$$I_1^n := \sum_{i=1}^n S_{t_{i-1}}|\widehat{H}_{t_i} - \widehat{H}_{t_{i-1}}| - \sum_{i=1}^n \sigma S_{t_{i-1}}^2 \widehat{h}_{t_{i-1}} Z_i^n n^{-1/2},$$

$$I_2^n := \sum_{i=1}^n \sigma S_{t_{i-1}}^2 \widehat{h}_{t_{i-1}} \big[Z_i^n - E(Z_i^n|\mathcal{F}_{t_{i-1}})\big] n^{-1/2},$$

$$I_3^n := \sum_{i=1}^n \sigma S_{t_{i-1}}^2 \widehat{h}_{t_{i-1}} E(Z_i^n|\mathcal{F}_{t_{i-1}}) n^{-1/2}.$$

Using the inequality $||a_1|-|a_2|| \le |a_1-a_2|$ and the representation (1.2.5), we estimate the summand I_1^n, regrouping the terms, as follows:

$$\begin{aligned} |I_1^n| \le & \sum_{i=1}^n S_{t_{i-1}}\big|M_{t_i} - M_{t_{i-1}} - \sigma S_{t_{i-1}}\widehat{h}_{t_{i-1}}(w_{t_i} - w_{t_{i-1}})\big| \\ & + \sum_{i=1}^n S_{t_{i-1}}\left|A_{t_i} - A_{t_{i-1}} - \sigma S_{t_{i-1}}\widehat{h}_{t_{i-1}}\left(\frac{\ln(S_{t_{i-1}}/K)}{2\sigma(1-t_{i-1})} - \frac{\widehat{\sigma}^2}{4\sigma}\right)\Delta t\right|. \end{aligned}$$

The first sum in the right-hand side coincides up to the constant k_0 with the majorant for $|L_3^n|$, which, as established in the proof of Lemma 1.2.7, converges to zero in probability. Consulting the table of derivatives, we observe that

$$x\widehat{C}_{xx}(t,x)\left(\frac{1}{2}\frac{1}{1-t}\ln(x/K) - \frac{1}{4}\widehat{\sigma}^2\right) = \widehat{C}_{xt}(t,x).$$

It follows that the second sum is dominated, up to a random but fixed multiplier, by

$$\int_0^1 |\widehat{C}_{xxx}(t,S_t)|\,dt + \sum_{i=1}^n \left|\int_{t_{i-1}}^{t_i} [\widehat{C}_{xt}(t,S_t)\,dt - \widehat{C}_{xt}(t_{i-1},S_{t_{i-1}})]\,dt\right|.$$

As already shown, the first integral converges to zero.

The convergence of the second term to zero (of course, outside the null-set $\{S_1 = K\}$) follows by our usual arguments based on the change of variables and dominating convergence.

Using the same consideration as in the proof of Lemma 1.2.5, we can show that $I_2^n \to 0$ in probability. Indeed, for every fixed n, the sequence of random variables $\tilde{Z}_i^n = Z_i^n - E(Z_i^n|\mathcal{F}_{t_{i-1}})$, $i = 1,\dots,n$, is a martingale difference with respect to the discrete-time filtration $(\mathcal{F}_{t_{i-1}})$. Easy calculations show that

$$\sum_{i=1}^n \sigma^2 S_{t_{i-1}}^4 \widehat{h}_{t_i}^2 E\big((\tilde{Z}_i^n)^2|\mathcal{F}_{t_{i-1}}\big) n^{-1} \to 0 \quad \text{a.s.},$$

implying by the Lenglart inequality the required convergence of I_2^n to zero in probability.

Taking into account that $\widehat{\sigma}/\sqrt{n} \sim \sqrt{8/\pi}k_0\sigma$, we obtain, substituting the expressions for $\widehat{h}$ and the conditional expectation, that I_3^n has the same limit as

$$\frac{1}{4}\sum_{i=1}^{n} \frac{S_{t_{i-1}}}{\widehat{\sigma}\sqrt{1-t_{i-1}}} G(S_{t_{i-1}}, \widehat{\sigma}^2(1-t_{i-1}))\varphi(d(t_{i-1}, S_{t_{i-1}}, \widehat{\sigma}))\Delta t,$$

which is obviously equal to $k_0^{-1}J_2(k_0)$. □

Remarks. 1. It is easy to see that for the discrepancy (i.e., for the limiting hedging error), we have the bounds

$$-\kappa k_0 \le J_1 - J_2(k_0) < 0, \tag{1.3.5}$$

where κ is a function of S_1 and K. They lead to the important conclusion that the option is always *underpriced* in the limit, though the hedging error is small for small values of k_0.

The proof of these bounds is easy. Inspecting the formula (1.3.3) and taking into account that $J_1 = J_2(0)$, we observe that they follow from the following bounds for a standard Gaussian random variable ξ:

$$-|c| \le E|\xi| - E|\xi - c| < 0$$

for any nonzero constant c. The first one is obvious since $|a|-|b| \le |a-b|$. The second holds because the function $f(x) := E|\xi - x| = 2\varphi(x) + x(2\Phi(x) - 1)$ with the derivative $f'(x) = 2\Phi(x) - 1$ attains its unique minimum at $x = 0$.

2. As was observed by Granditz and Schachinger in [85], the function F defined by (1.3.2) and (1.3.3) for $k_0 \ne 0$ has a removable discontinuity at $y = 0$. Since the random variable S_1 is not equal to K a.s., the value $F(1, k_0)$ does not matter and can be chosen arbitrarily, e.g., to make the function continuous. However, this may not be a reasonable idea because the mentioned formula appears in a natural way. The quality of a limiting approximation is deteriorating when S_1 is near the strike K, and the discontinuity of F (the point jumps upwards) indicates that it may be even worse for finite n. Simulations results confirmed this conjecture.

1.3.2 Discussion

As we have shown, the Leland theorem that the strategy based on the Black and Scholes formula with an enlarged volatility replicates in the limit the call option pay-off (inclusive transaction costs fail to be true when the transaction cost coefficient does not depend on the number of revisions (which is quite common). On the other hand, this strategy is used by traders routinely: the Leland pricing formula is an important tool to account market imperfections. How to solve this apparent contradiction between theoretical results and financial practice? The answer can be based on the Leland–Lott theorem and the usual methodology of using asymptotic results. The latter prescribes to

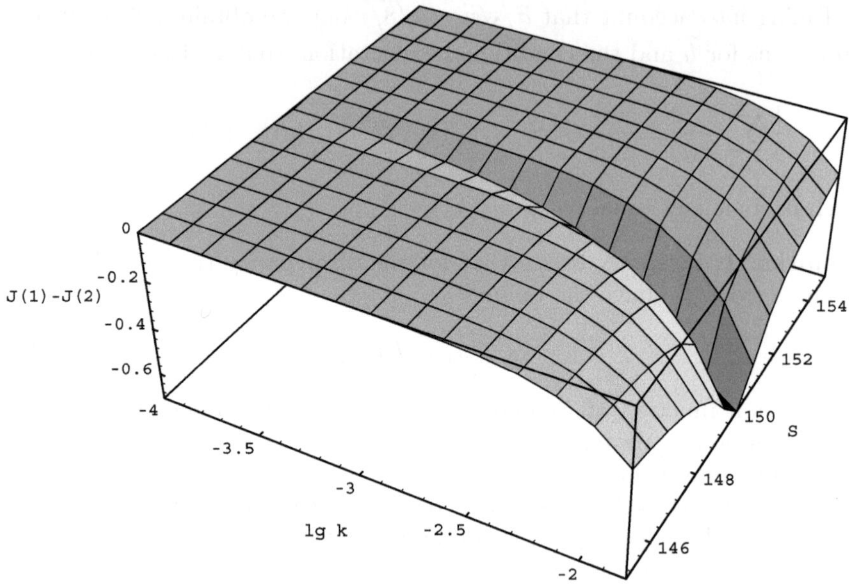

Picture 1. Dependence of $J_1 - J_2$ on $k = k_0$ and $S = S_1$ for $K = 150$.

imbed the model in a family of models parameterized by n (the "scheme of series"). So, if n revisions are planned by the investor, the transaction cost coefficient (e.g., provided by a trader) has to be interpreted as $k_n = k_0 n^{-\alpha}$ with certain artificial parameters k_0 and α needed to compute the modified volatility. In the realistic situations where n is several dozens and the transaction costs are fractions of percent, this interpretation seems to be legitimate and leads to satisfactory results.

So, though we provide arguments disclaiming the Leland theorem, they do not discard the approach based on the approximate hedging.

Denis [62] recently suggested a modification of the Leland strategy for which the convergence in probability to the pay-off of the call option holds for the constant transaction cost coefficient $k_n = k_0$. Namely, one can take

$$H^n_{t_{i-1}} = \widehat{C}_x(t_{i-1}, S_{t_{i-1}}) - \int_0^{t_{i-1}} \widehat{C}_{xt}(u, S_u)\, du.$$

Another improvement of the Leland strategy to ensure convergence was proposed in [179].

1.3.3 Pergamenshchikov Theorem

The convergence results presented below can be deepened. One of interesting questions is on the rate of convergence in Theorems 1.2.1 and 1.3.1. A result of Granditz and Schachinger [85] indicated that in the latter it should be $n^{-1/4}$. A complete answer is given by the following theorem of Pergamenshchikov [179].

Theorem 1.3.2 *Let $k = k_0 > 0$. Then the sequence of random variables*

$$\xi_n := n^{1/4}\big(V_1^n - (S_1 - K)^+ - J_1 + J_2(k_0)\big) \tag{1.3.6}$$

converges in law to a random variable ξ with a mixed Gaussian distribution.

The proof of this theorem is rather complicated, and we have no intention to give it here. In the next section we investigate the rate of convergence in Theorem 1.2.1, namely, the deviation in the L^2-norm of the terminal value of the portfolio process from the pay-off.

1.4 Rate of Convergence of the Replication Error

1.4.1 Formulation

In this section we consider the Lott case, where $\alpha = 1/2$, and therefore $\widehat{\sigma}_n$ does not depend on n. It is easy to prove that V_1^n converges also in L^2. We establish a much more delicate result giving the rate of convergence of the mean-square replication error. For simplicity, we continue to work under the martingale measure.

Theorem 1.4.1 *The mean-square approximation error of the Leland–Lott strategy for hedging the European call option with equidistant revision dates has the following asymptotics:*

$$E\big(V_1^n - V_1\big)^2 = A_1 n^{-1} + o(n^{-1}), \quad n \to \infty, \tag{1.4.1}$$

where the coefficient

$$A_1 = \int_0^1 \left[\frac{\sigma^4}{2} + \sigma^3 k_0 \sqrt{\frac{2}{\pi}} + k_0^2 \sigma^2 \left(1 - \frac{2}{\pi}\right)\right] \Lambda_t \, dt \tag{1.4.2}$$

with $\Lambda_t = E S_t^4 \widehat{C}_{xx}^2(t, S_t)$. Explicitly,

$$\Lambda_t = \frac{K^2}{2\pi\widehat{\sigma}\sqrt{1-t}\sqrt{2\sigma^2 t + \widehat{\sigma}^2(1-t)}} \exp\left\{-\frac{(\ln \frac{S_0}{K} - \frac{1}{2}\sigma^2 t - \frac{1}{2}\widehat{\sigma}^2(1-t))^2}{2\sigma^2 t + \widehat{\sigma}^2(1-t)}\right\}. \tag{1.4.3}$$

One can consider a slightly more general hedging strategy with a nonuniform revision grid defined by a smooth transformation of the uniform one.

Let f be a strictly increasing differentiable function on $[0, 1]$ such that $f(0) = 0$ and $f(1) = 1$, and let $g := f^{-1}$ denote its inverse. For each fixed n, we define the revision dates $t_i = t_i^n = g(i/n)$, $1, \dots, n$. The enlarged volatility now depends on t and is given by the formula

$$\widehat{\sigma}_t^2 = \sigma^2 + \sigma k_0 \sqrt{8/\pi}\sqrt{f'(t)}. \tag{1.4.4}$$

The pricing function

$$\widehat{C}(t,x) = E\big(xe^{\xi\rho_t - \frac{1}{2}\rho_t^2} - K\big)^+, \quad t \in [0,1],\ x > 0,$$

where

$$\rho_t^2 = \int_t^1 \widehat{\sigma}_s^2\, ds,$$

admits the explicit expression

$$\widehat{C}(t,x) = x\Phi\big(\rho_t^{-1}\ln(x/K) + \rho_t/2\big) - K\Phi\big(\rho_t^{-1}\ln(x/K) - \rho_t/2\big), \quad t < 1.$$

The function ρ_t decreases from ρ_0 to 0. The following bounds are obvious:

$$\sigma^2(1-t) \le \rho_t^2 \le \sigma^2(1-t) + \sigma k_0\sqrt{8/\pi}(1-t)^{1/2}\big(1-f(t)\big)^{1/2}.$$

Assumption 1. $g, f \in C^2([0,1])$.

Assumption 2. $g(t) = 1-(1-t)^\beta$, $\beta \ge 1$.

Note that in the second case where $f(t) = 1-(1-t)^{1/\beta}$, the derivative f' for $\beta > 1$ explodes at the maturity date, and so does the enlarged volatility. The notation $C^2([0,1])$ is used for the functions which are continuous with their two derivatives on the closed interval $[0,1]$.

Theorem 1.4.2 *Under any of the above assumptions, the mean-square approximation error for hedging the European call option has the following asymptotics:*

$$E\big(V_1^n - V_1\big)^2 = A_1(f)n^{-1} + o\big(n^{-1}\big), \quad n \to \infty, \tag{1.4.5}$$

where the coefficient

$$A_1(f) = \int_0^1 \left[\frac{\sigma^4}{2}\frac{1}{f'(t)} + k_0\sigma^3\sqrt{\frac{2}{\pi}}\frac{1}{\sqrt{f'(t)}} + k_0^2\sigma^2\left(1-\frac{2}{\pi}\right)\right]\Lambda_t\, dt \tag{1.4.6}$$

with $\Lambda_t = ES_t^4\widehat{C}_{xx}^2(t,S_t)$. *Explicitly,*

$$\Lambda_t = \frac{1}{2\pi\rho_t}\frac{K^2}{\sqrt{2\sigma^2 t + \rho_t^2}}\exp\left\{-\frac{(\ln\frac{S_0}{K} - \frac{1}{2}\sigma^2 t - \frac{1}{2}\rho_t^2)^2}{2\sigma^2 t + \rho_t^2}\right\}. \tag{1.4.7}$$

The case $f(t) = t$ corresponds to the model with the uniform grid and $A_1 = A_1(f)$.

We formulated Theorems 1.4.1 and 1.4.2 for convenience of references and because of their rather explicit formulae. Our main result is more general. It covers not only models with nonuniform grids but also gives the rate of convergence of the mean-square error in the problem of approximate hedging of options with pay-off function $G(x)$ satisfying the following hypothesis.

Assumption 3. *$G : \mathbf{R}_+ \to \mathbf{R}$ is a convex function such that $G|_{I_j} \in C^2(I_j)$, where the intervals $I_j := [K_{j-1}, K_j]$, $j = 1, \dots, N$, $I_{N+1} := [K_N, \infty[$ with $0 = K_0 < K_1 < \dots < K_N < \infty$, and $G''(x) \le \kappa(1 + x^m)$ for some constants $\kappa, M > 0$.*

The pricing function

$$\widehat{C}(t,x) = EG\big(xe^{\xi\rho_t - \frac{1}{2}\rho_t^2}\big), \quad t \in [0,1],\ x > 0, \tag{1.4.8}$$

solves the Cauchy problem

$$\widehat{C}_t(t,x) + \frac{1}{2}\widehat{\sigma}_t^2 x^2 \widehat{C}_{xx}(t,x) = 0, \qquad \widehat{C}(1,x) = G(x). \tag{1.4.9}$$

Theorem 1.4.3 *Suppose that for the scale function, one of Assumptions 1 or 2 is fulfilled and the pay-off function satisfies Assumption 3. Then*

$$E\big(V_1^n - V_1\big)^2 = A_1(f)n^{-1} + o(n^{-1}), \quad n \to \infty, \tag{1.4.10}$$

where $A_1(f)$ is given by the formula (1.4.6) with $\Lambda_t = ES_t^4\widehat{C}_{xx}^2(t,S_t)$.

This result makes plausible the conjecture that the normalized difference $n^{1/2}(V_1^n - V_1)$ converges in law. Indeed, this is the case; see further Theorem 1.5.1.

1.4.2 Preparatory Manipulations

First of all, we represent the deviation of the approximating portfolio from the pay-off in an integral form, which is instructive how to proceed further. This is an obvious extension of Lemma 1.2.2. There is only a minor difference: now $\widehat{\sigma}_t^2 - \sigma^2 = 2\sqrt{2/\pi}k_0\sqrt{f'(t)}$.

In the sequel we need to define a number of stochastic processes. Since the terminal date plays a particular role (we do not include the final transaction), they will be defined on the interval $[0,1[$ with an extension by continuity to its right extremity. With such a convention, the identity in the following lemma holds also for $s = 1$.

Lemma 1.4.4 *We have the representation $V_s^n - \widehat{V}_s = F_s^{1n} + F_s^{2n}$, $s \in [0,1[$, where $\widehat{V}_s = \widehat{C}(s,S_s)$,*

$$F_s^{1n} := \sigma \int_0^s S_t \sum_{i=1}^n \big(\widehat{C}_x(t_{i-1}, S_{t_{i-1}}) - \widehat{C}_x(t,S_t)\big) I_{[t_{i-1},t_i[}(t)\, dw_t,$$

$$F_s^{2n} := k_0\sqrt{\frac{2}{\pi}}\sigma \int_0^s S_t^2 \widehat{C}_{xx}(t,S_t)\sqrt{f'(t)}\, dt - \frac{k_0}{\sqrt{n}} \sum_{t_i \le s} |\Delta\widehat{C}_x(t_i)| S_{t_i},$$

with the abbreviation $\Delta\widehat{C}_x(t_i) := \widehat{C}_x(t_i, S_{t_i}) - \widehat{C}_x(t_{i-1}, S_{t_{i-1}})$.

Proof. Using the expression

$$V_s^n = \widehat{C}(0,S_0) + \int_0^s \sum_{i=1}^n \widehat{C}_x(t_{i-1},S_{t_{i-1}}) I_{]t_{i-1},t_i]}(u)\,dS_u - \frac{k_0}{\sqrt{n}} \sum_{t_i \le s} S_{t_i} |\Delta \widehat{C}_x(t_i)|$$

and applying the Itô formula to the increment $\widehat{C}(0,S_0) - \widehat{C}(s,S_s)$, we get that the difference $V_s^n - \widehat{V}_s$ is equal to

$$F_s^{1n} - \int_0^s \left(\widehat{C}_t(t,S_t) + \frac{1}{2}\sigma^2 S_t^2 \widehat{C}_{xx}(t,S_t) \right) dt - \frac{k_0}{\sqrt{n}} \sum_{t_i \le s} |\Delta \widehat{C}_x(t_i)| S_{t_i}.$$

Since $\widehat{C}(t,x)$ solves the Cauchy problem (1.4.9), the integrand above is equal to $(1/2)(\sigma^2 - \widehat{\sigma}_t^2) S_t^2 \widehat{C}_{xx}(t,S_t)$. We conclude by substituting the expression (1.4.4) for $\widehat{\sigma}_t^2$. □

Put (for $s \in [0,1[$)

$$M_s^{1n} := \frac{1}{2}\sigma^2 \sum_{t_i \le s} \widehat{C}_{xx}(t_{i-1},S_{t_{i-1}}) S_{t_{i-1}}^2 \left[\Delta t_i - (\Delta w_{t_i})^2\right],$$

$$M_s^{2n} := \sigma \frac{k_0}{\sqrt{n}} \sum_{t_i \le s} \widehat{C}_{xx}(t_{i-1},S_{t_{i-1}}) S_{t_{i-1}}^2 \left[\sqrt{2/\pi}\sqrt{\Delta t_i} - |\Delta w_{t_i}|\right],$$

where $\Delta t_i := t_i - t_{i-1}$ and $\Delta w_{t_i} := w_{t_i} - w_{t_{i-1}}$.

We introduce also two residual processes $R_s^{jn} := F_s^{jn} - M_s^{jn}$, $j = 1,2$.

Since $\widehat{V}_1 = G(S_1)$, Theorem 1.4.3 follows from the following assertions:

Proposition 1.4.5 $nE(M_1^{1n} + M_1^{2n} - A_1(f))^2 \to 0$ *as* $n \to \infty$.

Proposition 1.4.6 $nE\sup_{s \le 1} (R_s^{1n})^2 \to 0$ *as* $n \to \infty$.

Proposition 1.4.7 $nE\sup_{s \le 1} (R_s^{2n})^2 \to 0$ *as* $n \to \infty$.

Remark. In fact, to prove the theorem, it would be sufficient to show that $nE(R_1^{jn})^2 \to 0$. However, the stronger property claimed above happens to be useful in a study of more delicate results on the asymptotic behavior of the hedging error.

For a process $X = (X_t)$, we denote by X^* its maximal process. That is, $X_t^* = \sup_{u \le t} |X_u|$. In this (standard) notation the claims of Propositions 1.4.6 and 1.4.7 can be written as $n^{1/2} \|R_1^{jn*}\|_{L^2(\Omega)} \to 0$, $j = 1,2$.

Note that the sum in the expression for $F_1^{n2} = F_{1-}^{n2}$ does not include the term with $i = n$. Having in mind singularities of derivatives at the maturity, it is convenient to isolate the last summands also in other sums and treat them separately.

Now we analyze the expressions for F_s^{1n} and F_s^{1n} by applying the Taylor expansion of the first order to the differences $\widehat{C}_x(t_{i-1},S_{t_{i-1}}) - \widehat{C}_x(t,S_t)$ and

$\widehat{C}_x(t_i, S_{t_i}) - \widehat{C}_x(t_{i-1}, S_{t_{i-1}})$ at the point $(t_{i-1}, S_{t_{i-1}})$. A short inspection of the resulting formulae using the helpful heuristics $\Delta S_t \approx \sigma S_t \Delta w_t \approx \sigma S_t \sqrt{dt}$ reveals that the main contributions in the first-order Taylor approximations of increments originate from the derivatives of $\widehat{C}_x(t,x)$ in x. That is, the principal terms of asymptotics are

$$P_s^{1n} := \sigma \int_0^s \sum_{i=1}^{n-1} \widehat{C}_{xx}(t_{i-1}, S_{t_{i-1}}) S_{t_{i-1}}^2 (1 - S_t/S_{t_{i-1}}) S_t/S_{t_{i-1}} I_{[t_{i-1},t_i[}(t)\, dw_t,$$

$$P_s^{2n} := k_0 \sum_{t_i \le s} \widehat{C}_{xx}(t_{i-1}, S_{t_{i-1}}) S_{t_{i-1}}^2 \times \left[\sigma\sqrt{2/\pi}\sqrt{f'(t_{i-1})}\Delta t_i - |S_{t_i}/S_{t_{i-1}} - 1|/\sqrt{n}\right],$$

where $s \subset [0,1[$.

We write the first residual term $R_s^{1n} = (P_s^{1n} - M_s^{1n}) + (F_s^{1n} - P_s^{1n})$ in the following form:

$$R_s^{1n} := \left(R_s^{1Mn} + R_s^{1nn} - R_s^{1tn} - (1/2)\tilde{R}_s^{1n}\right)\sigma, \tag{1.4.11}$$

where

$$R_s^{1Mn} := (P_s^{1n} - M_s^{1n})/\sigma,$$

$$R_s^{1nn} := I_{[t_{n-1},t_n]}(s) \int_{t_{n-1}}^s \left(\widehat{C}_x(t_{n-1}, S_{t_{n-1}}) - \widehat{C}_x(t, S_t)\right) S_t\, dw_t,$$

$$R_s^{1tn} := \int_0^s \sum_{i=1}^{n-1} \widehat{C}_{xt}(t_{i-1}, S_{t_{i-1}})(t - t_{i-1}) S_t I_{[t_{i-1},t_i[}(t)\, dw_t,$$

$$\tilde{R}_s^{1n} := \int_0^s \sum_{i=1}^{n-1} \tilde{U}_t^i I_{[t_{i-1},t_i[}(t)\, dw_t$$

with

$$\tilde{U}_t^i = \widehat{C}_{xxx}(\tilde{t}_{i-1}, \tilde{S}_{t_{i-1}})(S_t - S_{t_{i-1}})^2 S_t + \widehat{C}_{xtt}(\tilde{t}_{i-1}, \tilde{S}_{t_{i-1}})(t - t_{i-1})^2 S_t + 2\widehat{C}_{xxt}(\tilde{t}_{i-1}, \tilde{S}_{t_{i-1}})(t - t_{i-1})(S_t - S_{t_{i-1}}) S_t.$$

The intermediate point $(\tilde{t}_{i-1}, \tilde{S}_{t_{i-1}})$ in the interval connecting $(t_{i-1}, S_{t_{i-1}})$ and (t_i, S_{t_i}) can be chosen in such a way that the mapping $\omega \mapsto (\tilde{t}_{i-1}, \tilde{S}_{t_{i-1}})$ is an $\mathcal{F}_{t_i}$-measurable random variable (for example, one cane take the first point on this interval for which the Taylor formula holds).

Notice that $\tilde{t}_{i-1} \in [t_{i-1}, t_i]$ and $\tilde{S}_{t_{i-1}} \in [S_{t_{i-1}}, S_t]$.

The structure of the above representation of R^{1n} is clear: the term R^{1nn} corresponds to the nth revision interval (it will be treated separately because of singularities at the left extremity of the time interval), the term R^{1tn} involving the first derivatives of $\widehat{C}_x$ in t at points $(t_{i-1}, S_{t_{i-1}})$ comes from the Taylor formula, and the "tilde" term is due to the remainder of the latter.

It is important to note that the integrals involving in the definition of P^{1n} depend only on the increments of the Wiener process on the intervals $[t_{i-1}, t_i]$ and, therefore, are independent on the σ-algebras $\mathcal{F}_{t_{i-1}}$. This helps to calculate the expectation of the squared sum: according to Lemma 1.4.16 below, it is the sum of expectations of the squared terms. We define P^{2n} in a way to enjoy the same property. The second residual term includes the term R^{2nn} corresponding to the last revision interval; the term R^{21n} represents the approximation error arising from replacement of the integral by the Riemann sum; we split the remaining part of the residual in a natural way into summands R^n_{22} and R^n_{23}. After these explanations we write the second residual term as follows:

We "telescope" the residual term $R^{2n}_s = (P^{2n}_s - M^{2n}_s) + (F^{2n}_s - P^{2n}_s)$ in the following way:

$$R^{2n}_s = \big(R^{2Mn}_s + \sigma\sqrt{2/\pi}R^{2nn}_s + \sigma\sqrt{2/\pi}R^{21n}_s + R^{22n}_s + R^{23n}_s + R^{24n}_s\big)k_0 \quad (1.4.12)$$

with

$$R^{2Mn}_s = \big(P^{2n}_s - M^{2n}_s\big)/k_0,$$

$$R^{2nn}_s = I_{[t_{n-1},t_n]}(s)\int_{t_{n-1}}^{s} S_t^2\widehat{C}_{xx}(t,S_t)\sqrt{f'(t)}\,dt,$$

$$R^{21n}_s = \sum_{i=1}^{n-1}\int_{t_{i-1}}^{t_i}\big(S_t^2\widehat{C}_{xx}(t,S_t)\sqrt{f'(t)} - S^2_{t_{i-1}}\widehat{C}_{xx}(t_{i-1},S_{t_{i-1}})\sqrt{f'(t_{i-1})}\big)\,dt,$$

$$R^{22n}_s = \frac{1}{\sqrt{n}}\sum_{t_i\le s}\widehat{C}_{xx}(t_{i-1},S_{t_{i-1}})|S_{t_{i-1}} - S_{t_i}|(S_{t_{i-1}} - S_{t_i}),$$

$$R^{23n}_s = \frac{1}{\sqrt{n}}\sum_{t_i\le s}[\ldots]_i(S_{t_i} - S_{t_{i-1}}),$$

$$R^{24n}_s = \frac{1}{\sqrt{n}}\sum_{t_i\le s}[\ldots]_i S_{t_{i-1}},$$

where

$$[\ldots]_i = \widehat{C}_{xx}(t_{i-1},S_{t_{i-1}})|S_{t_i} - S_{t_{i-1}}| - \big|\widehat{C}_x(t_i,S_{t_i}) - \widehat{C}_x(t_{i-1},S_{t_{i-1}})\big|. \quad (1.4.13)$$

1.4.3 Convenient Representations, Explicit Formulae, and Useful Bounds

1. Representations of derivatives in x. We consider the function $\widehat{C}(t,x)$ defined by the formula (1.4.8), i.e.,

$$\widehat{C}(t,x) = \int_{-\infty}^{\infty} G\big(xe^{\rho_t y - \frac{1}{2}\rho_t^2}\big)\varphi(y)\,dy. \quad (1.4.14)$$

To ensure that the integral is finite, we suppose that $G : \mathbf{R}_+ \to \mathbf{R}$ is of polynomial growth. We assume also that G is a convex function. Automatically, G, being locally Lipschitz, admits a positive Radon–Nikodym derivative G'. One can choose as G' the right derivative of G, which is increasing and has only a countable set of discontinuities.

Our aim is to get appropriate estimates of partial derivatives of $\widehat{C}(t,x)$. To this end we introduce the function

$$\bar{C}(x;\rho) = \int_{-\infty}^{\infty} G\big(xe^{\rho y - \frac{1}{2}\rho^2}\big)\varphi(y)\,dy. \tag{1.4.15}$$

Lemma 1.4.8 *Suppose that G is a convex function with Radon–Nikodym derivative G' of polynomial growth. Then for $n = 1,2,3,4$, we have the following representation:*

$$\frac{\partial^n}{\partial x^n}\bar{C}(x;\rho) = \frac{1}{\rho^{n-1}x^{n-1}}\int_{-\infty}^{\infty} G'\big(xe^{\rho y + \frac{1}{2}\rho^2}\big)P_{n-1}(y)\varphi(y)\,dy, \tag{1.4.16}$$

where

$$\begin{aligned}
P_0(y) &:= 1,\\
P_1(y) &:= y,\\
P_2(y) &:= y^2 - \rho y - 1,\\
P_3(y) &:= y^3 - 3\rho y^2 + \big(2\rho^2 - 3\big)y + 3\rho.
\end{aligned}$$

Proof. Let us introduce the function

$$\bar{G}(u;\rho) := G\big(e^{-\rho u - \frac{1}{2}\rho^2}\big), \quad u \in \mathbf{R},\ \rho > 0.$$

Recall that the convolution of $\bar{G}$ and φ is defined by the formula

$$\bar{G} * \varphi(z;\rho) := \int_{-\infty}^{\infty} \bar{G}(z-y;\rho)\varphi(y)\,dy.$$

The representation $\bar{C}(x;\rho) = \bar{G} * \varphi(-\rho^{-1}\ln x;\rho)$ allows us to calculate easily the derivatives in x.

Differentiating the convolution, we get that

$$\frac{\partial^n}{\partial z^n}\bar{G} * \varphi(z;\rho) = \bar{G} * \varphi^{(n)}(z;\rho) = \bar{G}' * \varphi^{(n-1)}(z;\rho).$$

Recalling that $\varphi^{(n)}(y) = (-1)^n H_n(y)\varphi(y)$, where H_n is the Hermite polynomial of order n, we obtain the representations

$$\frac{\partial^n}{\partial z^n}\bar{G} * \varphi(z;\rho) = (-1)^n\rho\int_{-\infty}^{\infty} G'\big(e^{-\rho(z-y)-\frac{1}{2}\rho^2}\big)e^{-\rho(z-y)-\frac{1}{2}\rho^2}H_{n-1}(y)\varphi(y)\,dy$$

and

$$\frac{\partial^n}{\partial z^n}\bar{G}*\varphi(-\rho^{-1}\ln x;\rho) = (-1)^n \rho x \int_{-\infty}^{\infty} G'\big(xe^{\rho y-\frac{1}{2}\rho^2}\big)e^{\rho y-\frac{1}{2}\rho^2}H_{n-1}(y)\varphi(y)\,dy.$$

Changing the variable, we rewrite the last formula as

$$\frac{\partial^n}{\partial z^n}\bar{G}*\varphi(-\rho^{-1}\ln x;\rho) = (-1)^n \rho x \int_{-\infty}^{\infty} G'\big(xe^{\rho y+\frac{1}{2}\rho^2}\big)H_{n-1}(y+\rho)\varphi(y)\,dy. \tag{1.4.17}$$

The first four derivatives of the function $f(g(.))$ at the point x are given by the formulae

$$\begin{aligned}
\frac{d}{dx}f(g(x)) &= f_g'(x)g'(x),\\
\frac{d^2}{dx^2}f(g(x)) &= f_g''(x)\big(g'(x)\big)^2 + f_g'(x)g''(x),\\
\frac{d^3}{dx^3}f(g(x)) &= f_g^{(3)}(x)\big(g'(x)\big)^3 + 3f_g''(x)g''(x)g'(x) + f_g'(x)g^{(3)}(x),\\
\frac{d^4}{dx^4}f(g(x)) &= f_g^{(4)}(x)\big(g'(x)\big)^4 + 6f_g^{(3)}(x)g''(x)\big(g'(x)\big)^2 + 3f_g''(x)\big(g''(x)\big)^2\\
&\quad + 4f_g''(x)g^{(3)}(x)g'(x) + f_g'(x)g^{(4)}(x),
\end{aligned}$$

where we use the abbreviations $f_g'(x) := f'(g(x))$, $f_g''(x) := f''(g(x))$, etc.

For the function $g(x) = -\rho^{-1}\ln x$, the mth derivative

$$g^{(m)}(x) = (-1)^m(m-1)!\rho^{-1}x^{-m}.$$

Applying the above formulae with $f = \bar{G}*\varphi$ and $f_g^{(n)}(x)$ given by the right-hand side of (1.4.17), we obtain the assertion of the lemma with

$$\begin{aligned}
P_0(y) &= H_0(y+\rho),\\
P_1(y) &= H_1(y+\rho) - \rho H_0(y+\rho),\\
P_2(y) &= H_2(y+\rho) - 3\rho H_1(y+\rho) + 2\rho^2 H_0(y+\rho),\\
P_3(y) &= H_3(y+\rho) - 6\rho H_2(y+\rho) + 11\rho^2 H_1(y+\rho) + 6\rho^3 H_0(y+\rho).
\end{aligned}$$

Since $H_0(y) = 1$, $H_1(y) = y$, $H_2(y) = y^2 - 1$, and $H_3(y) = y^3 - 3y$, these formulae can be rewritten as in the statement of the lemma. □

Remark. Using the well-known combinatorial formula for the nth derivative of $f(g(x))$ (see, e.g., Theorem III.21 in the textbook [208]), one can easily check that the representation (1.4.16) holds for each n with a certain polynomial P_{n-1} of two variables, y and ρ, of order $n-1$ and the coefficient at y^{n-1} equal to unity.

It follows from the above lemma and accompanying remark that in the case where $G'(x) \le \kappa(1+x^p)$,

$$\frac{\partial^n}{\partial x^n}\widehat{C}(t,x) \le \kappa_n \frac{(1+x^p)}{x^{n-1}(1-t)^{(n-1)/2}}. \tag{1.4.18}$$

In particular, if G' is bounded, we have that

$$\frac{\partial^n}{\partial x^n}\widehat{C}(t,x) \le \kappa_n \frac{1}{x^{n-1}(1-t)^{(n-1)/2}}. \tag{1.4.19}$$

Lemma 1.4.9 *Suppose that G is a convex function with Radon–Nikodym derivative G' of polynomial growth. Then $\widehat{C}_x(t,S_t) \to G'(S_1)$ as $t \to 1$ almost surely and in any $L^p(\Omega)$, $p < \infty$.*

Proof. From the representation (1.4.16) with $n = 1$ it follows that

$$\widehat{C}_x(t,S_t) = \int_{-\infty}^{\infty} G'\big(S_t e^{\rho_t y + \frac{1}{2}\rho_t^2}\big)\varphi(y)\,dy.$$

Since the distribution of S_1 is continuous, the set Ω_0 of ω for which $S_1(\omega)$ belongs to the (countable) set of discontinuities of G' has zero probability. Outside Ω_0 we apply to the integral the Lebesgue dominated convergence theorem using the assumption that G' has a polynomial growth. To get the convergence in $L^p(\Omega)$, we also apply the Lebesgue theorem but now to the expectation. Its condition holds because S_1^* is integrable in any power. □

2. Representations of mixed derivatives. Explicit formulae for derivatives involving the variable t are more cumbersome but also easy to obtain.

Let us define the operator T transforming the polynomial $P(y;\rho)$ into the polynomial

$$TP(y;\rho) = \big(yP(y;\rho) - P_y(y;\rho)\big)(y+\rho) + \rho P_\rho(y;\rho) - P(y;\rho).$$

Lemma 1.4.10 *Suppose that G is an increasing convex function with Radon–Nikodym derivative G' of polynomial growth. Then we have the following formulae:*

$$\begin{aligned}
\widehat{C}_t(t,x) &= -\frac{\widehat{\sigma}_t^2 x}{2\rho_t}\int_{-\infty}^{\infty} G'\big(xe^{\rho_t y + \frac{1}{2}\rho_t^2}\big)P_1(y)\varphi(y)\,dy,\\
\widehat{C}_{xt}(t,x) &= -\frac{\widehat{\sigma}_t^2}{2\rho_t^2}\int_{-\infty}^{\infty} G'\big(xe^{\rho_t y + \frac{1}{2}\rho_t^2}\big)TP_0(y;\rho_t)\varphi(y)\,dy,\\
\widehat{C}_{xxt}(t,x) &= -\frac{\widehat{\sigma}_t^2}{2\rho_t^3 x}\int_{-\infty}^{\infty} G'\big(xe^{\rho_t y + \frac{1}{2}\rho_t^3 x}\big)\big(TP_1(y;\rho_t) + P_1(y)\big)\varphi(y)\,dy,\\
\widehat{C}_{tt}(t,x) &= -\left(\frac{(\widehat{\sigma}_t^2)'}{2\rho_t} + \frac{\widehat{\sigma}_t^4}{4\rho_t^3}\right)x\int_{-\infty}^{\infty} G'\big(xe^{\rho_t y + \frac{1}{2}\rho_t^2}\big)P_1(y)\varphi(y)\,dy\\
&\quad + \frac{\widehat{\sigma}_t^4 x}{4\rho_t^3}\int_{-\infty}^{\infty} G'\big(xe^{\rho_t y + \frac{1}{2}\rho_t^2}\big)TP_1(y;\rho_t)(y)\varphi(y)\,dy,\\
\widehat{C}_{xtt}(t,x) &= -\left(\frac{(\widehat{\sigma}_t^2)'}{2\rho_t^2} + \frac{\widehat{\sigma}_t^4}{2\rho_t^4}\right)\int_{-\infty}^{\infty} G'\big(xe^{\rho_t y + \frac{1}{2}\rho_t^2}\big)TP_0(y;\rho_t)\varphi(y)\,dy\\
&\quad + \frac{\widehat{\sigma}_t^4}{2\rho_t^4}\int_{-\infty}^{\infty} G'\big(xe^{\rho_t y + \frac{1}{2}\rho_t^2}\big)T^2P_0(y;\rho_t)\varphi(y)\,dy,
\end{aligned}$$

where $P_j(y)$ are polynomials defined in Lemma 1.4.16. In accordance to the definition of the operator T,

$$TP_0(y) = y^2 + \rho y - 1,$$
$$TP_1(y) = y^3 + \rho y^2 - 2y - \rho.$$

Proof. Differentiating under the sign of integral in (1.4.15) and making a linear change of variables, we obtain the representation

$$\bar{C}_\rho(x;\rho) = x\int_{-\infty}^{\infty} G'\left(xe^{\rho y+\frac{1}{2}\rho^2}\right) y\varphi(y)\,dy.$$

Since $\rho_t' = -\sigma_t^2/(2\rho_t)$, the formula for $\widehat{C}_t(t,x)$ follows obviously.

Using the change of variable

$$z(y;x,\rho) = xe^{\rho y+\frac{1}{2}\rho^2}$$

with the inverse

$$y(z;x,\rho) = \frac{1}{\rho}\ln\frac{z}{x} - \frac{1}{2}\rho$$

and differentiating under the sign of integral, we get that

$$\frac{\partial}{\partial\rho}\int_{-\infty}^{\infty} G'\left(xe^{\rho y+\frac{1}{2}\rho^2}\right) P(y;\rho)\varphi(y)\,dy = \frac{1}{\rho}\int_{-\infty}^{\infty} G'\left(xe^{\rho y+\frac{1}{2}\rho^2}\right) TP(y;\rho)\varphi(y)\,dy.$$

This identity helps us to derive the formulae for $\widehat{C}_{xt}(t,x)$ and $\widehat{C}_{xxt}(t,x)$ from the representation (1.4.16) and also get the formulae for $\widehat{C}_{tt}(t,x)$ and $\widehat{C}_{xtt}(t,x)$ by the differentiation of those for $\widehat{C}_t(t,x)$ and $\widehat{C}_{xt}(t,x)$. □

From the above lemma we have the following bounds:

Lemma 1.4.11 *Suppose that one of Assumptions 1 or 2 is fulfilled and G' has a polynomial growth. Then*

$$\left|\widehat{C}_{xt}(t,x)\right| \le \kappa\frac{1}{1-t}\left(1+x^m\right), \tag{1.4.20}$$

$$\left|\widehat{C}_{xxt}(t,x)\right| \le \kappa\frac{1}{(1-t)^{3/2}}\frac{1}{x}\left(1+x^m\right), \tag{1.4.21}$$

$$\left|\widehat{C}_{xtt}(t,x)\right| \le \kappa\frac{1}{(1-t)^2}\left(1+x^m\right). \tag{1.4.22}$$

Proof. Under the Assumption 1, both $\widehat{\sigma}_t^2$ and $|(\widehat{\sigma}_t^2)'| = \kappa|f''(t)|/\sqrt{f'(t)}$ are bounded, and the statement is obvious. Under Assumption 2, i.e., when $f(t) = 1-(1-t)^{1/\beta}$, $\beta > 1$, direct calculations lead to the bounds

$$\frac{\widehat{\sigma}_t^2}{\rho_t^2} \le \kappa\frac{1}{1-t}, \qquad \frac{|(\widehat{\sigma}_t^2)'|}{\rho_t^2} \le \kappa\frac{1}{(1-t)^2}, \tag{1.4.23}$$

implying the required estimates. □

3. Sharper estimates of partial derivatives. For our analysis, we need also more precise estimates requiring further hypotheses on G.

Put

$$\Sigma_N(x,\rho) := \sum_{j=1}^{N} \exp\left\{-\frac{1}{4}\frac{\ln^2(K_j/x)}{\rho^2}\right\} \tag{1.4.24}$$

with the convention $\Sigma_0(x,\rho) := 0$.

Lemma 1.4.12 *Under Assumption 3, there is a constant κ such that, for any $\rho \in [0,\sigma]$,*

$$0 \le \bar{C}_{xx}(x,\rho) \le \kappa \frac{1}{\rho x^{3/2}} \Sigma_N(x,\rho) + \kappa(1+x^m). \tag{1.4.25}$$

Proof. Put

$$\delta_j := \frac{1}{\rho}\ln\frac{K_j}{x} - \frac{1}{2}\rho.$$

Integrating by parts on the closed intervals with the extremities δ_j, we obtain that

$$\int_0^\infty G'\big(xe^{\rho y+\frac{1}{2}\rho^2}\big) y\varphi(y)\,dy = -\sum_{j=0}^{N} G'\big(xe^{\rho y+\frac{1}{2}\rho^2}\big)\varphi(y)\Big|_{\delta_j+}^{\delta_{j+1}-}$$
$$+ \rho x \int_{-\infty}^{\infty} G''\big(xe^{\rho y+\frac{1}{2}\rho^2}\big) e^{\rho y+\frac{1}{2}\rho^2}\varphi(y)\,dy.$$

Clearly,

$$-\sum_{j=0}^{N} G'\big(xe^{\rho y+\frac{1}{2}\rho^2}\big)\varphi(y)\Big|_{\delta_j+}^{\delta_{j+1}-} = \sum_{j=1}^{N}\big(G'(K_j+) - G'(K_j-)\big)\varphi(\delta_j).$$

Due to the assumed convexity of G, the summands in the right-hand side are positive and dominated by

$$G'(K_N+)\varphi(\delta_j) = G'(K_N+)\frac{1}{\sqrt{2\pi}}e^{-\rho^2/8}\frac{K_j^{1/2}}{x^{1/2}}\exp\left\{-\frac{1}{2}\frac{\ln^2(K_j/x)}{\rho^2}\right\}.$$

Due to the polynomial-growth condition on G'' in Assumption 3,

$$\int_{-\infty}^{\infty} G''\big(xe^{\rho y+\frac{1}{2}\rho^2}\big)e^{\rho y+\frac{1}{2}\rho^2}\varphi(y)\,dy \le \kappa(1+x^m).$$

Combining the above estimates, we infer that

$$0 \le \int_0^\infty G'\big(xe^{\rho y+\frac{1}{2}\rho^2}\big)y\varphi(y)\,dy$$
$$\le \frac{\kappa}{x^{1/2}}\sum_{j=1}^{N}\exp\left\{-\frac{1}{2}\frac{\ln^2(K_j/x)}{\rho^2}\right\} + \kappa\rho x(1+x^m).$$

The claim follows now from the representation (1.4.16) for $n=2$. □

Lemma 1.4.13 *Under Assumption 3, there is a constant κ such that, for any $\rho \in]0,\sigma]$,*

$$\left|\bar{C}_{xxx}(x,\rho)\right| \le \kappa \frac{1}{\rho^2 x^{9/4}} \Sigma_N(x,\rho) + \kappa \frac{1}{\rho x}\left(1+x^m\right),$$
$$\left|\bar{C}_{xxxx}(x,\rho)\right| \le \kappa \frac{1}{\rho^3 x^{13/4}} \Sigma_N(x,\rho) + \kappa \frac{1}{\rho^2 x^2}\left(1+x^m\right).$$

Proof. Let $Q_n(y)$ be a polynomial the coefficients of which are functions of ρ bounded on $[0,\sigma]$. Then there exists a constant κ such that

$$\left|\int_0^\infty G'\left(xe^{\rho y+\frac{1}{2}\rho^2}\right)Q_n(y)\varphi(y)\,dy\right| \le \kappa \frac{1}{x^{1/4}}\Sigma_N(x,\rho) + \kappa\rho x\left(1+x^m\right). \quad (1.4.26)$$

By virtue of the representation (1.4.16), the bounds of the lemma immediately follow from the above inequality. To prove the latter, we first consider the case where $Q_n(y)$ is $H_n(y)$, the Hermite polynomial of order n. We argue in the same way as in the proof Lemma 1.4.12.

Taking into account that $\varphi^{(n)}(y) = (-1)^n H_n(y)\varphi(y)$, we obtain, using the integration by parts, that the left-hand side of (1.4.26) is dominated by

$$G'(K_N+)\sum_{j=1}^N \left|\varphi^{(n-1)}(\delta_j)\right| + \rho x \int_{-\infty}^\infty G''\left(xe^{\rho y+\frac{1}{2}\rho^2}\right)e^{\rho y+\frac{1}{2}\rho^2}\left|\varphi^{(n-1)}(y)\right|dy.$$

There is a constant κ_n such that $|\varphi^{(n-1)}(y)| \le \kappa_n\varphi(y/2)$ for all y. In particular,

$$\left|\varphi^{(n-1)}(\delta_j)\right| \le \kappa_n \frac{1}{\sqrt{2\pi}} e^{-\rho^2/16}\frac{{K_j}^{1/4}}{x^{1/4}}\exp\left\{-\frac{1}{4}\frac{\ln^2(K_j/x)}{\rho^2}\right\}.$$

This leads to the inequality (1.4.26) for the Hermite polynomials. The Hermite polynomials $H_n(y)$ form a basis in the linear space of polynomials in y. It follows that the inequality holds also for $Q_n(y) = y^n$ and, hence, for any polynomial the coefficients of which are functions of ρ bounded on $[0,\sigma]$. □

Using the estimate (1.4.26) we obtain from Lemma 1.4.10 the following:

Lemma 1.4.14 *Under Assumption 3 on the pay-off function G, there is a constant κ such that, for any $t \in [0,1[$,*

$$\left|\widehat{C}_t(t,x)\right| \le \kappa \frac{\widehat{\sigma}_t^2 x^{3/4}}{\rho_t}\Sigma_N(x,\rho_t) + \kappa\widehat{\sigma}_t^2 x^2\left(1+x^m\right),$$
$$\left|\widehat{C}_{xt}(t,x)\right| \le \kappa \frac{\widehat{\sigma}_t^2}{\rho_t^2 x^{1/4}}\Sigma_N(x,\rho_t) + \kappa\frac{\widehat{\sigma}_t^2}{\rho_t}x\left(1+x^m\right),$$
$$\left|\widehat{C}_{xxt}(t,x)\right| \le \kappa \frac{\widehat{\sigma}_t^2}{\rho_t^3 x^{7/4}}\Sigma_N(x,\rho_t) + \kappa\frac{\widehat{\sigma}_t^2}{\rho_t^2}\left(1+x^m\right),$$

$$|\widehat{C}_{tt}(t,x)| \le \kappa\left(\frac{(\widehat{\sigma}_t^2)'}{2\rho_t} + \frac{\widehat{\sigma}_t^4}{4\rho_t^3}\right)\left(x^{3/4}\Sigma_N(x,\rho_t) + \rho_t x^2(1+x^m)\right),$$
$$|\widehat{C}_{xtt}(t,x)| \le \kappa\left(\frac{(\widehat{\sigma}_t^2)'}{2\rho_t} + \frac{\widehat{\sigma}_t^4}{2\rho_t^4}\right)\left(\frac{1}{x^{1/4}}\Sigma_N(x,\rho_t) + \rho_t x(1+x^m)\right).$$

4. Call option: explicit formulae. For the classical call option with $G(x) = (x-K)^+$, the derivatives we need can be given explicitly. In particular,

$$\widehat{C}_x(t,x) = \Phi(\widehat{d}(t,x)),$$
$$\widehat{C}_{xx}(t,x) = \frac{1}{x\rho_t}\varphi(\widehat{d}(t,x)),$$

where

$$\widehat{d}(t,x) := \frac{1}{\rho_t}\ln\frac{x}{K} + \frac{1}{2}\rho_t. \tag{1.4.27}$$

To get the expression for the function $\Lambda_t = ES_t^4\widehat{C}_{xx}^2(t,S_t)$ from Theorem 1.4.5, we use the following easily verified formula.

Let $\xi \in \mathcal{N}(0,1)$, and let $a \ne 0$, b, c be arbitrary constants. Then

$$Ee^{c\xi}e^{-(a\xi+b)^2} = \frac{1}{\sqrt{2a^2+1}}\exp\left\{-\frac{\tilde{b}^2}{2a^2+1} + \tilde{b}^2 - b^2\right\}, \tag{1.4.28}$$

where $\tilde{b} := b - c/(2a)$.

The distribution of the random variable $2\pi S_t^p\widehat{C}_{xx}^2(t,S_t)$ is the same as that of

$$S_0^{p-2}e^{-\frac{1}{2}(p-2)\sigma^2 t}\rho_t^{-2}e^{c_t\xi}e^{-(a_t\xi+b_t)^2},$$

where $c_t = (p-2)\sigma t^{1/2}$, $a_t = \frac{1}{\rho_t}\sigma t^{1/2}$,

$$b_t = \frac{1}{\rho_t}\left(\ln\frac{S_0}{K} - \frac{1}{2}\sigma^2 t\right) + \frac{1}{2}\rho_t, \qquad \tilde{b}_t = b_t - \frac{1}{2}(p-2)\rho_t.$$

Since

$$\tilde{b}_t^2 - b_t^2 = -(p-2)\left[\left(\ln\frac{S_0}{K} - \frac{1}{2}\sigma^2 t\right) + \rho_t^2 - \frac{1}{4}p\rho_t^2\right],$$

we obtain from the above that

$$ES_t^p\widehat{C}_{xx}^2(t,S_t) = \frac{1}{2\pi\rho_t}\frac{K^{p-2}}{\sqrt{2\sigma^2 t + \rho_t^2}}e^{-B_t}, \tag{1.4.29}$$

where

$$B_t := \frac{(\ln\frac{S_0}{K} - \frac{1}{2}\sigma^2 t - \frac{1}{2}(p-3)\rho_t^2)^2}{2\sigma^2 t + \rho_t^2} - \frac{(p-2)(p-4)}{4}\rho_t^2. \tag{1.4.30}$$

In particular, with $p = 4$, we have

$$\Lambda_t = \frac{1}{2\pi\rho_t} \frac{K^2}{\sqrt{2\sigma^2 t + \rho_t^2}} \exp\left\{ -\frac{(\ln \frac{S_0}{K} - \frac{1}{2}\sigma^2 t - \frac{1}{2}\rho_t^2)^2}{2\sigma^2 t + \rho_t^2} \right\}. \tag{1.4.31}$$

5. Bounds for expectations. Using (1.4.28), we obtain from Lemmas 1.4.12–1.4.14 and (1.4.23) the bounds which will be used in the proof of Theorem 1.4.3.

Lemma 1.4.15 *Suppose that one of Assumptions 1 or 2 is fulfilled and G satisfies Assumption 3. Then*

$$ES_t^p \widehat{C}_{xx}^{2m}(t, S_t) \le \kappa \frac{1}{(1-t)^{m-1/2}}, \tag{1.4.32}$$

$$ES_t^p \widehat{C}_{xt}^{2m}(t, S_t) \le \kappa \frac{1}{(1-t)^{2m-1/2}}, \tag{1.4.33}$$

$$ES_t^p \widehat{C}_{xxx}^{2m}(t, S_t) \le \kappa \frac{1}{(1-t)^{2m-1/2}}, \tag{1.4.34}$$

$$ES_t^p \widehat{C}_{xxt}^{2m}(t, S_t) \le \kappa \frac{1}{(1-t)^{3m-1/2}}, \tag{1.4.35}$$

$$ES_t^p \widehat{C}_{xxxx}^{2m}(t, S_t) \le \kappa \frac{1}{(1-t)^{3m-1/2}}, \tag{1.4.36}$$

where the constant κ depends on p and m. In particular,

$$\Lambda_t \le \kappa \frac{1}{\sqrt{1-t}}. \tag{1.4.37}$$

1.4.4 Tools

In our computations we shall frequently use the following two assertions. The first one is a standard fact on discrete-time square-integrable martingales.

Lemma 1.4.16 *Let $M = (M_i)$ be a square-integrable martingale with respect to a filtration $(\mathcal{G}_i)$, $i = 0, \dots, k$, and let $X = (X_i)$ be a predictable process with $EX^2 \cdot \langle M \rangle_k < \infty$. Then*

$$E(X \cdot M_k)^2 = EX^2 \cdot \langle M \rangle_k = \sum_{i=1}^{k} EX_i^2 (\Delta M_i)^2,$$

where, as usual, $\Delta \langle M \rangle_i := E((\Delta M_i)^2 | \mathcal{G}_{i-1})$,

$$X \cdot M_k := \sum_{i=1}^{k} X_i \Delta M_i, \qquad X^2 \cdot \langle M \rangle_k := \sum_{i=1}^{k} X_i^2 \langle M \rangle_i.$$

Lemma 1.4.17 *Suppose that $g', f' \in C([0,1])$. Let $p > 0$ and $a \geq 0$. Then*

$$\sum_{i=1}^{n-1} \frac{(\Delta t_i)^{p+a}}{(1-t_i)^p} = \begin{cases} O(n^{1-p-a}), & p < 1, \\ O(n^{-a} \ln n), & p = 1, \\ O(n^{-a}), & p > 1. \end{cases}$$

If $g(t) = 1 - (1-t)^\beta$, $\beta \geq 1$, then

$$\sum_{i=1}^{n-1} \frac{(\Delta t_i)^{p+a}}{(1-t_i)^p} = \begin{cases} O(n^{1-p-a}), & p < 1 + a(\beta - 1), \\ O(n^{-a\beta} \ln n), & p = 1 + a(\beta - 1), \\ O(n^{-a\beta}), & p > 1 + a(\beta - 1). \end{cases}$$

Proof. We first consider the case where $g', f' \in C([0,1])$, i.e., g' is not only bounded but also bounded away from zero. By the finite increment formula $\Delta t_i = g'(x_i)n^{-1}$, where $x_i \in [(i-1)/n, i/n]$, and, hence, $\Delta t_i \leq \text{const}\, n^{-1}$. Applying again the finite increment formula and taking into account that $\min g'(t) > 0$, it is easy to check that there is a constant c such that

$$\frac{1 - t_{i-1}}{1 - t_i} \leq c, \quad 1 \leq i \leq n-1.$$

Thus,

$$\sum_{i=1}^{n-1} \frac{\Delta t_i}{(1-t_i)^p} \leq c \sum_{i=1}^{n-1} \frac{\Delta t_i}{(1-t_{i-1})^p} \leq c \int_0^{t_{n-1}} \frac{dt}{(1-t)^p}.$$

Since

$$n^{-1} \min g'(t) \leq 1 - g(1 - 1/n) \leq n^{-1} \max g'(t),$$

the asymptotics of the last integral is $O(1)$ if $p < 1$ (the integral converges), $O(\ln n)$ if $p = 1$, and $O(n^{p-1})$ if $p > 1$. This implies the claimed property.

In the second case where $g(t) = 1 - (1-t)^\beta$, $\beta \geq 1$, we have

$$\sum_{i=1}^{n-1} \frac{(\Delta t_i)^{p+a}}{(1-t_i)^p} = \frac{\beta^{p+a}}{n^{p-1+a}} \sum_{i=1}^{n-1} \frac{(1-x_i)^{(\beta-1)(p+a)}}{(1-i/n)^{\beta p}} \frac{1}{n}.$$

The sum in the right-hand side is dominated, up to a multiplicative constant, by

$$\sum_{i=1}^{n-1} \frac{1}{(1-(i-1)/n)^{p+a-\beta a}} \frac{1}{n} \leq \int_0^{1-1/n} \frac{dt}{(1-t)^{p+a-\beta a}}.$$

Using the explicit formulae for the integral, we infer that the required property holds whatever are the parameters $p > 0$, $a \geq 0$, and $\beta \geq 1$. □

1.4.5 Analysis of the Principal Terms: Proof of Proposition 1.4.5

Since $E(M_1^{1n} + M_1^{2n}) = 0$, we need to verify that $nE(M_1^{1n} + M_1^{2n})^2 \to A_1(f)$ as $n \to \infty$.

Recall that $E(\xi^2-1)^2=2$ and $E|\xi|^3=2E|\xi|=2\sqrt{2/\pi}$ for $\xi\in\mathcal{N}(0,1)$. Using Lemma 1.4.16, we obtain the representation

$$nE\big(M_1^{1n}+M_1^{2n}\big)^2=\frac{\sigma^4}{2}n\sum_{i=1}^{n-1}\Lambda_{t_{i-1}}(\Delta t_i)^2+k_0\sigma^3\sqrt{\frac{2}{\pi}}n^{1/2}\sum_{i=1}^{n-1}\Lambda_{t_{i-1}}(\Delta t_i)^{3/2}$$
$$+k_0^2\sigma^2\left(1-\frac{2}{\pi}\right)\sum_{i=1}^{n-1}\Lambda_{t_{i-1}}\Delta t_i.$$

By the finite increment formula $\Delta t_i=g(i/n)-g((i-1)/n)=g'(x_i)/n$, where $x_i\in[(i-1)/n,i/n]$. We substitute this expression into the sums above. Let us introduce the function F_n (depending on p) by the formula

$$F_n(t):=\sum_{i=1}^{n-1}\Lambda_{g((i-1)/n)}\big[g'(x_i)\big]^pI_{[(i-1)/n,i/n[}(t).$$

For $p\ge 1$, we have

$$\sum_{i=1}^{n-1}\Lambda_{g((i-1)/n)}\big[g'(x_i)\big]^p\frac{1}{n}=\int_0^1F_n(t)\,dt\to\int_0^1\Lambda_{g(t)}\big[g'(t)\big]^p\,dt.$$

The needed uniform integrability of the sequence $\{F_n\}$ with respect to the Lebesgue measure follows from the de la Vallée-Poussin criterion because the estimate $\Lambda_t\le\kappa(1-t)^{-1/2}$ and the boundedness of g' imply that

$$\int_0^1F_n^{3/2}(t)\,dt\le\text{const}\int_0^1\frac{dg(t)}{(1-g(t))^{3/4}}=\text{const}\int_0^1\frac{ds}{(1-s)^{3/4}}<\infty.$$

By the change of variable, taking into account that $g'(t)=1/f'(g(t))$, we transform the limiting integral into the form used in the formulations of the theorem:

$$\int_0^1\Lambda_{g(t)}\big[g'(t)\big]^p\,dt=\int_0^1\Lambda_{g(t)}\big[g'(t)\big]^{p-1}\,dg(t)=\int_0^1\Lambda_t\big[f'(t)\big]^{1-p}\,dt.$$

The claimed property on the convergence of $n^{1/2}(M_1^{1n}+M_1^{2n})$ to $A_1(f)$ in L^2-norm is verified.

1.4.6 Analysis of the Residual R^{1n}

In this subsection we give a proof of Proposition 1.4.6.

1. To check the convergence of the sequence $n^{1/2}R_1^{1Mn*}$ to zero in L^2, it is convenient to introduce the "intermediate" process

$$\bar M_s^{1n}:=\sigma^2\int_0^s\sum_{i=1}^{n-1}\widehat C_{xx}(t_{i-1},S_{t_{i-1}})S_{t_{i-1}}^2(w_{t_{i-1}}-w_t)I_{[t_{i-1},t_i[}(t)\,dw_t.$$

The difference $P^{1n} - \bar{M}^{1n}$ is a square-integrable martingale, and

$$E(P_1^{1n} - \bar{M}_1^{1n})^2 = \sigma^2 \sum_{i=1}^{n-1} \Lambda_{t_{i-1}} \int_{t_{i-1}}^{t_i} E\left[\left(\frac{S_t}{S_{t_{i-1}}} - 1\right)\frac{S_t}{S_{t_{i-1}}} - \sigma(w_t - w_{t_{i-1}})\right]^2 dt.$$

It is a simple exercise to check that

$$E\big((e^{u\xi - \frac{1}{2}u^2} - 1)e^{u\xi - \frac{1}{2}u^2} - u\xi\big)^2 = O(u^4), \quad u \to 0.$$

Hence, we can dominate the expectations in the integrals by a quadratic function and obtain that

$$nE(P_1^{1n} - \bar{M}_1^{1n})^2 \le \text{const}\, n \sum_{i=1}^{n-1} \Lambda_{t_{i-1}} (\Delta t_i)^3 \to 0, \quad n \to \infty.$$

By virtue of the Doob inequality, also $nE\sup_s(P_s^{1n} - \bar{M}_s^{1n})^2 \to 0$.

Note that $\bar{M}_{t_{i-1}}^{1n} = M_{t_{i-1}}^{1n}$, the process M^{1n} is constant on the interval $[t_{i-1}, t_i[$, while

$$\bar{M}_s^{1n} - \bar{M}_{t_{i-1}}^{1n} = \sigma^2 \widehat{C}_{xx}(t_{i-1}, S_{t_{i-1}}) S_{t_{i-1}}^2 \int_{t_{i-1}}^{s} (w_{t_{i-1}} - w_t)\, dw_t.$$

It follows that

$$n \sup_s (M_s^{1n} - \bar{M}_s^{1n})^2 = n\sigma^4 \max_{i \le n-1} \widehat{C}_{xx}^2(t_{i-1}, S_{t_{i-1}}) S_{t_{i-1}}^4 \eta_i^2,$$

where

$$\frac{1}{2} \sup_{s \in [t_{i-1}, t_i]} \big|(w_s - w_{t_{i-1}})^2 - (s - t_{i-i})\big|.$$

Let $m \in]1, 3/2[$. Using the elementary inequality $\max_i |a_i| \le \sum_i |a_i|$, the independence of increments of the Wiener process from the past, the bound (1.4.32), and the estimate $E|\eta_i|^{2m} \le \kappa(\Delta t_i)^{2m}$, we obtain that

$$n^m E \sup_s (M_s^{1n} - \bar{M}_s^{1n})^{2m} \le \kappa n^m \sum_{i=1}^{n-1} \frac{(\Delta t_i)^{2m}}{(1 - t_{i-1})^{m-1/2}} = O(n^{1-m}), \quad n \to \infty.$$

The sequence $n\sup_s(M_s^{1n} - \bar{M}_s^{1n})^2$ converges to zero in L^m and, hence, in L^1.

Summarizing, we conclude that $n^{1/2}\|R_1^{1Mn*}\|_{L^2} \to 0$ as $n \to \infty$.

2. The residual process R^{1nn} is a martingale, and by the Doob inequality $E(R_1^{1nn*})^2 \le 4E(R_1^{1nn})^2$. We have

$$E(R_1^{1nn})^2 = \int_{t_{n-1}}^{1} E\big(\widehat{C}_x(t_{n-1}, S_{t_{n-1}}) - \widehat{C}_x(t, S_t)\big)^2 S_t^2\, dt \le \kappa_n (1 - t_{n-1}),$$

where κ_n is the supremum of the integrand over $[t_{n-1}, 1]$. By virtue of Lemma 1.4.9, $\kappa_n \to 0$. Since $1 - t_{n-1} \le \kappa n^{-1}$ (due to the boundedness of g'), we conclude that $nE(R_1^{1nn})^2 \to 0$.

3. By the Doob inequality asymptotic analysis of the sequence R_1^{1tn*} can be reduced to that of

$$R_1^{1tn} = \sum_{i=1}^{n-1} \widehat{C}_{xt}(t_{i-1}, S_{t_{i-1}}) \int_{t_{i-1}}^{t_i} (t - t_{i-1}) S_t \, dw_t.$$

According to (1.4.33),

$$E\widehat{C}_{xt}^2(t, S_t) S_t^2 \le \kappa \frac{1}{(1-t)^{3/2}}.$$

Therefore,

$$\begin{aligned} E\big(R_1^{1tn}\big)^2 &= \sum_{i=1}^{n-1} E\widehat{C}_{xt}^2(t_{i-1}, S_{t_{i-1}}) S_{t_{i-1}}^2 \int_{t_{i-1}}^{t_i} (t - t_{i-1})^2 E(S_t/S_{t_{i-1}})^2 \, dt \\ &\le \text{const} \sum_{i=1}^{n-1} \frac{(\Delta t_i)^3}{(1 - t_{i-1})^{3/2}} = O\big(n^{-3/2}\big), \quad n \to \infty, \end{aligned}$$

in virtue of Lemma 1.4.17. Hence, $E(R_1^{1tn})^2 \to 0$.

4. Now we estimate the expectation $E(\tilde{R}_1^{1n})^2$ corresponding to the terminal value of the martingale arising from the remainder term in the Taylor formula for $\widehat{C}_x$. We have

$$E\big(\tilde{R}_1^{1n}\big)^2 = \sum_{i=1}^{n-1} \int_{t_{i-1}}^{t_i} E\big(\tilde{U}_t^i\big)^2 \, dt.$$

Since $(a+b+c)^2 \le 3(a^2+b^2+c^2)$, it is sufficient to check that each of the following sums converge to zero as $o(n^{-1})$:

$$\begin{aligned} \Sigma_1^n &= \sum_{i=1}^{n-1} \int_{t_{i-1}}^{t_i} E\widehat{C}_{xxx}^2(\tilde{t}_{i-1}, \tilde{S}_{t_{i-1}}) (S_t - S_{t_{i-1}})^4 S_t^2 \, dt, \\ \Sigma_2^n &:= \sum_{i=1}^{n-1} \int_{t_{i-1}}^{t_i} E\widehat{C}_{xtt}^2(\tilde{t}_{i-1}, \tilde{S}_{t_{i-1}}) (t - t_{i-1})^4 S_t^2 \, dt, \\ \Sigma_3^n &:= \sum_{i=1}^{n-1} \int_{t_{i-1}}^{t_i} E\widehat{C}_{xxt}^2(\tilde{t}_{i-1}, \tilde{S}_{t_{i-1}}) (t - t_{i-1})^2 (S_t - S_{t_{i-1}})^2 S_t^2 \, dt. \end{aligned}$$

Using the continuity of the process S_t, we obtain from the formula (1.4.24) that

$$\lim_{t \to 1} \sup_{r \ge t} \Sigma_N(S_r, \rho_t) = 0 \quad \text{a.s.}$$

Applying Lemma 1.4.13, we infer that for any $\varepsilon > 0$ and $m \ge 1$, there exists $a \in]0,1[$ such that

$$E\big|\widehat{C}_{xxx}(\tilde{t}_{i-1}, \tilde{S}_{t_{i-1}})\big|^{2m} \le \varepsilon \frac{1}{(1-t_i)^{2m}} \tag{1.4.38}$$

for every $t_{i-1} \geq a$. For $t_{i-1} < a$, the above expectation is bounded by a constant which does not depend on n.

Let $\xi \sim N(0,1)$, and let $b \in [0,1]$. Using the elementary bound

$$\left|e^{bx} - 1\right| \leq b\left(e^{|x|} - 1\right),$$

which follows from the Taylor expansion, we obtain, for $m \geq 1$, the estimate

$$E\left(e^{u\sigma\xi-(1/2)\sigma^2u^2} - 1\right)^{2m} \leq \kappa u^{2m},$$

where the constant κ depends on m and σ. Applying the Cauchy–Schwarz inequality and this estimate, we get that

$$E(S_t - S_{t_{i-1}})^{2m} S_t^p \leq \kappa(t - t_{i-1})^m.$$

Manipulating again with the Cauchy–Schwarz inequality, we obtain with the help of the above bounds that

$$\Sigma_1^n \leq \kappa \sum_{t_{i-1}<a} (\Delta t_i)^3 + \kappa\varepsilon \sum_{i=1}^{n-1} \frac{(\Delta t_i)^3}{(1-t_i)^2}.$$

The first sum in the right-hand side is of order $O(n^{-2})$. According to Lemma 1.4.17, the second one is of order $O(n^{-1})$. Since $\varepsilon > 0$ is arbitrary, it follows that $\lim_n n\Sigma_1^n = 0$.

Similarly to the bound (1.4.38) but referring now to Lemma 1.4.14, we can establish that for any $\varepsilon > 0$, there is a threshold $a \in]0,1]$ such that for any $t_{i-1} \geq a$, the following inequalities hold:

$$E\left|\widehat{C}_{xxt}(\tilde{t}_{i-1}, \tilde{S}_{t_{i-1}})\right|^{2m} \leq \varepsilon \frac{1}{(1-t_i)^{3m}} \tag{1.4.39}$$

and

$$E\left|\widehat{C}_{xtt}(\tilde{t}_{i-1}, \tilde{S}_{t_{i-1}})\right|^{2m} \leq \varepsilon \frac{1}{(1-t_i)^{4m}}. \tag{1.4.40}$$

With these bounds, we prove, making obvious changes in arguments, that $\lim_n n\Sigma_2^n = 0$ and $\lim_n n\Sigma_3^n = 0$. Thus, $nE(\tilde{R}_1^{1n*})^2 \to 0$.

1.4.7 Analysis of the Residual R^{2n}

Now we give a proof of Proposition 1.4.7.

1. Put (for $s < 1$)

$$\bar{P}_s^{2n} = k_0 \frac{1}{\sqrt{n}} \sum_{t_i \leq s} \widehat{C}_{xx}(t_{i-1}, S_{t_{i-1}}) S_{t_{i-1}}^2 \left[E|S_{t_i}/S_{t_{i-1}} - 1| - |S_{t_i}/S_{t_{i-1}} - 1|\right].$$

The processes P^{2n}, $\bar{P}^{2n}$, and M^{2n} have piecewise constant trajectories jumping at the moments t_i, $i \le n-1$. Thus,

$$\Big\|\sup_s \big|P^{2n}_s - M^{2n}_s\big|\Big\|_{L^2} \le \Big\|\sup_i \big|P^{2n}_{t_i} - \bar{P}^{2n}_{t_i}\big|\Big\|_{L^2} + \Big\|\sup_i \big|\bar{P}^{2n}_{t_i} - M^{2n}_{t_i}\big|\Big\|_{L^2}.$$

We have

$$n^{1/2}\Big\|\sup_i \big|P^{2n}_{t_i} - \bar{P}^{2n}_{t_i}\big|\Big\|_{L^2} \le k_0 \sum_{i=1}^{n-1} \Lambda^{1/2}_{t_{i-1}} B_i,$$

where

$$B_i := \big|\sigma\sqrt{2/\pi}\sqrt{nf'(t_{i-1})}\Delta t_i - E|S_{t_i}/S_{t_{i-1}} - 1|\big|.$$

Using the Taylor formula, it is easy to verify that, for $u > 0$,

$$E\big|e^{u\xi - \frac{1}{2}u^2} - 1\big| = 2\big[\Phi(u/2) - \Phi(-u/2)\big] = \sqrt{2/\pi}u + O\big(u^3\big), \quad u \to 0.$$

It follows that

$$B_i = \sigma\sqrt{2/\pi}(\Delta t_i)^{1/2}\big|\sqrt{nf'(t_{i-1})\Delta t_i} - 1\big| + O\big((\Delta t_i)^{3/2}\big).$$

By the Taylor formula,

$$\Delta t_i = g(i/n) - g\big((i-1)/n\big) = g'\big((i-1)/n\big)\frac{1}{n} + \frac{1}{2}g''(y_i)\frac{1}{n^2},$$

where the point $y_i \in [(i-1)/n, i/n]$. Since f is the inverse of g, we have $f'(t_{i-1}) = 1/g'((i-1)/n)$. Using these identities and the elementary inequality $|\sqrt{1+a} - 1| \le |a|$ for $a \ge -1$, we obtain that

$$B_i \le \text{const}\,\frac{|g''(y_i)|}{g'((i-1)/n)}(\Delta t_i)^{1/2}\frac{1}{n} + O\big((\Delta t_i)^{3/2}\big).$$

Fix $\varepsilon \in]0, 1/4[$. Substituting the finite increment formula $\Delta t_i = g'(x_i)/n$ with an intermediate point x_i in $[(i-1)/n, i/n]$, we infer that

$$B_i \le \text{const}\, a_n \frac{g'(x_i)}{[1 - g((i-1)/n)]^{3/4-\varepsilon}}\frac{1}{n} + O\big((\Delta t_i)^{3/2}\big),$$

where

$$a_n = \frac{1}{n^{1/2}} \sup_{i \le n-1} \sup_{x_i, y_i} \frac{|g''(y_i)|[1 - g((i-1)/n)]^{3/4-\varepsilon}}{g'((i-1)/n)(g'(x_i))^{1/2}}.$$

Recall that

$$\sum_{i=1}^{n-1} \frac{g'(x_i)}{[1 - g((i-1)/n)]^{1-\varepsilon}}\frac{1}{n} \to \int_0^1 \frac{dg(t)}{[1 - g(t)]^{1-\varepsilon}} = \int_0^1 \frac{dt}{(1-t)^{1-\varepsilon}} < \infty$$

and $a_n \to 0$ under each of our assumptions. These observations lead to the conclusion that

$$\sum_{i=1}^{n-1} \Lambda_{t_{i-1}}^{1/2} B_i \to 0.$$

Noticing that

$$E\big(\big|e^{u\xi-\frac{1}{2}u^2}-1\big|-u|\xi|\big)^2 = O(u^4), \quad u \to 0,$$

we infer that

$$E\big[\big(E|S_{t_i}/S_{t_{i-1}}-1|-|S_{t_i}/S_{t_{i-1}}-1|\big)-\sigma\big(E|\Delta w_{t_i}|-|\Delta w_{t_i}|\big)\big]^2 = O((\Delta t_i)^2).$$

Applying Lemma 1.4.16 and the Doob inequality to the discrete-time square-integrable martingale $(\bar{P}_{t_i}^{2n}-M_{t_i}^{2n},\mathcal{F}_{t_i})$, we get that

$$nE\sup_i\big|\bar{P}_{t_i}^{2n}-M_{t_i}^{2n}\big|^2 \le \text{const}\sum_{i=1}^{n-1}\Lambda_{t_{i-1}}(\Delta t_i)^2 \to 0, \quad n\to\infty.$$

We conclude that $n^{1/2}\|R_1^{2Mn*}\|_{L^2} \to 0$ as $n \to \infty$.

2. Noting that $\|S_t^2\widehat{C}_{xx}(t,S_t)\|_{L^2} = \Lambda_t^{1/2}$, we have

$$\|R_1^{2nn}\|_{L^2} \le \int_{t_{n-1}}^1 \Lambda_t^{1/2}\sqrt{f'(t)}\,dt \le \bigg(\int_{t_{n-1}}^1 \Lambda_t\,dt\bigg)^{1/2}\big(1-f(t_{n-1})\big)^{1/2}.$$

Since $f(t_{n-1}) = f(g((n-1)/n)) = 1-1/n$ and the function Λ is integrable, it follows that $nE(R_1^{2nn})^2 \to 0$.

3. The process R^{21n} describes the error in approximation of an integral by the Riemann sums. To analyze the approximation rate, we need the following auxiliary result.

Lemma 1.4.18 *Let $X = (X_t)_{t\in[0,T]}$ be a process with*

$$dX_t = \mu_t\,dt + \vartheta_t\,dw_t, \qquad X_0 = 0,$$

where $\mu = (\mu_t)_{t\in[0,T]}$ and $\vartheta = (\vartheta_t)_{t\in[0,T]}$ are predictable processes such that

$$\int_0^T (|\mu_t| + \vartheta_t^2)\,dt < \infty.$$

Let $X_t^n := \sum_{i=1}^n X_{t_{i-1}} I_{]t_{i-1},t_i]}(t)$. Then

$$\begin{aligned} &E\sup_{s\in[0,T]}\bigg(\int_0^s (X_t - X_t^n)\,dt\bigg)^2 \\ &\le 8\int_0^T \sum_{i=1}^n (t_i-u)^2 I_{]t_{i-1},t_i]}(u) E\vartheta_u^2\,du \\ &\quad + 2\bigg(\int_0^T \sum_{i=1}^n (t_i-u) I_{]t_{i-1},t_i]}(u)\big(E\mu_u^2\big)^{1/2}\,du\bigg)^2. \end{aligned}$$

Proof. It is sufficient to work assuming that the right-hand side of the inequality is finite. Having in mind that $(a+b)^2 \le 2a^2 + 2b^2$, we may consider separately the cases where one of the coefficients is zero. Let us start with the case where $\mu = 0$. For $v \in [t_{i-1}, t_i[$, we have, using the stochastic Fubini theorem,

$$\int_{t_{i-1}}^{v} (X_t - X_{t_{i-1}})\, dt = \int_{t_{i-1}}^{v} \int_{t_{i-1}}^{v} \vartheta_u I_{]t_{i-1},t]}(u)\, dw_u\, dt = \int_{t_{i-1}}^{v} (v-u)\vartheta_u\, dw_u.$$

Thus,

$$\int_0^s (X_t - X_t^n)\, dt = \int_0^s \sum_{i=1}^n (t_i - u) I_{]t_{i-1},t_i]}(u)\vartheta_u\, dw_u.$$

The right-hand side is a local martingale, and by the Doob inequality we obtain

$$E \sup_{s\in[0,T]} \left(\int_0^s (X_t - X_t^n)\, dt \right)^2 \le 4 \int_0^T \sum_{i=1}^n (t_i - u)^2 I_{]t_{i-1},t_i]}(u) E\vartheta_u^2\, du.$$

In the case where $\vartheta = 0$, we have, this time by the ordinary Fubini theorem, that

$$\int_{t_{i-1}}^{v} (X_t - X_{t_{i-1}})\, dt = \int_{t_{i-1}}^{v} (v-u)\mu_u\, du, \quad v \in [t_{i-1}, t_i[,$$

and this representation allows us to transform the squared process of interest to the following form:

$$\int_0^s \int_0^s \sum_{i,j=1}^n (t_i - u)(t_j - r) I_{]t_{i-1},t_i]}(u) I_{]t_{j-1},t_j]}(r) \mu_u \mu_r\, du\, dr.$$

Its expectation can be dominated by

$$\int_0^s \int_0^s \sum_{i,j=1}^n (t_i - u)(t_j - r) I_{]t_{i-1},t_i]}(u) I_{]t_{j-1},t_j]}(r) E|\mu_u \mu_r|\, du\, dr.$$

Using the Cauchy–Schwarz inequality $E|\mu_u\mu_r| \le (E\mu_u^2)^{1/2}(E\mu_r^2)^{1/2}$ and once again the Fubini theorem, we obtain the needed bound. □

Let $X_t := S_t^2 \widehat{C}_{xx}(t, S_t)\sqrt{f'(t)} I_{[0,t_{n-1}]}$. Then

$$R_s^{21n} = Y_s^n + \int_0^s X_t^n\, dt - \sum_{t_i \le s} \widehat{C}_{xx}(t_{i-1}, S_{t_{i-1}}) S_{t_{i-1}}^2 \sqrt{f'(t_{i-1})}\Delta t_i, \quad s < 1,$$

where

$$Y_s^n := \int_0^s (X_t - X_t^n)\, dt.$$

The process X on the interval $[0,t_{n-1}]$ admits the representation of the above lemma with the coefficients

$$\begin{aligned}\vartheta_t &= \left[2S_t\widehat{C}_{xx}(t,S_t)+S_t^2\widehat{C}_{xxx}(t,S_t)\right]\sqrt{f'(t)}\sigma S_t,\\ \mu_t &= \frac{1}{2}\left[2\widehat{C}_{xx}(t,S_t)+4S_t\widehat{C}_{xxx}(t,S_t)+S_t^2\widehat{C}_{xxxx}(t,S_t)\right]\sqrt{f'(t)}\sigma^2 S_t^2\\ &\quad+\frac{1}{2}S_t^2\widehat{C}_{xx}(t,S_t)\frac{f''(t)}{\sqrt{f'(t)}}+S_t^2\widehat{C}_{xxt}(t,S_t)\sqrt{f'(t)}.\end{aligned}$$

In the case where g' is bounded away from zero (hence, f' is bounded), estimates (1.4.29) and (1.4.34) imply that $E\vartheta_t^2\le\kappa/(1-t)^{3/2}$. If also f'' is bounded, then estimates (1.4.29) and (1.4.34)–(1.4.36) ensure that $E\mu_t^2\le\kappa/(1-t)^{5/2}$.

Applying the lemma, we have

$$E\sup_{s\in[0,1]}\left|Y_s^n\right|^2\le\kappa\sum_{i=1}^{n-1}\frac{(\Delta t_i)^3}{(1-t_i)^{3/2}}+\kappa\left(\sum_{i=1}^{n-1}\frac{(\Delta t_i)^2}{(1-t_i)^{5/4}}\right)^2.$$

According to Lemma 1.4.17, the right-hand side is $O(n^{-3/2})$ as $n\to\infty$.

In the case where $g(t)=1-(1-t)^\beta$, $\beta>1$, we obtain in the same way that $E\vartheta_t^2\le\kappa/(1-t)^{5/2-1/\beta}$, $E\mu_t^2\le\kappa/(1-t)^{7/2-1/\beta}$, and

$$E\sup_{s\in[0,1]}\left|Y_s^n\right|^2\le\kappa\sum_{i=1}^{n-1}\frac{(\Delta t_i)^3}{(1-t_i)^{5/2-1/\beta}}+\kappa\left(\sum_{i=1}^{n-1}\frac{(\Delta t_i)^2}{(1-t_i)^{7/4-1/(2\beta)}}\right)^2.$$

By Lemma 1.4.17 the first sum in the right-hand side can be of order $O(n^{-2})$, $O(n^{-2}\ln n)$, or $O(n^{-(\beta/2+1)})$, that is, $o(n^{-1})$ as $n\to\infty$. The second sum can be $O(n^{-1})$, $O(n^{-1}\ln n)$, or $O(n^{-(\beta/4+1/2)})$, i.e., $o(n^{-1/2})$. In all cases $nE\sup_s|Y_s^n|^2\to 0$.

The process $R^{21n}-Y^n$ vanishes in the revision dates, and

$$\sup_{s\in[0,1]}\left|R_s^{21n}-Y_s^n\right|\le\kappa\max_{i\le n-1}\int_{t_{i-1}}^{t_i}\widehat{C}_{xx}(t,S_t)S_t^2\sqrt{f'(t)}\,dt.$$

By the Cauchy–Schwarz inequality,

$$\int_{t_{i-1}}^{t_i}\widehat{C}_{xx}(t,S_t)S_t^2\sqrt{f'(t)}\,dt\le\left(\int_{t_{i-1}}^{t_i}\widehat{C}_{xx}^2(t,S_t)S_t^4\,dt\right)^{1/2}\left(\int_{t_{i-1}}^{t_i}f'(t)\,dt\right)^{1/2}.$$

Note that the second integral in the right-hand side is equal to $1/n$.

Using the bound $\max_i|a_i|\le\sum_i|a_i|$, the Jensen inequality, and estimate (1.4.32) we obtain from here that, for $m\ge 3/2$,

$$n^m E \sup_{s\in[0,1]} \left|R_s^{21n} - Y_s^n\right|^{2m} \le \kappa E \sum_{i\le n-1} \left(\int_{t_{i-1}}^{t_i} \widehat{C}_{xx}^2(t, S_t) S_t^4 \, dt\right)^m$$
$$\le \kappa \sum_{i\le n-1} (\Delta t_i)^{m-1} \int_{t_{i-1}}^{t_i} E\widehat{C}_{xx}^{2m}(t, S_t) S_t^{4m} \, dt$$
$$\le \kappa \sum_{i\le n-1} \frac{(\Delta t_i)^m}{(1-t_i)^{m-1/2}} = O\big(n^{-1/2}\big)$$

by virtue of Lemma 1.4.17. That is, $n \sup_{s\in[0,1]} |R_s^{21n} - Y_s^n|^2$ tends to zero in L^m and, hence, in L^1.

4. The residual processes R_s^{22n} have piecewise constant trajectories, and the analysis of the asymptotic behavior is reduced to the discrete-time scheme.

Let

$$\xi_i^n := (S_{t_i}/S_{t_{i-1}} - 1)^2 \operatorname{sign}(S_{t_i}/S_{t_{i-1}} - 1),$$

$\Delta M_i^n := \xi_i^n - E\xi_i^n$, and $X_i^n := \widehat{C}_{xx}(t_{i-1}, S_{t_{i-1}}) S_{t_{i-1}}^2$. With this notation, we have the representation

$$n^{1/2} R_{t_k}^{22n} = X^n \cdot M_k^n + A_k^n, \quad k \le n-1,$$

where

$$A_k^n := \sum_{i\le k} X_i^n E\xi_i^n.$$

Note that

$$E\big(e^{u\xi - \frac{1}{2}u^2} - 1\big)^4 = O\big(u^4\big), \quad u \to 0.$$

Applying the Doob inequality and Lemma 1.4.16, we obtain that

$$E \sup_{i\le n-1} \big(X^n \cdot M_{n-1}^n\big)^2 \le 4E\big(X^n \cdot M_{n-1}^n\big)^2 \le 4 \sum_{i\le n-1} E\Lambda_{t_{i-1}} (S_{t_i}/S_{t_{i-1}} - 1)^4$$
$$\le \kappa \sum_{i\le n-1} \frac{(\Delta t_i)^2}{(1-t_{i-1})^{1/2}} = O\big(n^{-1}\big)$$

according to Lemma 1.4.17.

By virtue of Lemma 1.4.20 given below in the section on asymptotics of Gaussian integrals, for sufficiently large n, we have the inequalities

$$0 \le E\xi_i^n \le \kappa (\Delta t_i)^{3/2},$$

implying that the discrete-time process A is increasing and

$$\big\|A_{n-1}^n\big\|_{L^2} \le \sum_{i\le n-1} \big\|X_i^n\big\|_{L^2} E\xi_i^n \le \kappa \sum_{i\le n-1} \Lambda_{t_{i-1}}^{1/2} (\Delta t_i)^{3/2}$$
$$\le \kappa \sum_{i\le n-1} \frac{(\Delta t_i)^{3/2}}{(1-t_{i-1})^{1/4}} = O\big(n^{-1/2}\big),$$

again according to Lemma 1.4.17.

It follows that $nE(R_1^{22n*})^2 \to 0$.

5. We verify now that $nE(R_1^{23n*})^2 \to 0$. Recall that

$$E(S_{t_i} - S_{t_{i-1}})^{2m} \le c_m (\Delta t_i)^m.$$

Using (1.4.33), we obtain the bound

$$E\widehat{C}^2_{xt}(t_{i-1}, S_{t_{i-1}})(\Delta t_i)^2 (S_{t_i} - S_{t_{i-1}})^2 \le \kappa \frac{(\Delta t_i)^3}{(1-t_{i-1})^{3/2}}.$$

To estimate the terms coming from the residual term of the Taylor expansion, we use the Cauchy–Schwarz inequality and the bounds (1.4.18), (1.4.21), and (1.4.22). This yields the following:

$$E\widehat{C}^2_{xxx}(\tilde{t}_{i-1}, \tilde{S}_{t_{i-1}})(S_{t_i} - S_{t_{i-1}})^6 \le \kappa \frac{(\Delta t_i)^3}{(1-t_i)^2},$$

$$E\widehat{C}^2_{xxt}(\tilde{t}_{i-1}, \tilde{S}_{t_{i-1}})(S_{t_i} - S_{t_{i-1}})^4 (\Delta t_i)^2 \le \kappa \frac{(\Delta t_i)^4}{(1-t_i)^3},$$

$$E\widehat{C}^2_{xtt}(\tilde{t}_{i-1}, \tilde{S}_{t_{i-1}})(\Delta t_i)^4 (S_{t_i} - S_{t_{i-1}})^2 \le \kappa \frac{(\Delta t_i)^5}{(1-t_i)^4}.$$

Obviously,

$$n^{1/2} \|R_1^{23n*}\|_{L^2} \le \sum_{i \le n-1} \big\| [\ldots]_i (S_{t_i} - S_{t_{i-1}}) \big\|_{L^2},$$

where $[\ldots]_i$ is defined in (1.4.13). Taking into account that $\widehat{C}_{xx}(t,x) \ge 0$ and using the inequality $\big||a| - |b|\big| \le |a-b|$, we can write that

$$\big\|[\ldots]_i (S_{t_i} - S_{t_{i-1}})\big\|_{L^2} \le \kappa \big(\|\widehat{C}_{xt}(t_{i-1}, S_{t_{i-1}})(t_i - t_{i-1})(S_{t_i} - S_{t_{i-1}})\|_{L^2} + \cdots \big),$$

where we denote by dots the L^2-norms of the residual term in the first-order Taylor expansion of the difference $\widehat{C}_x(t_i, S_{t_i}) - \widehat{C}_x(t_{i-1}, S_{t_{i-1}})$. Summing up and using the above estimates, we conclude, applying Lemma 1.4.17, that the right-hand side of the above inequality tends to zero as $n \to \infty$, and we conclude.

6. It remains to check that $nE(R_1^{24n*})^2 \to 0$, and this happens to be the most delicate part of the proof. Again the analysis can be reduced to the discrete-time case. We note that

$$nE\big(R_1^{24n*}\big)^2 \le \sum_{i \le n-1} ES^2_{t_{i-1}}[\ldots]_i^2 + 2\sum_{i<j} E\big|S_{t_{i-1}}[\ldots]_i S_{t_{j-1}}[\ldots]_j\big|.$$

The estimation of the first sum is rather straightforward. Applying the Itô formula to the function $\widehat{C}_x(t,x)$ and using the positivity of $\widehat{C}_{xx}(t,x)$ and the inequality $\big||a| - |b|\big| \le |a-b|$, we dominate the absolute value of random variable denoted by $[\dots]_i$, see the formula (1.4.13), by the absolute value of

$$\int_{t_{i-1}}^{t_i} (\widehat{C}_{xx}(t_{i-1}, S_{t_{i-1}}) - \widehat{C}_{xx}(t, S_t))\, dS_t - \int_{t_{i-1}}^{t_i} \left(\widehat{C}_{xt}(t, S_t) + \frac{\sigma^2}{2} S_t^2 \widehat{C}_{xxx}(t, S_t)\right) dt.$$

We check that

$$\sum_{i=1}^{n-1} ES_{t_{i-1}}^2 \int_{t_{i-1}}^{t_i} (\widehat{C}_{xx}(t_{i-1}, S_{t_{i-1}}) - \widehat{C}_{xx}(t, S_t))^2 S_t^2\, dt = O(n^{-1/4}), \quad (1.4.41)$$

$$\sum_{i=1}^{n-1} \Delta t_i ES_{t_{i-1}}^2 \int_{t_{i-1}}^{t_i} (\widehat{C}_{xt}^2(t, S_t) + S_t^4 \widehat{C}_{xxx}^2(t, S_t))\, dt = O(n^{-1/2}). \quad (1.4.42)$$

A generic term of the first sum is dominated by

$$\Delta t_i E \sup_{t \le 1} S_t^4 \sup_{t_{i-1} \le t \le t_i} (\widehat{C}_{xx}(t, S_t) - \widehat{C}_{xx}(t_{i-1}, S_{t_{i-1}}))^2.$$

The Cauchy–Schwarz inequality allows us to separate the terms under the sign of expectation and reduce the problem to the estimation of the fourth power of the difference $\widehat{C}_{xx}(t, S_t) - \widehat{C}_{xx}(t_{i-1}, S_{t_{i-1}})$. The Itô formula transforms this difference into the sum of a stochastic integral and an ordinary integral. Using consecutively the Burkholder and Cauchy–Schwarz inequalities and the bound (1.4.34), we have

$$\begin{aligned} E \sup_{t \in [t_{i-1}, t_i]} \left[\int_{t_{i-1}}^{t} \widehat{C}_{xxx}(u, S_u) S_u\, dS_u\right]^4 &\le c_4 E\left[\int_{t_{i-1}}^{t_i} \widehat{C}_{xxx}^2(u, S_u) S_u^4\, du\right]^2 \\ &\le c_4 \Delta t_i E \int_{t_{i-1}}^{t_i} \widehat{C}_{xxx}^4(u, S_u) S_u^8\, du \\ &\le \kappa \frac{(\Delta t_i)^2}{(1 - t_i)^{7/2}}. \end{aligned}$$

To estimate the ordinary integral, we use the Jensen inequality for $f(x) = x^4$ and the bounds (1.4.35) and (1.4.36) and get that

$$E \sup_{t \in [t_{i-1}, t_i]} \left[\int_{t_{i-1}}^{t} \left(\widehat{C}_{xxt}(u, S_u) + \frac{1}{2}\sigma^2 S_u^2 \widehat{C}_{xxxx}(u, S_u)\right) du\right]^4 \le \kappa \frac{(\Delta t_i)^4}{(1 - t_i)^{11/2}}.$$

Using these estimates, we obtain that the sum in (1.4.41) is dominated, up to a multiplicative constant, by

$$\sum_{i=1}^{n-1} \left[\frac{(\Delta t_i)^2}{(1 - t_i)^{7/4}} + \frac{(\Delta t_i)^3}{(1 - t_i)^{11/4}}\right],$$

and the claimed asymptotics follows from Lemma 1.4.17.

Similar arguments, but using the inequalities (1.4.33) and (1.4.34), give us the second asymptotic formula.

From the same estimates we obtain that

$$\sum_{i=1}^{n-1} \left(ES^2_{t_{i-1}}[\ldots]^2_i\right)^{1/2} \le \kappa \sum_{i=1}^{n-1} \frac{\Delta t_i}{(1-t_i)^{7/8}} + \kappa \sum_{i=1}^{n-1} \frac{(\Delta t_i)^{3/2}}{(1-t_i)^{11/8}}.$$

The second sum in the right-hand side converges to zero, while for the first one, we can say only that it is dominated by a convergent integral. Using this observation, we conclude that the sum of expectations of cross terms over indices i, j with $i < j$ and $t_j > a$ also can be done arbitrarily small by choosing a sufficiently close to one.

Unexpectedly, the most difficult part of the proof is in establishing the convergence to zero of the sum of cross terms corresponding to the dates of revisions before $a < 1$, i.e., bounded away from the singularity.

To formulate the claim, we introduce "reasonable" notations. Put

$$\alpha_i := \widehat{C}_{xx}(S_{t_{i-1}}, t_{i-1}) S^2_{t_{i-1}} \left(\frac{S_{t_i}}{S_{t_{i-1}}} - 1\right),$$

$$\beta_i := S_{t_{i-1}} \widehat{C}_{xt}(S_{t_{i-1}}, t_{i-1}) \Delta t_i + \frac{1}{2} S^3_{t_{i-1}} \widehat{C}_{xxx}(S_{t_{i-1}}, t_{i-1}) \left(\frac{S_{t_i}}{S_{t_{i-1}}} - 1\right)^2,$$

$\gamma_i := |\alpha_i + \beta_i| - |\alpha_i|$. Let us also define the random variable $\chi_i := \operatorname{sign} \alpha_i \beta_i$ and the set $A_i := \{|\beta_i| < |\alpha_i|\}$, which will be used in the lemma below.

Now we have the identity

$$S_{t_{i-1}}[\ldots]_i = -\gamma_i + \zeta_i,$$

where the expression $[\ldots]_i$ is given in (1.4.13), and

$$\zeta_i := |\alpha_i + \beta_i| - \left|\widehat{C}_x(t_i, S_{t_i}) - \widehat{C}_x(t_{i-1}, S_{t_{i-1}})\right| S_{t_{i-1}}.$$

Using the second-order Taylor expansion of $\widehat{C}_x(t, x)$ at the point $(t_{i-1}, S_{t_{i-1}})$ and the elementary inequality $||x| - |y|| \le |x - y|$, we obtain that

$$|\zeta_i| \le \frac{1}{2} S_{t_{i-1}} \left|\widehat{C}_{xtt}(t_{i-1}, S_{t_{i-1}})(\Delta t_i)^2 + 2\widehat{C}_{xxt}(t_{i-1}, \tilde{S}_{t_{i-1}}) \Delta t_i \Delta S_{t_i}\right| + \frac{1}{6} S_{t_{i-1}} |r_i|,$$

where the residual term is of the form

$$\begin{aligned} r_i := {} & \widehat{C}_{xttt}(\tilde{t}_{i-1}, \tilde{S}_{t_{i-1}})(\Delta t_i)^3 + 3\widehat{C}_{xxtt}(\tilde{t}_{i-1}, \tilde{S}_{t_{i-1}})(\Delta t_i)^2 \Delta S_{t_i} \\ & + 3\widehat{C}_{xxxt}(\tilde{t}_{i-1}, \tilde{S}_{t_{i-1}}) \Delta t_i (\Delta S_{t_i})^2 + \widehat{C}_{xxxx}(\tilde{t}_{i-1}, \tilde{S}_{t_{i-1}})(\Delta S_{t_i})^3. \end{aligned}$$

For t bounded away from unit, the derivatives of $\widehat{C}$ can be estimated by polynomials of x and x^{-1}. This allows us to obtain the bound

$$|\zeta_i| \le \kappa_a \eta \left(|\Delta S_{t_i}|^3 + |\Delta S_{t_i}|^2 \Delta t_i + |\Delta S_{t_i}| \Delta t_i + (\Delta t_i)^2\right),$$

where $\eta = 1 + \exp\{\kappa w_1^*\}$ is a random variable having moments of any order. It follows that

$$E|\zeta_i|^2 \le \kappa(\Delta t_i)^3.$$

By the Cauchy–Schwarz inequality we infer from here that

$$\sum_{t_i,t_j\le a} E|\zeta_i\zeta_j| \le \Big(\sum_{t_i\le a}(E|\zeta_i|^2)^{1/2}\Big)^2 = O(n^{-1}), \quad n\to\infty.$$

Taking into account that $|\gamma_i| \le |\beta_i|$, we get the estimate $E|\gamma_i|^2 \le \kappa(\Delta t_i)^2$. Therefore,

$$\sum_{t_i,t_j\le a} E|\gamma_i\zeta_j| \le \Big(\sum_{t_i\le a}(E|\gamma_i|^2)^{1/2}\Big)\Big(\sum_{t_i\le a}(E|\zeta_i|^2)^{1/2}\Big) = O(n^{-1/2}).$$

The assertion needed to conclude is the lemma below. It is based on asymptotic analysis of expectations of some Gaussian integrals given in the next section and the following identities:

$$\begin{aligned}|\alpha+\beta| - |\alpha| &= |\beta|\chi I_A + |\beta| I_{\{\chi>0\}} I_{A^c} + (|\beta| - 2|\alpha|) I_{\{\chi\le 0\}} I_{A^c}\\ &= |\beta|\chi + 2(|\beta| - |\alpha|) I_{\{\chi\le 0\}} I_{A^c} - |\beta| I_{\{\chi=0\}} I_{A^c},\end{aligned}$$

where α,β are arbitrary random variables, $\chi := \operatorname{sign}(\alpha\beta)$, and $A := \{|\beta| < |\alpha|\}$.

Lemma 1.4.19 *For every fixed* $a \in]0,1[$,

$$\Big|\sum_{i<j,\,t_j\le a} E\gamma_i\gamma_j\Big| = o(1), \quad n\to\infty.$$

Proof. The routine estimation $|E\gamma_i\gamma_j| \le E|\gamma_i||\gamma_j|$ does not work in our case. However, for $i<j$,

$$|E\gamma_i\gamma_j| = \big|E(\gamma_i E(\gamma_j|\mathcal{F}_{t_{j-1}}))\big| \le E(|\gamma_i||E(\gamma_j|\mathcal{F}_{t_{j-1}})|) \le E(|\beta_i||E(\gamma_j|\mathcal{F}_{t_{j-1}})|).$$

According to the above identity,

$$|E(\gamma_j|\mathcal{F}_{t_{j-1}})| \le |E(|\beta_j|\chi_j|\mathcal{F}_{t_{j-1}})| + 2E(|\beta_j| I_{A_j^c}|\mathcal{F}_{t_{j-1}}).$$

Using Lemma 1.4.21 of the next section with $\eta_u = S_{t_j}/S_{t_{j-1}} - 1$, $u = (\Delta t_j)^{1/2}$, and $A = -S_{t_{i-1}}^2 \widehat{C}_{xxx}(S_{t_{i-1}}, t_{i-1})/\widehat{C}_{xt}(S_{t_{i-1}}, t_{i-1})$, we dominate the first term in the right-hand side of the above inequality by

$$\kappa\big(S_{t_{i-1}}|\widehat{C}_{xt}(S_{t_{i-1}}, t_{i-1})| + S_{t_{i-1}}^3|\widehat{C}_{xxx}(S_{t_{i-1}}, t_{i-1})|\big)(\Delta t_j)^{3/2}.$$

Applying again the bounds for the derivatives $\widehat{C}(t,x)$ when $t_j \le a < 1$, we infer that the coefficients can be dominated by $\kappa_a(1+\exp\{\kappa w_1^*\})$, i.e., by a random variable having all moments. In the same range of indices we have also the bound $E(\beta_i^2|\mathcal{F}_{t_{i-1}}) \le \zeta_a(\Delta t_i)^2$, where ζ_a is a random variable having all moments. It follows from here that

$$\sum_{i<j, t_j \le a} E\big(|\beta_i|\big|E\big(|\beta_j|\chi_j|\mathcal{F}_{t_{j-1}}\big)\big|\big) = O\big(n^{-1/2}\big).$$

We estimate $P(A_j^c|\mathcal{F}_{t_{j-1}})$ applying Lemma 1.4.22 of the next section with

$$c_1(t_{j-1}) := \frac{S_{t_{j-1}}^3 \widehat{C}_{xxx}(S_{t_{j-1}}, t_{j-1})}{S_{t_{j-1}}^2 \widehat{C}_{xx}(S_{t_{j-1}}, t_{j-1})}, \qquad c_2(t_{j-1}) := \frac{S_{t_{j-1}} \widehat{C}_{xt}(S_{t_{j-1}}, t_{j-1})}{S_{t_{j-1}}^2 \widehat{C}_{xx}(S_{t_{j-1}}, t_{j-1})},$$

and $c(t_{j-1}) := 2\big(|c_1(t_{j-1})| + |c_2(t_{j-1})| + 1\big)$. On the interval $[0,a]$ the continuous process $c(t)$ can be dominated by a random variable ξ_a. Fix $\varepsilon > 0$ and choose N such that $P(\xi_a > N) < \varepsilon$. Lemma 1.4.22 implies that

$$P\big(A_j^c|\mathcal{F}_{t_{j-1}}\big) \le L_N(\Delta t_j)^{1/2} I_{\{c(t_{j-1}\})\le N} + I_{\{c(t_{j-1})>N\}},$$

and, therefore, $P(A_j^c) \le L_N(\Delta t_j)^{1/2} + \varepsilon \le 2\varepsilon$ when n is large enough. Using the Cauchy–Schwarz and Jensen inequalities, we get that

$$\begin{aligned}\sum_{i<j, t_j \le a} E\big(|\beta_i|\big|E\big(|\beta_j|I_{A_j^c}|\mathcal{F}_{t_{j-1}}\big)\big|\big) &\le \sum_{t_i \le a}\big(E\beta_i^2\big)^{1/2}\sum_{t_j\le a}\big(E\beta_i^4\big)^{1/4}\big(P(A_j^c)\big)^{1/4} \\ &\le (2\varepsilon)^{1/4}\sum_{t_i \le a}\big(E\beta_i^2\big)^{1/2}\sum_{t_j\le a}\big(E\beta_j^4\big)^{1/4}.\end{aligned}$$

Both sums in the right-hand side are bounded because $E\beta_j^2 \le \kappa(\Delta t_i)^2$ and $E\beta_j^4 \le \kappa(\Delta t_i)^4$. By the choice of ε the right-hand side can be made arbitrarily small. □

Thus, $nE(R_1^{24n*})^2 \to 0$.

1.4.8 Asymptotics of Gaussian Integrals

Let $\xi \in \mathcal{N}(0,1)$, and let $\eta_u := e^{u\xi - \frac{1}{2}u^2} - 1$, $u \in [0,1]$.

Lemma 1.4.20 *The following asymptotical formulae hold as $u \to 0$:*

$$\begin{aligned}E\big[\eta_u^2 - \eta_{-u}^2\big]I_{\{\eta_u>0\}} &= \frac{2}{\sqrt{2\pi}}u^3 + O\big(u^4\big), \\ E\eta_u^2 \operatorname{sign}\eta_u &= \frac{2}{\sqrt{2\pi}}u^3 + O\big(u^4\big), \\ E\operatorname{sign}\eta_u &= -\frac{1}{\sqrt{2\pi}}u + O\big(u^3\big).\end{aligned}$$

Proof. Put

$$Z(u) := \left(e^{u\xi - \frac{1}{2}u^2} - 1\right)^2 - \left(e^{-u\xi - \frac{1}{2}u^2} - 1\right)^2.$$

Then $Z(0) = Z'(0) = Z''(0) = 0$, $Z'''(0) = 12(\xi^3 - \xi)$, and the function $Z^{(4)}(u)$ is bounded by a random variable having moments of any order. Using the Taylor formula, we obtain that

$$EZ(u)I_{\{\xi \geq \frac{1}{2}u\}} = 2u^3 E\left(\xi^3 - \xi\right)I_{\{\xi \geq \frac{1}{2}u\}} + O\left(u^4\right), \quad u \to 0,$$

and we obtain the first formula. The second formula is a corollary of the first one since

$$E\eta_u^2 \operatorname{sign} \eta_u = EZ(u)I_{\{\xi \geq \frac{1}{2}u\}} - E\eta_u^2 I_{\{|\xi| \leq \frac{1}{2}u\}},$$

and the last term is $O(u^4)$ as $u \to 0$. Finally,

$$\begin{aligned} E \operatorname{sign} \eta_u = P(\xi > u/2) - P(\xi < u/2) &= 2\big(\Phi(0) - \Phi(u/2)\big) \\ &= -\frac{1}{\sqrt{2\pi}}u + \frac{1}{4}\varphi(\tilde{u})\tilde{u}u^2, \end{aligned}$$

where $\tilde{u} \in [0, u/2]$. □

Lemma 1.4.21 *There exists a constant κ such that, for any $A \in \mathbf{R}$,*

$$\left|E|\eta_u^2 - Au^2| \operatorname{sign}\left(\eta_u^2 - Au^2\right)\eta_u\right| \leq \kappa\left(1 + |A|\right)u^3. \tag{1.4.43}$$

Proof. Note that $|x| \operatorname{sign} xy = x \operatorname{sign} y$. Therefore, the left-hand side of (1.4.43) is dominated by

$$\left|E\eta_u^2 \operatorname{sign} \eta_u\right| + |A|u^2|E \operatorname{sign} \eta_u|,$$

and the result holds by virtue of the previous lemma. □

Lemma 1.4.22 *For every $N > 0$, there is a constant L_N such that, for all $u \in [0, 1]$,*

$$P\left(\left|c_1\eta_u^2 + c_2u^2\right| > |\eta_u|\right) \leq L_N I_{\{c \leq N\}}u + I_{\{c > N\}}$$

for any constants c_1, c_2 and $c := 2(|c_1| + |c_2| + 1)$.

Proof. Suppose that $N \geq c > 2$, the only case where the work is needed. It is easy to see that

$$\begin{aligned} P\left(\left|c_1\eta_u^2 + c_2u^2\right| > |\eta_u|\right) &\leq P\left((c/2)\eta_u^2 + (c/2)u^2 > |\eta_u|\right) \\ &\leq P\left(c|\eta_u| > 1\right) + P\left(|\eta_u| < cu^2\right). \end{aligned}$$

The probabilities in the right-hand side as functions of c are increasing, and it remains to dominate their values at the point $c = N$. The required bound holds for the first probability in the right-hand side (and even with a constant which does not depend on N). Indeed, using the Chebyshev inequality, finite increment formula, and the bound $\varphi(x) \le 1/\sqrt{2\pi}$, we have

$$P(N|\eta_u| > 1) \le \frac{1}{N} E|\eta_u| \le \frac{1}{2} E|\eta_u| = \Phi(u/2) - \Phi(-u/2) \le \frac{1}{\sqrt{2\pi}} u.$$

For $u \ge 1/\sqrt{2N}$, the second probability is dominated by linear functions with $L_N \ge \sqrt{2N}$. For $u < 1/\sqrt{2N}$, we write it as

$$P(u/2 \le \xi < (1/u)\ln(1+Nu^2)+u/2)+P((1/u)\ln(1-Nu^2)+u/2 < \xi < u/2).$$

Using again the finite increment formula, we obtain that

$$P(u/2 \le \xi < (1/u)\ln(1+Nu^2)+u/2) \le \frac{1}{\sqrt{2\pi}} Nu.$$

On the interval $]0, 1/\sqrt{2N}[$, we have the bound $(1/u)\ln(1-Nu^2) \ge -\kappa N u$, where $\kappa > 0$ is the maximum of the function $-\ln(1-x)/x$ on the interval $]0, 1/2]$. It follows that

$$P((1/u)\ln(1-Nu^2)+u/2 < \xi < u/2) \le \frac{1}{\sqrt{2\pi}} \kappa N u.$$

Thus, the second probability also admits a linear majorant on the whole interval $[0,1]$. □

1.5 Functional Limit Theorem for $\alpha = 1/2$

1.5.1 Formulation

The asymptotic behavior of the hedging error is mathematically interesting and practically important issue. It is well known that, in finance, gains and losses have quite different weights. That is why measuring hedging errors using L^2-norm is strongly criticized. Of course, the limiting distribution of the hedging errors is much more informative. However, the problem to find it is nontrivial even for models without transaction costs, i.e., for the case where $k_0 = 0$.

The exact rate of the L^2-convergence for $\alpha = 1/2$ provided by Theorem 1.4.3 indicates that in this case the approximation errors multiplied by the amplifying factor growing as $n^{1/2}$ also should converge in law. Our aim is to show this property: the sequence of random variables $n^{1/2}(V_1^n - V_1)$ converges in law. In fact, we prove a more general result on the Markov diffusion approximation which claims that the process $n^{1/2}(V^n - \widehat{V})$, where $\widehat{V}_t = \widehat{C}(t, S_t)$, converges in law in the Skorokhod space to a continuous process and calculate the limit.

Theorem 1.5.1 *Suppose that $\alpha = 1/2$ and that Assumption 1 or 2 holds. Suppose also that Assumption 3 on the pay-off function is fulfilled. Then the distributions of the processes $Y^n := n^{1/2}(V^n - \widehat{V})$ in the Skorokhod space*

$\mathcal{D}[0,1]$ *converge weakly to the distribution of the process*

$$Y_t = \int_0^t F(t, S_t)\, dw'_t, \tag{1.5.1}$$

where w' *is an independent Wiener process, and*

$$F(t,x) = \left[\frac{\sigma^4}{2}\frac{1}{f'(t)} + k_0\sqrt{\frac{2}{\pi}}\frac{\sigma^3}{\sqrt{f'(t)}} + k_0^2\sigma^2\left(1-\frac{2}{\pi}\right)\right]^{1/2} \widehat{C}_{xx}(t,x)x^2.$$

It is important to note that the limiting process Y itself is not a Markov diffusion, but it is a component of the Markov diffusion process (S,Y). Though the problem looks as a scalar one, its natural framework is two-dimensional, and it can be treated as a problem of diffusion approximation in a general semimartingale setting. In fact, we prove a more general result on the convergence in distribution of the two-dimensional processes (S, Y^n) to (S,Y). The strategy of the proof is a standard one: to reduce the problem to the well-known semimartingale scheme and then use an appropriate result from a large variety of limit theorems. Our choice is Theorem IX.3.39 (in fact, its simplified version) from the fundamental monograph [102]. However, a straightforward application of this theorem is not possible because its conditions, rather delicate, are not fulfilled. The problem is due to the singularity of the coefficients at the maturity date. To circumvent the difficulty, we first check the C-tightness of distributions using the relatively simple classical sufficient condition from the mentioned book. As the second step, we identify the limit by applying to this aim Theorem IX.3.39 on a shorter intervals $[0,T]$, $T<1$.

1.5.2 Limit Theorem for Semimartingale Scheme

For the reader's convenience, we formulate a functional limit theorem for a weakly convergent sequence of semimartingales X^n. This theorem is just a simplified version of Theorem IX.3.39 from [102] adapted to our purposes. Namely, we assume that the triplet (B,C,ν) of predictable characteristics of the limit is assumed to be $(0,C,0)$, while X^n have no continuous martingale component. We shall use it to identify the limit.

Let $(D[0,T],\mathcal{D},\mathbf{D}=(\mathcal{D}_t)_{t\le T})$ be the Skorokhod space of d-dimensional càdlàg functions on $[0,T]$ with its natural filtration, $\mathcal{D}=\mathcal{D}_T$, and let $C[0,T]$ be its subspace formed by continuous functions. We denote by α a generic point of $D[0,T]$ and consider the canonical process X with $X_t(\alpha)=\alpha_t$.

On this stochastic basis, we are given a $d\times d$-dimensional continuous adapted process C of finite variation with $C_0=0$ and increments $C_t - C_s$, $s\le t$, taking values in the set of nonnegative definite matrices.

Put $\tau_a(\alpha)=\inf\{t>0:\ |\alpha_t|\vee|\alpha_{t-}|>a\}$, $a>0$.

For the formulation of the theorem, we need the following conditions:

(H_1) Continuity hypothesis: the function $\alpha \mapsto C_t(\alpha)$ is continuous (in the Skorokhod topology).

(H_2) Local strong majoration hypothesis: for each $a > 0$, there is a continuous deterministic increasing function $s \mapsto F_s^a$ strongly dominating[4] the stopped process $(\sum_{i \le d} C^{ii}_{s \wedge \tau_a})$.

(H_3) Local uniqueness property holds for the martingale problem with the triplet $(0, C, 0)$ (see details in [102]); we denote by Q the unique solution.

We do not want to discuss the latter hypothesis here: as we shall see later, it is always fulfilled in the case of interest.

In conformity with [102], we consider a fixed continuous truncation function $h(x) = x\delta(x)$, where $\delta : \mathbf{R}^d \to [0, 1]$ is a function such that $\delta(x) = 1$ for $|x| \le 1$ and $\delta(x) = 0$ for $|x| \ge 2$.

Let $X^n = (X_t^n)_{t \le 1}$ be a d-dimensional semimartingale, $X_0^n = X_0$, defined on a stochastic basis $(\Omega, \mathcal{F}, \mathbf{F}^n, P)$. Let μ^n and $(B^n, 0, \nu^n)$ be the jump measure and the triplet of predictable characteristics of X^n corresponding to the truncation function h. Let $\mathcal{L}(X^n)$ be the distribution of X^n.

Put $\tau_a^n := \tau_a \circ X^n = \inf\{t > 0 : \ |X_t^n| \vee |X_{t-}^n| > a\}$ and

$$\tilde{C}_t^{n,i,j} := (h^i h^j) * \nu_t^n - \sum_{s \le t} \Delta B_s^{n,i} \Delta B_s^{n,j}.$$

With this notation, we write the simplified version of Theorem IX.3.39 as follows:

Theorem 1.5.2 *Assume that hypotheses (H_1)–(H_3) are fulfilled and the following conditions hold for all $t \in [0, T]$ and $a > 0$:*

(a) $\sup_{s \le T} |B^n_{s \wedge \tau_a^n}| \to_P 0$;

(b) $\tilde{C}^n_{t \wedge \tau_a^n} - C_{t \wedge \tau_a} \circ X^n \to_P 0$;

(c) $g * \nu^n_{t \wedge \tau_a^n} \to_P 0$ *for each bounded function $g \ge 0$ vanishing in a neighborhood of the origin.*

Then $\mathcal{L}(X^n)$ weakly converge to Q.

1.5.3 Problem Reformulation

First, we formulate an assertion following immediately from the results above and which reduces the asymptotic analysis of the processes $V^n - \widehat{V}$ to the analysis of processes of more simple structure.

The processes in the formulation of the lemma below are defined on $[0, 1[$; we extend them to the closed interval $[0, 1]$ by continuity.

Put

$$\lambda_t := \widehat{C}^2_{xx}(t, S_t) S_t^4. \tag{1.5.2}$$

[4] By definition, $\tilde{A}$ strongly dominates A if $A_r - A_s \le \tilde{A}_r - \tilde{A}_s$ for $r \ge s$.

Lemma 1.5.3 *The error process admits the representation*

$$V_s^n - \widehat{V}_s = M_s^{1n} + M_s^{2n} + R_s^n,$$

where

$$M_s^{1n} := \frac{1}{2}\sigma^2 \sum_{t_i \le s} \lambda_{t_{i-1}} \big[\Delta t_i - (\Delta w_{t_i})^2\big],$$

$$M_s^{2n} := k_0 \sigma n^{-1/2} \sum_{t_i \le s} \lambda_{t_{i-1}} \big[\sqrt{2/\pi}\sqrt{\Delta t_i} - |\Delta w_{t_i}|\big],$$

and $\sqrt{n}\|\sup_{s\le 1} R_s^n\| \to 0$ *as* $n \to \infty$.

Since $R_s^n = R_s^{1n} + R_s^{2n}$, the claimed property of the residual is a direct corollary of Lemma 1.4.4 and Propositions 1.4.6 and 1.4.7. It implies, according to Lemma IX.3.31 in [102], that it is sufficient to establish the functional limit theorem for the processes $n^{1/2}(M^{1n} + M^{2n})$.

To apply the general semimartingale scheme, we need to reformulate our problem appropriately and adjust the notation. We consider the stochastic basis $(\Omega, \mathcal{F}, \mathbf{F}^n = (\mathcal{F}^n), P)$ with $\mathcal{F}_t^n = \mathcal{F}_{t_{i-1}^n}$ for $t \in [t_{i-1}^n, t_i^n[$ supporting the two-dimensional square-integrable martingale $X^n = (X^{1n}, X^{2n})$ with the components defined, for $t < 1$, by the formulae

$$X_t^{1n} := \sum_{i=1}^{n} S_{t_{i-1}} I_{[t_{i-1}, t_i[}(t) = \sum_{t_i \le t} \Delta S_{t_i},$$

$$X_t^{2n} := n^{1/2}\big(M_t^{1n} + M_t^{2n}\big) = \sum_{t_i \le t} U_i^n,$$

where $U_i^n := Y_i^n + Z_i^n$,

$$Y_i^n := \frac{1}{2}\sigma^2 n^{1/2} \lambda_{t_{i-1}} \big[\Delta t_i - (\Delta w_{t_i})^2\big],$$

$$Z_i^n := k_0 \sigma \lambda_{t_{i-1}} \left[\sqrt{\frac{2}{\pi}}\sqrt{\Delta t_i} - |\Delta w_{t_i}|\right],$$

and $\Delta S_{t_i} = S_{t_i} - S_{t_{i-1}}$, $\Delta w_{t_i} = w_{t_i} - w_{t_{i-1}}$ (as S and w are continuous, we hope that the reader will not be confused by this notation). We put $X_1^n = X_{t_{n-1}}^n$.

Note that

$$\langle X^{1n} \rangle_t = \sum_{t_i \le t} E\big((\Delta S_{t_i})^2 | \mathcal{F}_{t_{i-1}}\big),$$

$$\langle X^{2n} \rangle_t = \sum_{t_i \le t} E\big(U_i^{n2} | \mathcal{F}_{t_{i-1}}\big).$$

It is easily seen that the triplet of predictable characteristics of X^n (with respect to the truncation function h) is $(B^n, 0, \nu^n)$, where

$$B_t^n = \sum_{t_i \le t} E\big(h(\Delta X_{t_i}^n)|\mathcal{F}_{t_{i-1}}\big),$$

$$\nu^n\big([0,t] \times \Gamma\big) = \sum_{t_i \le t} E\big(I_\Gamma(\Delta X_{t_i}^n)|\mathcal{F}_{t_{i-1}}\big).$$

The components of the matrix-valued process $\tilde{C}_t^n$ are as follows:

$$\tilde{C}_t^{n,1,1} = \sum_{t_i \le t} \big[E\big((\Delta S_{t_i})^2 \delta^2(\Delta X_{t_i}^n)|\mathcal{F}_{t_{i-1}}\big) - \big(E\big(\Delta S_{t_i}\delta(\Delta X_{t_i}^n)|\mathcal{F}_{t_{i-1}}\big)\big)^2\big],$$

$$\tilde{C}_t^{n,1,2} = \tilde{C}_t^{n,2,1} = \sum_{t_i \le t} \big[E\big(\Delta S_{t_i} U_i^n \delta^2(\Delta X_{t_i}^n)|\mathcal{F}_{t_{i-1}}\big) - E\big(\Delta S_{t_i}\delta(\Delta X_{t_i}^n)|\mathcal{F}_{t_{i-1}}\big)E\big(U_i^n\delta(\Delta X_{t_i}^n)|\mathcal{F}_{t_{i-1}}\big)\big],$$

$$\tilde{C}_t^{n,2,2} = \sum_{t_i \le t} \big[E\big(U_i^{n2}\delta^2(\Delta X_{t_i}^n)|\mathcal{F}_{t_{i-1}}\big) - \big(E\big(U_i^n\delta(\Delta X_{t_i}^n)|\mathcal{F}_{t_{i-1}}\big)\big)^2\big].$$

1.5.4 Tightness

We consider $G^n = \langle X^{1n}\rangle + \langle X^{2n}\rangle$. According to [102, Theorem VI.4.13], the sequence $\mathcal{L}(X^n)$ is tight if the sequence $\mathcal{L}(G^n)$ is C-tight. In fact, we have a much stronger property: G^n converges uniformly a.s. to a certain continuous process. Since the pointwise convergence of increasing functions to a continuous function implies the uniform convergence, it is sufficient to verify that

$$\langle X^{1n}\rangle_t \to \langle S\rangle_t = \sigma^2 \int_0^t S_r^2\, dr$$

and

$$\langle X^{2n}\rangle_t \to \int_0^t F^2(r, S_r)\, dr$$

a.s. for $t \in [0,1]$.

Using the independence of increments of the Wiener process, we have

$$\langle X^{1n}\rangle_t = \sum_{t_i \le t} S_{t_{i-1}}^2 E(S_{t_i}/S_{t_{i-1}} - 1)^2 = \sigma^2 \sum_{t_i \le t} S_{t_{i-1}}^2 \Delta t_i + o_n(1),$$

where $o_n(1)$ is a sequence of random variables converging to zero almost surely. This representation implies the first claimed relation.

For $\xi \sim N(0,1)$, we have $E(\xi^2 - 1)^2 = 2$, $E(\xi^2 - 1)(|\xi| - E|\xi|) = \sqrt{2/\pi}$, and $E(|\xi| - E|\xi|)^2 = 1 - 2/\pi$.

Using these formulae and the independence of increments, we get that

$$E(U_i^{n2}|\mathcal{F}_{t_{i-1}}) = F_n^2(t_{i-1})\Delta t_i,$$

where

$$F_n^2(t_{i-1}) := \left[\frac{1}{2}\sigma^4 n\Delta t_i + k_0\sqrt{\frac{2}{\pi}}\sigma^3 (n\Delta t_i)^{1/2} + k_0^2\sigma^2\Big(1-\frac{2}{\pi}\Big)\right]\lambda_{t_{i-1}}$$

with $\lambda_t := \widehat{C}_{xx}^2(t,S_t)S_t^4$.

Defining the process $F_n(t)$ with trajectories that are constant on each interval $[t_{i-1},t_i[$, where they take values $F_n(t_{i-1})$, we obtain that

$$\langle X^{2n}\rangle_t = \sum_{t_i\le t} E(U_i^{n2}|\mathcal{F}_{t_{i-1}}) = \int_0^t F_n^2(r)\,dr.$$

By virtue of Lemma 1.4.12, $(\lambda_t)_{t\in[0,1[}$ is a process with trajectories that are bounded continuous functions for all ω except a P-null set. Recall that $n\Delta t_i = g'(y_i)$, where y_i is a point between $(i-1)/n$ and i/n, and also the differentiation rule for the inverse function, $g'((i-1)/n) = 1/f'(t_{i-1})$. Under any of our two assumptions on the scale transform, g' is bounded. It follows immediately that $F_n^2(r)\to F^2(r,S_r)$ for all $r<1$ and that the limit and the integral are interchangeable in virtue of the Lebesgue dominated convergence theorem (except a P-null set).

So, the second claimed relation holds.

Noting that, for $\xi\sim N(0,1)$,

$$E\big(e^{u\xi-u^2/2}-1\big)\big(\xi^2-1\big) = E\big((\xi+u)^2-1\big) = u^2$$

and

$$\big|E\big(e^{u\xi-u^2/2}-1\big)\big(|\xi|-E|\xi|\big)\big| = \big|E|\xi+u|-E|\xi|\big| \le \kappa u^2,$$

we obtain that $\langle X^{1n},X^{2n}\rangle_t\to 0$ a.s. for all $t\le 1$.

It is easy to check that $\sup_t|\Delta X_t^n|\to_P 0$. Indeed, $\sup_i|\Delta S_{t_i}|\to 0$ and $\sup_i|Z_i^n|\to 0$ almost surely due to the continuity of S and w. Using the Chebyshev inequality, we have

$$\begin{aligned} P\Big(n^{1/2}\sup_i|(\Delta w_{t_i})^2-\Delta t_i|>\varepsilon\Big) &\le \sum_i P(n^{1/2}|(\Delta w_{t_i})^2-\Delta t_i|>\varepsilon) \\ &\le \varepsilon^{-4}n^2\sum_i E\big|(\Delta w_{t_i})^2-\Delta t_i\big|^4 \\ &\le \kappa\varepsilon^{-4}n^2\sum_i(\Delta t_i)^4\to 0, \quad n\to\infty. \end{aligned}$$

Recall that $\sup_t\lambda_t$ is finite a.s. It follows that $\sup_i|Y_i^n|\to_P 0$, and, therefore, $\sup_t|\Delta X_t^n|\to_P 0$. So, the sequence $\mathcal{L}(X^n)$ is C-tight.

The standard method to prove the limit theorem is to show that any convergent subsequence of $\mathcal{L}(X^n)$ has as the limit the measure defined in the formulation of the theorem, i.e., the distribution of the diffusion process

$X = (X_t^1, X_t^2)$ with

$$dX_t^1 = \sigma X_t^1 \, dw_t^1, \qquad X_0^1 = 1,$$
$$dX_t^2 = F(t, X_t^1) \, dw_t^2, \qquad X_0^2 = 0,$$

where (w^1, w^2) is a standard Wiener process in $\mathbf{R}^2$.

Recall that measures coinciding on an algebra also coincide on the σ-algebra generated by the latter. Also, $\mathcal{C}_1 = \sigma\{\bigcup_{T<1} \mathcal{C}_T\}$, where $(\mathcal{C}_t)$ is the natural filtration in the space of continuous functions. Thus, it is sufficient to identify the limit by considering the restrictions of the processes to each interval $[0, T]$, $T < 1$.

1.5.5 Limit Measure

We fix $T \in]0, 1[$ and define on the canonical space $D[0, T]$ (with $d = 2$) the matrix-valued process

$$C = C(t, \alpha) = \int_0^t c(s, \alpha_s) \, ds, \qquad \alpha_s = (\alpha_s^1, \alpha_s^2),$$

where $c(s, x) = a(s, x)a^*(s, x)$, and $a(s, x)$ is the diagonal 2×2-matrix with

$$a^{11}(s, x) = \sigma x^1, \qquad a^{22}(s, x) = F(s, x^1).$$

Notice that the function $(t, x) \mapsto c(t, x)$ is continuous on $[0, T] \times \mathbf{R}^2$. It follows that $C(t, \alpha)$ is continuous in α on the whole space $D[0, T]$ (see, e.g., [161], Chap. 6.2) and the condition (H_1) is fulfilled. Note that we cannot claim this property with $T = 1$.

The local boundedness hypothesis (H_2) holds obviously with $F_t^a = \bar{c}_a t$, where the constant $\bar{c}_a$ is the maximal value of the function $c(t, x)$ on the set $[0, T] \times \{x : \ |x| \leq a\}$.

The "triangle" structure of C allows us to solve the martingale problem recursively. The martingale problem corresponding to the triplet of predictable characteristics $(0, C, 0)$ with the initial condition $x \in \mathbf{R}^2$ has a unique solution Q_x. It is the distribution of the Markov diffusion process X^x with

$$dX_t^{x1} = \sigma X_t^{x1} \, dw_t^1, \qquad X_0^{x1} = x^1,$$
$$dX_t^{x2} = F(t, X_t^{x1}) \, dw_t^2, \qquad X_0^{x2} = x^2.$$

By virtue of the representation given by Lemma 1.4.8 and Assumption 3, the mapping $\Psi : C[0, T] \to C[0, T]$ with $(\Psi y)_t = \widehat{C}_{xx}(t, y_t) y_t^2$ is locally Lipschitz. Using this, we infer that $\sup_{t \leq T} |X_t^{x_n} - X_t^x| \to 0$ as $x_n \to x$. It follows that, for a bounded continuous function ψ on $D[0, T]$, the function $x \mapsto E\psi(X^x)$ is continuous and, hence, Borel. By monotone class arguments we obtain that the function $x \mapsto Q_x(A)$ is Borel whatever is $A \in \mathcal{D}_T$. According to Lemma IX.4.4, hypothesis (H_3) on local uniqueness of the martingale problem is also fulfilled.

1.5.6 Identification of the Limit

To identify the limit, it remains to verify conditions (a), (b), and (c) of Theorem 1.5.2.

For $\varepsilon > 0$, by the Cauchy–Schwarz and Chebyshev inequalities we have

$$E|\Delta X^n_{t_i}|I_{\{|\Delta X^n_{t_i}|>\varepsilon\}} \le \big(E|\Delta X^n_{t_i}|^2\big)^{1/2}\big(P(|\Delta X^n_{t_i}|>\varepsilon)\big)^{1/2} \le \frac{1}{\varepsilon^2}\big(E|\Delta X^n_{t_i}|^2\big)^{1/2}\big(E|\Delta X^n_{t_i}|^4\big)^{1/2}.$$

Similarly,

$$E|\Delta X^n_{t_i}|^2 I_{\{|\Delta X^n_{t_i}|>\varepsilon\}} \le \frac{1}{\varepsilon^2}E|\Delta X_{t_i}|^4.$$

Recall that $E|\Delta S_{t_i}|^{2p} \le \kappa_p n^{-p}$ for $p \ge 1$. By virtue of Lemma 1.4.12, the random variable $\sup_{t\le T}\widehat{C}^{2p}_{xx}(t,S_t)S_t^{4p}$ for $T<1$ is dominated, up to a multiplicative constant depending on T, by a power of the random variable $\sup_{t\le 1} S_t$. The latter has moments of any order, and we easily infer that

$$\max_i E|\Delta U^n_i|^{2p} \le \kappa_{p,T} n^{-p},$$

where the maximum is taken over the set of indices i for which $t_i \le T$.

Put $\bar h(x) := x - h(x)$. Since $E(\Delta X^n_{t_i}|\mathcal{F}_{t_{i-1}}) = 0$, we have

$$\sup_{t\le T}|B^n_t| \le \sum_{t_i\le T}\big|E\big(h(\Delta X^n_{t_i})|\mathcal{F}_{t_{i-1}}\big)\big| = \sum_{t_i\le T}\big|E\big(\bar h(\Delta X^n_{t_i})|\mathcal{F}_{t_{i-1}}\big)\big|. \tag{1.5.3}$$

Using the inequality $|\bar h(x)| \le |x|I_{\{|x|>1\}}$, from the above bounds we obtain that

$$E\sup_{t\le T}|B^n_t| \le \sum_{t_i\le T} E|\Delta X^n_{t_i}|I_{\{|\Delta X^n_{t_i}|>1\}} = O\big(n^{-1/2}\big), \qquad n\to\infty,$$

and this property is stronger than (a).

In a similar way we get that, for any $\varepsilon > 0$,

$$\sum_{t_i\le T} E|\Delta X^n_{t_i}|^2 I_{\{|\Delta X^n_{t_i}|>\varepsilon\}} = O\big(n^{-1}\big), \qquad n\to\infty. \tag{1.5.4}$$

The latter property implies (c).

Since $1-\delta^2(x) \le I_{\{|x|>1\}}$ and $1-\delta(x)\le I_{\{|x|>1\}}$, we have that

$$\sum_{t_i\le t} E\big(|\Delta X^n_{t_i}|^2(1-\delta^2(\Delta X^n_{t_i}))|\mathcal{F}_{t_{i-1}}\big) \le \sum_{t_i\le t} E\big(|\Delta X^n_{t_i}|^2 I_{\{|\Delta X^n_{t_i}|>1\}}|\mathcal{F}_{t_{i-1}}\big),$$

$$\Big|\sum_{t_i\le t} E\big(\Delta X^n_{t_i}(1-\delta(\Delta X^n_{t_i}))|\mathcal{F}_{t_{i-1}}\big)\Big| \le \sum_{t_i\le t} E\big(|\Delta X^n_{t_i}|I_{\{|\Delta X^n_{t_i}|>1\}}|\mathcal{F}_{t_{i-1}}\big).$$

For every $t \in [0,T]$, the right-hand sides of the above inequalities tend to zero in L^1 and, hence, in probability as $n \to \infty$. Thus, the sequence $\tilde{C}_t^n$ has the same limit as $\langle X^n \rangle_t$, i.e., $\langle X \rangle_t = \int_0^t c(s,S_s)\,ds$. On the other hand,

$$\int_0^t c^{11}(s,X_s^n)\,ds = \sigma^2 \int_0^t \sum_i S_{t_{i-1}}^2 I_{[t_{i-1},t_i[}(s)\,ds \to \sigma^2 \int_0^t S_s^2\,ds$$

and

$$\int_0^t c^{22}(s,X_s^n)\,ds = \sigma^2 \int_0^t \sum_i F^2(s,S_{t_{i-1}}) I_{[t_{i-1},t_i[}(s)\,ds \to \sigma^2 \int_0^t F^2(s,S_s)\,ds.$$

That is, $\tilde{C}_t^n - C_t(X^n) \to_P 0$, and this implies (b).

1.6 Superhedging by Buy-and-Hold

1.6.1 Levental–Skorokhod Theorem

Let $S_t = e^{M_t}$, where M is a continuous semimartingale on $[0,T]$ starting from zero. Recall that any adapted right-continuous process L of bounded variation admits the Hahn decomposition $L = L^+ - L^-$ into a difference of two adapted right-continuous increasing processes with $L_0^+ = L_0$ and $L_0^- = 0$ such that the total variation process $|L| = L^+ + L^-$.

The main objects of our interest will be the processes of the form

$$Y = L_- \cdot S - \lambda S \cdot |L|, \tag{1.6.1}$$

where $\lambda > 0$ is a fixed number. If L is piecewise constant, L can be interpreted as a "simple" portfolio strategy and Y as the corresponding value process (starting from zero) in a market with proportional transaction costs λ.

We immediately formulate the assumption on M needed for the central result. To this aim, we define $\varepsilon > 0$ as the solution of the equation $\lambda := e^{2\varepsilon} - 1$ and introduce the sequence of stopping times (τ_n) forming the ε-chain for M, i.e., in this sequence, $\tau_0 := 0$, and the other terms are defined recursively by

$$\tau_n := \inf\{t \ge \tau_{n-1} : \ |M_t - M_{\tau_{n-1}}| = \varepsilon\} \wedge T, \quad n \ge 1.$$

Using the ε-chain, we define the (discrete-time) "imbedded" process, or "skeleton", of M by putting $\tilde{M}_n := M_{\tau_n}$ on the set $\tau_n < T$ and $\tilde{M}_n := \tilde{M}_{n-1}$ on its complement. Its increments $\Delta\tilde{M}_n := \tilde{M}_n - \tilde{M}_{n-1}$ take three values: $-\varepsilon$, ε, and 0. We shall also consider the processes $\tilde{S}_n := S_{\tau_n}$, $\tilde{L}_n := L_{\tau_n}$, and $\tilde{Y}_n := Y_{\tau_n}$ and the filtration $\tilde{\mathcal{F}}_n$ with $\tilde{\mathcal{F}}_n := \mathcal{F}_{\tau_n}$.

We further assume that M satisfies the following hypothesis:

$\mathbf{H}^\varepsilon$. For every $n \geq 0$,

$$P(\tilde{M}_{\tau_{n+1}} - \tilde{M}_{\tau_n} = \varepsilon | \mathcal{F}_{\tau_n}) > 0 \quad \text{on } \{\tau_n < T\}, \tag{1.6.2}$$

$$P(\tilde{M}_{\tau_{n+1}} - \tilde{M}_{\tau_n} = -\varepsilon | \mathcal{F}_{\tau_n}) > 0 \quad \text{on } \{\tau_n < T\}. \tag{1.6.3}$$

Clearly, the Wiener process with drift satisfies $\mathbf{H}^\varepsilon$ for every $\varepsilon > 0$, and so the standard model with a geometric Brownian motion S as the price process falls in the scope of the present setting. The above conditions can be rewritten as

$$P(\Delta\tilde{M}_{n+1} = \varepsilon | \tilde{\mathcal{F}}_n) > 0 \quad \text{on } \{\Delta\tilde{M}_n \neq 0\}, \tag{1.6.4}$$

$$P(\Delta\tilde{M}_{n+1} = -\varepsilon | \tilde{\mathcal{F}}_n) > 0 \quad \text{on } \{\Delta\tilde{M}_n \neq 0\}. \tag{1.6.5}$$

Let $\mathcal{G}$ be the set of convex functions $g : [0, \infty[\to \mathbf{R}$ with $g(0) = 0$ and $g(y) \sim y$ as $y \to \infty$. An important example: $g(y) = (y - K)^+$, where $K > 0$.

Theorem 1.6.1 *Let $g \in \mathcal{G}$. If $x + Y \geq g(S)$, then $x \geq 1$.*

This theorem due to Levental and Skorokhod is the principal result of the section. Its proof is nontrivial and will be given later, after some preliminary work. Contrary to this, the assertion below is its easy consequence, and we establish it immediately.

Corollary 1.6.2 *Let $g \in \mathcal{G}$. Assume that $S \in \mathcal{M}$ and $Y \geq -\kappa(1 + S)$ for some constant $\kappa > 0$. If $x + Y_T \geq g(S_T)$, then $x \geq 1$.*

Proof. In virtue of the assumed lower bound for Y, we have

$$\lambda S \cdot |L| =:= L \cdot S - Y \leq L \cdot S + \kappa(1 + S).$$

Thus, $\lambda S \cdot |L|$ is locally integrable increasing process, and, therefore, Y is a local supermartingale. Due to the convexity of g, so is the process $-g(S)$ and also the process $X := x + Y - g(S)$. The latter, being bounded from below by a martingale, is a true supermartingale (by the Fatou lemma). But a supermartingale, nonnegative at T, is nonnegative on the whole interval $[0, T]$, i.e., the hypothesis of the theorem is fulfilled. □

The main assumption of the theorem is the inequality between two processes. It corresponds to the hedging of an American option when the investor's portfolio should dominate the pay-off process to meet at any time the eventual exercise of the option. Recall that we choose the unit of stock at time zero to be equal to the unit of money. Thus, the conclusion of the theorem means that the minimal initial endowment for the seller which guarantees to be always on the safe side allows one to buy the unit share of stock and hold it until the exercise date. This trivial strategy is always feasible. The corollary covering the situation with the European options yields the similar conclusion, but in this case one needs an additional admissibility assumption on the strategy which is fulfilled automatically for American options.

Notice that in the formulation of the above assertions one can replace the class G by the wider class F of the Borel functions f which are bounded from below and such that $f(0+) = 0$ and $\liminf_{y\to\infty} f(y)/y \geq 1$. Indeed, if the inequality $x + Y \geq f(S)$ holds for $f \in F$, then the inequality $x + Y \geq g(S)$ holds for $g \in G$. To see this, it is sufficient to consider the convex envelope of the function $f(y) \wedge y$.

1.6.2 Proof

The structure of the proof is very simple. It is based on a reduction to the discrete-time processes by noting that, for every n, the following inequality holds:

$$g(\tilde{S}_n) - \tilde{Y}_n \geq g(\tilde{S}_n) - \sum_{k=0}^{n-1} \tilde{L}_k(\tilde{S}_{k+1} - \tilde{S}_k), \tag{1.6.6}$$

see Lemma 1.6.4 below. With this observation, it remains to show that, for every $c < 1$, the supremum over n of the right-hand side is greater than c with positive probability. For this, it is sufficient to construct an *arbitrary* discrete-time $(\mathcal{F}_n)$-adapted process $\bar{S}$ such that the increments of its logarithm $\bar{M} := \ln \bar{S}$ take only two values ε and $-\varepsilon$ and

$$\sup_n \left[g(\bar{S}_n) - \sum_{k=0}^{n-1} \tilde{L}_k(\bar{S}_{k+1} - \bar{S}_k) \right] > c \quad \text{(a.s.)}. \tag{1.6.7}$$

Indeed, the property of increments alone and the hypothesis $\mathbf{H}^\varepsilon$ imply that $P(\bar{S}_n = \tilde{S}_n,\ \forall n \leq N) > 0$ for every finite integer N (Lemma 1.6.3), and the claimed property follows.

Lemma 1.6.3 *Let w_n be a sequence of $\tilde{\mathcal{F}}_{n-1}$-measurable r.v.'s such that $w_0 = 0$ and $|\Delta w_n| = \varepsilon$, $n \geq 1$. Let $A_N := \{\tilde{M}_n = w_n,\ n \leq N\}$. Then $P(A_N) > 0$.*

Proof. For $N = 0$, the claim is trivial. Assume that $P(A_k) > 0$. At least one of the sets $A_k \cap \{\Delta w_{k+1} = \varepsilon\}$ or $A_k \cap \{\Delta w_{k+1} = -\varepsilon\}$, say, the first one, has a positive probability. Then

$$P(A_{k+1}) \geq P(A_k,\ \Delta w_{k+1} = \varepsilon,\ \Delta\tilde{M}_{k+1} = \varepsilon) > 0$$

in virtue of $\mathbf{H}^\varepsilon$. □

Lemma 1.6.4 *Let σ, τ be stopping times such that $\sigma \leq \tau \leq T$. If $|M_t - M_\sigma| \leq \varepsilon$ for all $t \in]\sigma, \tau]$, then*

$$Y_\tau - Y_\sigma \leq L_\sigma(S_\tau - S_\sigma).$$

Proof. With L of bounded variation and S continuous, the Itô formula for the product coincides with the classical one, and, hence, $L\,dS = d(LS) - S\,dL$. It follows that

$$\int_\sigma^\tau L_t\,dS_t = L_\sigma(S_\tau - S_\sigma) + \int_\sigma^\tau (S_\tau - S_t)\,dL_t \le L_\sigma(S_\tau - S_\sigma) + \int_\sigma^\tau |S_\tau - S_t|\,d|L|_t.$$

For $t \in]\sigma, \tau]$, due to the assumed bounds $e^{-\varepsilon} \le S_t/S_\sigma \le e^{\varepsilon}$, we have

$$|S_\tau - S_t| \le (e^\varepsilon - e^{-\varepsilon})S_\sigma \le (e^{2\varepsilon} - 1)S_t = \lambda S_t$$

according to the definition of ε. Using this, we get the inequality

$$\int_\sigma^\tau L_t\,dS_t \le L_\sigma(S_\tau - S_\sigma) + \lambda \int_\sigma^\tau S_t\,d|L|_t,$$

equivalent to the claimed one. □

By "telescoping" and by the above lemma we infer that

$$\tilde{Y}_n = Y_{\tau_n} = Y_0 + \sum_{k=0}^{n-1}(Y_{\tau_{k+1}} - Y_{\tau_k}) \le \sum_{k=0}^{n-1} L_{\tau_k}(S_{\tau_{k+1}} - S_{\tau_k}) = \sum_{k=0}^{n-1} \tilde{L}_k(\tilde{S}_{k+1} - \tilde{S}_k).$$

This inequality is equivalent to (1.6.6).

The most delicate part of the proof is constructing a process $\bar{S} = (\bar{S}_n)$ satisfying the indicated properties. This can be done as follows.

Let $c \in]0,1[$, and let $f : \mathbf{R} \to [0,1]$ be a strictly increasing function with

$$c < f(0) < f(\infty) < 1.$$

For an arbitrary sequence of reals $l = (l_k)_{k\ge 0}$, we define recursively the integer-valued sequence $m = (m_k)$ and the sequence $s = (s_k)$ with $s_k := e^{\varepsilon m_k}$ by putting $m_0 := 0$ and

$$m_{k+1} := m_k + I_{\{l_k < f(s_k)\}} - I_{\{l_k \ge f(s_k)\}}, \quad k \ge 0.$$

We also put $\Sigma_0 := 0$ and

$$\Sigma_n := \sum_{k=0}^{n-1} l_k(s_{k+1} - s_k).$$

Obviously, $s_k = w_k(l_0, \dots, l_{k-1})$ for some Borel function $w_k : \mathbf{R}^k \to \mathbf{R}$.

To accomplish the proof of Theorem 1.6.1, it remains to show that the $(\mathcal{F}_n)$-predictable process $\bar{S}$ with $\bar{S}_n := s_n(\tilde{L}_0, \dots, \tilde{L}_{n-1})$ satisfies inequality (1.6.7). This is implied by the following assertion, which provides a property of the introduced sequences which is even stronger than the needed one.

Proposition 1.6.5 *Let $g \in \mathcal{G}$. There exists $N = N(c, \varepsilon)$ such that*

$$\sup_{n \le N} \big(g(s_n) - \Sigma_n\big) > c \tag{1.6.8}$$

whatsoever is the sequence (l_k).

Proof. We start with some estimates.

Lemma 1.6.6 *Let $n \ge k$. Then*

$$\Sigma_n - \Sigma_k < f(\infty)(s_n - s_k)I_{\{s_n > s_k\}} + f(0)(s_n - s_k)I_{\{s_n < s_k\}}. \tag{1.6.9}$$

Let $p \in \mathbf{N}$ and $n > 2p$. If $|m_k| < p$ for all $k \le n$, then

$$\Sigma_n \le f(\infty)\big(e^{p\varepsilon} - e^{-p\varepsilon}\big) - (n - 2p)\theta/2, \tag{1.6.10}$$

where

$$\theta := \inf_{|k| \le p} \big(f\big(e^{(k+1)\varepsilon}\big) - f\big(e^{k\varepsilon}\big)\big)\big(e^{(k+1)\varepsilon} - e^{k\varepsilon}\big).$$

Proof. By definition,

$$\Sigma_{k+1} - \Sigma_k = l_k(s_{k+1} - s_k) < f(s_k)(s_{k+1} - s_k). \tag{1.6.11}$$

Since f is increasing, inequality (1.6.9) holds for $n = k + 1$. Notice also that if $m_i = m_{j+1}$ and $m_{i+1} = m_j$, $i \ne j$, then

$$(\Sigma_{i+1} - \Sigma_i) + (\Sigma_{j+1} - \Sigma_j) < -\big(f(s_{i+1}) - f(s_i)\big)(s_{i+1} - s_i) \tag{1.6.12}$$

because (1.6.11) implies that the left-hand side above is dominated by

$$f(s_i)(s_{i+1} - s_i) + f(s_j)(s_{j+1} - s_j) = f(s_i)(s_{i+1} - s_i) + f(s_{i+1})(s_i - s_{i+1}).$$

The function f being increasing, the right-hand side of (1.6.12) is always negative. With this important observation, we easily complete the proof. Indeed,

$$\Sigma_n - \Sigma_k = \sum_{q=k}^{n-1} (\Sigma_{q+1} - \Sigma_q).$$

In the sum of one-step increments we can couple, for each $r \in \mathbf{Z}$, pairs of indices corresponding to up-crossings and down-crossings by the sequence (m_q) of each interval $[r, r+1[$. These pairs give negative contributions to the sum. For the uncoupled pairs, all the increments $s_{q+1} - s_q$ are either positive or negative in dependence of the sign of $s_n - s_k$, and their sum equals to $s_n - s_k$. Thus, inequality (1.6.9) follows from the one-step bound and monotonicity of f. With these arguments, the bound (1.6.10) is also obvious: the first term of the majorant is the upper bound for uncoupled summands, while the second expresses that every coupled pair of summands is dominated, in accordance with (1.6.12), by $-\theta$. □

Since $g(0+) = 0$ and $g(y) \sim y$ as $y \to \infty$, there is an integer p such that for $a := e^{-p\varepsilon}$ and $b := e^{p\varepsilon}$, we have

$$g(a) - f(0)a > c - f(0), \qquad b\Big(\frac{g(b)}{b} - f(\infty)\Big) > c.$$

Take $N > 2p$ such that

$$\kappa - f(\infty)b + (N - 2p)\theta/2 > c,$$

where κ is the lower bound for g.

If $s_n \in]a, b[$ for all $n \le N$, then by (1.6.10)

$$g(s_N) - \Sigma_N \ge \kappa - f(\infty)b + (N - 2p)\theta/2 > c.$$

Otherwise, either $s_n = a$ or $s_n = b$ for some $n \le N$, and in virtue of (1.6.9) we have, respectively,

$$g(s_n) - \Sigma_n \ge g(a) - f(0)(a - 1) = f(0) + \big(g(a) - f(0)a\big) > c$$

or

$$g(s_n) - \Sigma_n \ge g(b) - f(\infty)(b - 1) > b\Big(\frac{g(b)}{b} - f(\infty)\Big) > c.$$

In all cases (1.6.8) holds, and Proposition 1.6.5 is proven. □

1.6.3 Extensions for One-Side Transaction Costs

It is clear that the previous results hold true with obvious modifications for the processes of the form

$$Y = L \cdot S - \mu_1 S \cdot L^+ - \mu_2 S \cdot L^- \tag{1.6.13}$$

with both μ_i strictly positive (because $Y \le L \cdot S - \lambda S \cdot |L|$, where $\lambda = \mu_1 \wedge \mu_2$). It is less evident that it will be the case where one of the coefficients vanishes. Nevertheless, the arguments can be modified and extended in an appropriate way to produce the same conclusions.

First, let us consider the model where

$$Y = L \cdot S - \lambda S \cdot L^+ \tag{1.6.14}$$

where $\lambda > 0$, that is, only buying of stocks is charged. The analysis of the proof is based again on a reduction to the imbedded process.

We start with a revision of the basic inequalities. Choose $\varepsilon > 0$ small enough to ensure the inequalities $\lambda > 2\alpha/(1 - \alpha) > 0$, where $\alpha := e^{2\varepsilon} - 1$.

Lemma 1.6.7 *Let σ, τ be stopping times such that $\sigma \le \tau \le T$. If $|M_t - M_\sigma| \le \varepsilon$ for all $t \in]\sigma, \tau]$, then*

$$Y_\tau - Y_\sigma \le \frac{1}{1-\alpha} L_\sigma (S_\tau - S_\sigma) - \frac{\alpha}{1-\alpha}(L_\tau S_\tau - L_\sigma S_\sigma). \tag{1.6.15}$$

On the set $\{M_\tau = M_\sigma + \varepsilon\}$ we have the inequality

$$Y_\tau - Y_\sigma \le L_\sigma (S_\tau - S_\sigma) - \lambda' S_\sigma (L_\tau^+ - L_\sigma^+), \tag{1.6.16}$$

where $\lambda' := (\lambda - \alpha)e^{-\varepsilon}$.

Proof. Repeating the proof of Lemma 1.6.4, we use the integration by parts to replace the integral with respect to S by the integral with respect to L and, estimating the latter, arrive at the inequality

$$\int_\sigma^\tau L_t\, dS_t \le L_\sigma (S_\tau - S_\sigma) + \alpha \int_\sigma^\tau S_t\, d|L|_t.$$

Using the identity $|L| = 2L^+ - L$, we replace the integral with respect to $|L|$ by the integrals with respect to L^+ and L. Applying again the integration by parts formula to the second one (to be back with the integral with respect to S), we rewrite the above inequality as

$$\int_\sigma^\tau L_t\, dS_t \le L_\sigma (S_\tau - S_\sigma) + 2\alpha \int_\sigma^\tau S_t\, dL_t^+ - \alpha (L_\tau S_\tau - L_\sigma S_\sigma) + \alpha \int_\sigma^\tau L_t\, dS_t.$$

"Solving" it, we get that

$$\int_\sigma^\tau L_t\, dS_t \le \frac{1}{1-\alpha} L_\sigma (S_\tau - S_\sigma) + \frac{2\alpha}{1-\alpha} \int_\sigma^\tau S_t\, dL_t^+ - \frac{\alpha}{1-\alpha}(L_\tau S_\tau - L_\sigma S_\sigma).$$

Due to the assumed inequality $\lambda > 2\alpha/(1-\alpha)$, this implies (1.6.15).

Since $0 \le S_\tau - Z_t \le \alpha Z_t$ for $t \in [\sigma, \tau]$, we have that

$$\int_\sigma^\tau (S_\tau - S_t)\, dL_t \le \alpha \int_\sigma^\tau S_t\, dL_t^+.$$

Using this, we infer from the integration by parts formula that

$$\int_\sigma^\tau L_t\, dS_t \le L_\sigma (S_\tau - S_\sigma) + \alpha \int_\sigma^\tau S_t\, dL_t^+.$$

So,

$$\int_\sigma^\tau L_t\, dS_t - \lambda \int_\sigma^\tau S_t\, dL_t^+ \le L_\sigma (S_\tau - S_\sigma) - (\lambda - \alpha) \int_\sigma^\tau S_t\, dL_t^+,$$

and the result follows because $e^{-\varepsilon} S_\sigma \le S_t$ for $t \in [\sigma, \tau]$. □

The first inequality of the above lemma will be used to show that, when the hedge is nontrivial, there is a strictly positive probability that it will not work at one of the stopping times of the ε-chain (for suitably chosen ε) unless the number of shares at that time is negative. The second inequality, giving the bound on the gain when the stock price goes up, will be used on the set where the latter event happened. Roughly speaking, the bad things always may arrive if the hedge has a short position and the stock price (along the ε-chain) goes up and up.

Theorem 1.6.8 *Suppose that* $\mathbf{H}^\varepsilon$ *holds for values of* $\varepsilon > 0$ *arbitrarily close to zero. Let* $g \in \mathcal{G}$*, and let* Y *be given by (1.6.14). If* $x+Y \geq g(S)$*, then* $x \geq 1$.

Proof. Fix $\eta > 0$. Choose $\varepsilon > 0$ small enough not only to apply the above lemma (i.e., to satisfy the bounds $\lambda > 2\alpha/(1-\alpha) > 0$, where $\alpha := e^{2\varepsilon} - 1$) but also to ensure the following inequalities:

$$\frac{\alpha}{1-\alpha} L_0 < \eta, \qquad \frac{\alpha}{1-\alpha} < e^{3\varepsilon} - 1, \qquad \lambda' = (\lambda - \alpha)e^{-\varepsilon} > e^{3\varepsilon} - e^{\varepsilon}. \tag{1.6.17}$$

It follows from (1.6.15) that

$$\tilde{Y}_n \leq \frac{1}{1-\alpha} \sum_{k=0}^{n-1} \tilde{L}_k(\tilde{S}_{k+1} - \tilde{S}_k) - \frac{\alpha}{1-\alpha} \tilde{L}_n \tilde{S}_n + \frac{\alpha}{1-\alpha} L_0. \tag{1.6.18}$$

Taking into account (1.6.17), from this we immediately obtain the following inequality, which will replace (1.6.6):

$$g(\tilde{S}_n) - \tilde{Y}_n + \eta \geq g(\tilde{S}_n) - \frac{1}{1-\alpha} \sum_{k=0}^{n-1} \tilde{L}_k(\tilde{S}_{k+1} - \tilde{S}_k) + \frac{\alpha}{1-\alpha} \tilde{L}_n \tilde{S}_n. \tag{1.6.19}$$

Since $\eta > 0$ is arbitrary, it remains to check that, for any constant $c \in]0,1[$, the supremum of the right-hand side dominates c with positive probability. To this aim, we define the auxiliary function $h(z) := g(z) \wedge g(e^{3\varepsilon} z)$ and the sets

$$B_n := \left\{ h(\tilde{S}_n) - \frac{1}{1-\alpha} \sum_{k=0}^{n-1} \tilde{L}_k(\tilde{S}_{k+1} - \tilde{S}_k) > c \right\}.$$

Applying the previous results for the function $h \in G$ and the strategy $L/(1-\alpha)$ (namely, Proposition 1.6.5 and Lemma 1.6.3), we note that there is n such that $P(B_n,\ \tau_n < 1) > 0$. Thus, either $P(\{\tilde{L}_n \geq 0, \tau_n < 1\} \cap B_n) > 0$ (in this case the claimed property is obvious) or

$$P(\{\tilde{L}_n < 0, \tau_n < 1\} \cap B_n) > 0. \tag{1.6.20}$$

To complete the proof, it is sufficient to verify that in the latter case

$$P\big(g(\tilde{S}_{n+3}) - \tilde{Y}_{n+3} \geq c - \eta\big) > 0. \tag{1.6.21}$$

Let us consider the set $\Gamma_n := \{\Delta\tilde{M}_{n+k} = \varepsilon,\ 1 \le k \le 3\}$, where the imbedded process has three consecutive moves upwards. Due to the hypothesis on the process fluctuations, (1.6.20) implies that the set $\{\tilde{L}_n < 0,\ \tau_n < 1\} \cap B_n \cap \Gamma_n$ has strictly positive probability. According to the second inequality of Lemma 1.6.7, on this set we have

$$\tilde{Y}_{n+3} - \tilde{Y}_n \le \sum_{k=n}^{n+2} \tilde{L}_k(\tilde{S}_{k+1} - \tilde{S}_k) - \lambda' \sum_{k=n}^{n+2} \tilde{S}_k(\tilde{L}^+_{k+1} - \tilde{L}^+_k) \tag{1.6.22}$$

and, hence, due to the bound (1.6.18) and the definition of h, we have that

$$g(\tilde{S}_{n+3}) - \tilde{Y}_{n+3} + \frac{\alpha}{1-\alpha}L_0 \ge h(\tilde{S}_n) - \frac{1}{1-\alpha}\sum_{k=0}^{n-1} \tilde{L}_k(\tilde{S}_{k+1} - \tilde{S}_k) + R_n,$$

where

$$R_n := \frac{\alpha}{1-\alpha}\tilde{L}_n\tilde{S}_n - \sum_{k=n}^{n+2} \tilde{L}_k(\tilde{S}_{k+1} - \tilde{S}_k) + \lambda' \sum_{k=n}^{n+2} \tilde{S}_k(\tilde{L}^+_{k+1} - \tilde{L}^+_k)$$

on the considered set $R_n \ge 0$. Indeed,

$$\begin{aligned}
R_n &\ge \frac{\alpha}{1-\alpha}\tilde{L}_n\tilde{S}_n - \sum_{k=n}^{n+2} (\tilde{L}_n + \tilde{L}^+_k - \tilde{L}^+_n)(\tilde{S}_{k+1} - \tilde{S}_k) + \lambda' \sum_{k=n}^{n+2} \tilde{S}_k(\tilde{L}^+_{k+1} - \tilde{L}^+_k) \\
&\ge \left[\frac{\alpha}{1-\alpha} - (e^{3\varepsilon} - 1)\right]\tilde{L}_n\tilde{S}_n - (e^{3\varepsilon} - e^{\varepsilon})\tilde{S}_n(\tilde{L}^+_{n+2} - \tilde{L}^+_n) \\
&\quad + \lambda'\tilde{S}_n(\tilde{L}^+_{n+2} - \tilde{L}^+_n) \\
&\ge \left[\frac{\alpha}{1-\alpha} - (e^{3\varepsilon} - 1)\right]\tilde{L}_n\tilde{S}_n + [\lambda' - (e^{3\varepsilon} - e^{\varepsilon})]\tilde{S}_n(\tilde{L}^+_{n+2} - \tilde{L}^+_n) \ge 0
\end{aligned}$$

because of our choice of ε and the negativity of $\tilde{L}_n$.

The claimed property (1.6.21) follows now from definition of B_n. □

The result for the model

$$Y = L \cdot S - \lambda S \cdot L^-, \tag{1.6.23}$$

$\lambda > 0$, describing the situation where only selling of stocks is charged, is similar.

Theorem 1.6.9 *Suppose that* $\mathbf{H}^\varepsilon$ *holds for values of* $\varepsilon > 0$ *arbitrarily close to zero. Let* $g \in \mathcal{G}$, *and let* Y *be given by (1.6.23). If* $x + Y \ge g(S)$, *then* $x \ge 1$.

The reasoning is based on the following lemma the proof of which is analogous to that of Lemma 1.6.7 and left to the reader.

Lemma 1.6.10 *Let σ, τ be stopping times such that s $\sigma \le \tau \le T$. Suppose that $\lambda > 2\alpha/(1+\alpha)$, where $\alpha := e^{2\varepsilon} - 1$. If $|M_t - M_\sigma| \le \varepsilon$ for all $t \in]\sigma, \tau]$, then*

$$Y_\tau - Y_\sigma \le \frac{1}{1+\alpha} L_\sigma (S_\tau - S_\sigma) - \frac{\alpha}{1+\alpha}(L_\tau S_\tau - L_\sigma S_\sigma). \tag{1.6.24}$$

On the set $\{M_\tau = M_\sigma - \varepsilon\}$ we have the inequality

$$Y_\tau - Y_\sigma \le L_\sigma (S_\tau - S_\sigma) - \lambda' S_\sigma (L_\tau^- - L_\sigma^-), \tag{1.6.25}$$

where $\lambda' := (\lambda - \alpha)e^{-\varepsilon}$.

The auxiliary function needed in the proof of the theorem can be given as $h(z) = g(z) \wedge g(ze^{-3\varepsilon})$. Note that $h(z)/z \to e^{-3\varepsilon}$ (and not to 1) as $z \to \infty$. Nevertheless, Proposition 1.6.5 holds for such a function but with $c < e^{-3\varepsilon}$. Since $e^{-3\varepsilon}$ can be chosen arbitrarily close to 1, this does not matter, and the arguments work well with obvious changes (e.g., moves upwards should be replaced by moves downwards).

At last, the proofs of corresponding corollaries for the European options in the case of one-side transaction costs are exactly the same as that of Corollary 1.6.2.

1.6.4 Hedging of Vector-Valued Contingent Claims

Until now we considered hedging problems for models where the initial and final transactions were charge-free and the accounting was on the lump sum of the asset values. However, the final adjustment may not be exempted from transaction costs. In such a case, e.g., the hedging of a call option will depend on the clause of the contract whether the stock is delivered to the buyer or the latter receives only its money equivalent equal to the market price. More elaborate models, admitting transaction costs for any portfolio revisions and separating the accounting for each asset, allow us to incorporate the mentioned specificity. A general theory of such models will be developed in the next chapters. Here we present, in a rather sketchy way, a two-asset model of this kind assuming that the first one is the *numéraire*, the price process of the second is a continuous martingale $S > 0$, $S_0 = 1$, and at least one of the transaction cost coefficients λ^{12}, λ^{21} is strictly positive. The notation with superscripts in the two-dimensional case looks rather awkward, but it is chosen to be coherent with that we shall use later for multi-asset models.

We suppose that the process $M := \ln S$ satisfies condition $\mathbf{H}^\varepsilon$ for all sufficiently small $\varepsilon > 0$.

Now the value process $V = V^{v,L}$ of a portfolio is two-dimensional, and its components are

$$V^1 = v^1 + L^{21} - (1 + \lambda^{12})L^{12},$$
$$V^2 = v^2 + \widehat{V}^2 \cdot S + L^{12} - (1 + \lambda^{21})L^{21},$$

where L^{12}, L^{21} are increasing adapted right-continuous processes representing the accumulated net wealth transferred to a corresponding position, V_t^1 is the wealth on the bank account, V_t^2 is the wealth on the stock account, while $\widehat{V}_t^2 := V_t^2/S_t$ is the number of stock units held at time t.

An important concept of the model is the solvency cone K defined as the set of points $x = (x^1, x^2)$ for which there exist $a^{12}, a^{21} \geq 0$ such that

$$x^1 + a^{21} - (1 + \lambda^{12})a^{12} \geq 0,$$
$$x^2 + a^{12} - (1 + \lambda^{21})a^{21} \geq 0.$$

The closed convex cone K generates a partial ordering in $\mathbf{R}^2$; namely, $x \succeq y$ if $x - y \in K$.

It is easy to check that the vectors $w_1 := (1, 1+\lambda^{12})$ and $w_2 := (1+\lambda^{21}, 1)$ are generators of the dual positive cone $K^* := \{w : \ wx \geq 0 \ \forall x \in K\}$. Recall that $K^{**} = K$ and, hence, $x \succeq y$ if and only if $wx \geq wy$ for all $w \in K^*$. Of course, the inequality for all $w \in K^*$ holds if and only if it holds for the generators: the symbol $x \succeq y$ replaces comfortably the system of two linear inequalities $w_i x \geq w_i y$, $i = 1, 2$.

It is easy to see that the vectors $(1 + \lambda^{12}, -1)$ and $(-1, 1 + \lambda^{21})$ generate the solvency cone K.

The set of hedging endowment for the contingent claim $C = (C^1, C^2)$ is

$$\Gamma := \{v \in \mathbf{R}^2 : \ \exists L \text{ such that } V_T^{v,L} \succeq C\}.$$

The agent strategy L in the above definition should be admissible. This means that $V \succeq -\kappa S$ for some $\kappa > 0$.

Assume that $C = g(S_T)$, where $g :]0, \infty[\to \mathbf{R}^2$ is such that the functions $w_i g(y)$ are bounded from below with $w_i g(0+) = 0$, $i = 1, 2$,

$$\liminf_{y \to \infty} w_1 g(y)/y = 1 + \lambda^{12},$$
$$\liminf_{y \to \infty} w_2 g(y)/y = 1.$$

Examples of the pay-off functions for the call option:

(1) the stock is delivered $g^1(y) := -KI_{\{y \geq 0\}}$, $g^2(y) := yI_{\{y \geq 0\}}$;
(2) the stock is not delivered $g^1(y) := (y - K)^+ I_{\{y \geq 0\}}$, $g^2(y) := 0$.

Theorem 1.6.11 *If a point v is in Γ, then the following two linear equations are satisfied:*

$$v^2 \geq 1 - (1 + \lambda^{21})v^1,$$
$$v^2 \geq 1 - \frac{1}{1 + \lambda^{12}} v^1.$$

Proof. Let $v \in \Gamma$. The relation $V_T^{v,L} \succeq C$ holds if and only if, for all $\alpha, \beta \in]0,1[$ such that $\alpha + \beta = 1$,

$$(\alpha w_1 + \beta w_2) V_T^{v,L} \geq (\alpha w_1 + \beta w_2) C.$$

Substituting the expression for $V_T^{v,L}$ and setting $\kappa_0 := (1+\lambda^{12})(1+\lambda^{21}) - 1$ and $\tilde{L}^{12} = (1/S) \cdot L^{12}$, $\tilde{L}^{21} = (1/S) \cdot L^{21}$, we can rewrite this inequality as

$$(\alpha w_1 + \beta w_2)v + (1 + \alpha\lambda^{12})\widehat{V}^2 \cdot S_t - \kappa_0\big(\alpha S \cdot \tilde{L}^{21} + \beta S \cdot \tilde{L}^{12}\big) \geq (\alpha w_1 + \beta w_2)C.$$

Notice also that the process

$$\widehat{V}^2 = \widehat{V}^2_{0-} + \tilde{L}^{12} - (1+\lambda^{21})\tilde{L}^{21}$$

is of bounded variation. It follows that the left-hand side of the above inequality is dominated by $(\alpha w_1 + \beta w_2)v + Y$, where Y has the structure demanded in Theorem 1.6.2. The assumptions on the function $g = (g^1, g^2)$ ensure that this theorem can be applied (after an appropriate normalization) yielding the inequality

$$(\alpha w_1 + \beta w_2)v \geq \alpha(1+\lambda^{12}) + \beta.$$

By continuity it holds also for α equal to 0 and 1, and we get the system

$$\begin{aligned} (1+\lambda^{21})v^1 + v^2 &\geq 1, \\ v^1 + (1+\lambda^{12})v^2 &\geq 1 + \lambda^{12}. \end{aligned}$$

The theorem is proven. □

2

Arbitrage Theory for Frictionless Markets

2.1 Models without Friction

2.1.1 DMW Theorem

The classical result by Dalang–Morton–Willinger, usually abbreviated as DMW and sometimes referred to as the Fundamental Theory of Asset (or Arbitrage) Pricing (FTAP) for the discrete finite-time model of a frictionless financial market, says:

There is no arbitrage if and only if there is an equivalent martingale measure.

This formulation is due to Harrison and Pliska, who established it for a model with finite number of states of the nature, i.e., for finite Ω. Retrospectively, one can insinuate that in this case it is mainly a "linguistic" exercise: the result expressed in geometric language was known a long time ago as the Stiemke lemma. This is, to large extent, true. However, a remarkable fact is that, contrarily to its predecessors, exactly this formulation of a no-arbitrage criterion, involving an important probability concept, a martingale measure, opens a way to numerous generalizations of great theoretical and practical value.

Loosely speaking, the result can be viewed as a partial converse to the assertion that one cannot win (in finite time) by betting on a martingale: if one cannot win betting on a process, the latter is a martingale with respect to an equivalent martingale measure.

We start our presentation here with a detailed analysis of the Dalang–Morton–Willinger theorem. The assertion in italics is, in fact, a *grand public* formulation which hides a profound difference between these two results, and the authors of advanced textbooks prefer to give a longer list of *NA* criteria. We follow this tradition.

The model is given by a complete probability space $(\Omega, \mathcal{F}, P)$ with a discrete-time filtration $\mathbf{F} = (\mathcal{F}_t)_{t=0,1,\ldots,T}$ and an adapted d-dimensional price

Yu. Kabanov, M. Safarian, *Markets with Transaction Costs*,
Springer Finance, DOI 10.1007/978-3-540-68121-2_2,

process $S = (S_t)$ with the constant first component. It is convenient to assume that $\mathcal{F}_0$ is trivial and $\mathcal{F}_T = \mathcal{F}$.

The set of "results" R_T (obtained from zero starting value) consists of the terminal values of discrete-time integrals

$$H \cdot S_T := \sum_{t=1}^{T} H_t \Delta S_t,$$

where $\Delta S_t := S_t - S_{t-1}$, and H runs over the linear space $\mathcal{P}$ of predictable processes, i.e., $H_t \in L^0(\mathbf{R}^d, \mathcal{F}_{t-1})$ (the first component of S plays no role here because $\Delta S_t^1 = 0$).

The common terminology: H is a (portfolio) *strategy*, while $H \cdot S$ is called a *value process*. The larger set $A_T := R_T - L^0_+$ can be interpreted as the set of *hedgeable claims*; it is the set of random variables $H \cdot S_T - h$ where the r.v. $h \geq 0$.

By definition, the *NA property* of the model means that $R_T \cap L^0_+ = \{0\}$ (or, equivalently, $A_T \cap L^0_+ = \{0\}$). We prefer to use from the very beginning these mathematically convenient definitions in terms of intersections of certain sets rather than a popular form like this: the property $H \cdot S_T \geq 0$ implies that $H \cdot S_T = 0$.

Theorem 2.1.1 *The following properties are equivalent:*

(a) $A_T \cap L^0_+ = \{0\}$ *(NA condition);*
(b) $A_T \cap L^0_+ = \{0\}$ *and* $A_T = \bar{A}_T$ *(closure in* L^0*);*
(c) $\bar{A}_T \cap L^0_+ = \{0\}$*;*
(d) *there is a strictly positive process* $\rho \in \mathcal{M}$ *such that* $\rho S \in \mathcal{M}$*;*
(e) *there is a bounded strictly positive process* $\rho \in \mathcal{M}$ *such tat* $\rho S \in \mathcal{M}$*.*

As usual, $\mathcal{M}$ is the space of martingales (if necessary, we shall also use a more complicated notation showing the probability, time range, etc.).

Of course, the last two properties are usually formulated as:

(d′) *there is a probability* $\tilde{P} \sim P$ *such that* $S \in \mathcal{M}(\tilde{P})$*;*
(e′) *there is a probability* $\tilde{P} \sim P$ *with* $d\tilde{P}/dP \in L^\infty$ *such that* $S \in \mathcal{M}(\tilde{P})$*.*

However, the chosen versions have more direct analogs in the model with transaction costs. Their equivalences are obvious due to the following elementary fact about martingales with respect to a probability measure $\tilde{P} \ll P$ with density ρ_T:

$$S \in \mathcal{M}(\tilde{P}) \text{ if and only if } \rho S \in \mathcal{M}(P) \text{ where } \rho_t = E(\rho_T | \mathcal{F}_t),\ t \leq T.$$

Collecting conditions in the single theorem is useful because one can clearly see that in numerous generalizations and ramifications certain properties remain equivalent (of course, appropriately modified), but others do not. Note also that, in the case of finite Ω, the set A_T is always closed. Indeed, it is the arithmetic sum of a linear space and the polyhedral cone $-L^0_+$ in the finite-dimensional linear space L^0. Thus, it is a polyhedral cone. So, we have no difference between the first three properties, while the last two coincide trivially. The situation is completely different for arbitrary Ω. Though the linear space R_T is always closed (we show this later), the set A_T may be not closed even for $T = 1$ and a countable Ω. To see this, let $\mathcal{F}_0$ be trivial and take $\mathcal{F}_1 = \sigma\{\xi\}$ and $\Delta S_1 = \xi$, where ξ is a strictly positive finite random variable such that $P(\xi > \varepsilon) > 0$ whatever is $\varepsilon > 0$. The set $A_1 = \mathbf{R}\xi - L^0_+$ does not contain any strictly positive constant, but each constant $c > 0$ belongs to its closure $\bar{A}_1$ because $(n\xi) \wedge c \to c$ as $n \to \infty$.

To the already long list, one can add several other equivalent conditions:

(f) *there is a strictly positive process* $\rho \in \mathcal{M}$ *such that* $\rho S \in \mathcal{M}_{\text{loc}}$;
(f′) *there is a probability* $\tilde{P} \sim P$ *such that* $S \in \mathcal{M}_{\text{loc}}(\tilde{P})$;
(g) $\{\eta \Delta S_t : \ \eta \in L^0(\mathcal{F}_{t-1})\} \cap L^0_+ = \{0\}$ *for all* $t \le T$ *(NA for one-step models).*

With other conditions already established, the above addendum poses no problems. Indeed, (f′) is obviously implied by (e′). On the other hand, if $S \in \mathcal{M}_{\text{loc}}(\tilde{P})$, then $\tilde{H} \cdot S \in \mathcal{M}(\tilde{P})$ with $\tilde{H}_t := 1/(1 + \tilde{E}(|\Delta S_t| \| \mathcal{F}_{t-1}))$. So, we know that *NA* holds for the process $\tilde{H} \cdot S$; hence, it holds also for S as both processes have the same set of hedgeable claims, i.e., (f′) implies a property from the "main" list of equivalent conditions. Suppose now that the implication (g) $\Rightarrow$ (a) fails. Take the smallest $t \le T$ such that $A_t \cap L^0_+ \ne \{0\}$ (the set of such dates is nonempty: it contains, at least, T). We have a strategy $H = (H_s)_{s \le T}$ such that $H \cdot S_t \ge 0$ and $P(H \cdot S_t > 0) > 0$. Due to the choice of t, either the set $\Gamma' := \{H \cdot S_{t-1} < 0\}$ is of strictly positive probability (and (g) is violated by $\eta := I_{\Gamma'} H_t$), or the set $\Gamma'' := \{H \cdot S_{t-1} = 0\}$ is of full measure (and (g) is violated by $\eta := I_{\Gamma''} H_t$), a contradiction.

Remark. The *NA* property for the class of all strategies, as defined above, is equivalent to the *NA* property in the narrower class of bounded strategies H. Indeed, if there is an arbitrage opportunity, then, in virtue of the condition (g), there is an arbitrage opportunity η for a certain one-step model. Clearly, when n is sufficiently large, $\eta I_{\{|\eta| \le n\}}$ will be an arbitrage opportunity for this one-step model. Note that the presence of (g) in the list of equivalent conditions is crucial in this reasoning.

Similarly, *NA* is equivalent to the absence of arbitrage in the class of so-called *admissible* strategies for which the value processes are bounded from below by constants (depending on the strategy). Moreover, if H is an arbitrage opportunity generating the value process $V = H \cdot S$, one can find another arbitrage opportunity $\tilde{H}$ such that the value process $\tilde{H} \cdot S \ge 0$. To see this, we consider the sets $\Gamma_t := \{H \cdot S_t < 0\}$ and the last instant r for which the probability of such a set is strictly positive; $0 < r < T$ since H is an arbitrage opportunity. Let us check that the strategy $\tilde{H} := I_{\Gamma_r} I_{]r,T]} H$ has the claimed property. Indeed, the process $\tilde{V} := \tilde{H} \cdot S$ is zero for all $t \le r$ and remains zero outside the set Γ_r until T. On the set Γ_r, the increments $\Delta\tilde{V}_t = \Delta V_t$ for

$t \geq r+1$, and hence the trajectories of $\tilde{V}$ are the trajectories of V shifted upwards on the value $-V_r > 0$.

Before the proof of Theorem 2.1.1, we give in the following subsection several elementary results which will be also useful in obtaining NA criteria in models with transaction costs.

2.1.2 Auxiliary Results: Measurable Subsequences and the Kreps–Yan Theorem

Lemma 2.1.2 *Let $\eta^n \in L^0(\mathbf{R}^d)$ be such that $\underline{\eta} := \liminf |\eta^n| < \infty$. Then there are $\tilde{\eta}^k \in L^0(\mathbf{R}^d)$ such that for all ω, the sequence of $\tilde{\eta}^k(\omega)$ is a convergent subsequence of the sequence of $\eta^n(\omega)$.*

Proof. Define the random variables $\tau_k := \inf\{n > \tau_{k-1} : \ ||\eta^n| - \underline{\eta}| \leq k^{-1}\}$ starting with $\tau_0 := 0$. Then $\tilde{\eta}_0^k := \eta^{\tau_k}$ is in $L^0(\mathbf{R}^d)$, and $\sup_k |\tilde{\eta}_0^k| < \infty$. Working further with the sequence of $\tilde{\eta}_0^n$, we construct, applying the above procedure to the first component and its lim inf, a sequence of $\tilde{\eta}_1^k$ with convergent first component and such that for all ω, the sequence of $\tilde{\eta}_1^k(\omega)$ is a subsequence of the sequence of $\tilde{\eta}_0^n(\omega)$. Passing on each step to the newly created sequence of random variables and to the next component, we arrive at a sequence with the desired properties. □

Remark. The claim can be formulated as follows: there exists a (strictly) increasing sequence of integer-valued random variables σ_k such that η^{σ_k} converges a.s.

Lemma 2.1.3 *Let $\mathcal{G} = \{\Gamma_\alpha\}$ be a family of measurable sets such that any nonnull set Γ has a nonnull intersection with an element of $\mathcal{G}$. Then there is an at most countable subfamily of sets $\{\Gamma_{\alpha_i}\}$ the union of which is of full measure.*

Proof. Suppose that $\mathcal{G}$ is closed under countable unions. Then $\sup_\alpha P(\Gamma_\alpha)$ is attained on some $\tilde{\Gamma} \in \mathcal{G}$. The subfamily consisting of a single $\tilde{\Gamma}$ gives the answer. Indeed, $P(\tilde{\Gamma}) = 1$: otherwise we could enlarge the supremum by adding a set from $\mathcal{G}$ having a nonnull intersection with $\tilde{\Gamma}^c$. The general case follows by considering the family formed by countable unions of sets from $\mathcal{G}$. □

The following result is referred to as the Kreps–Yan theorem. It holds for arbitrary $p \in [1, \infty]$, $p^{-1} + q^{-1} = 1$, but the cases $p = 1$ and $p = \infty$ are the most important. Recall that for $p \neq \infty$, the norm closure of a convex set in L^p coincides with the closure in $\sigma\{L^p, L^q\}$.

Theorem 2.1.4 *Let $\mathcal{C}$ be a convex cone in L^p closed in $\sigma\{L^p, L^q\}$, containing $-L^p_+$ and such that $\mathcal{C} \cap L^p_+ = \{0\}$. Then there is $\tilde{P} \sim P$ with $d\tilde{P}/dP \in L^q$ such that $\tilde{E}\xi \leq 0$ for all $\xi \in \mathcal{C}$.*

Proof. By the Hahn–Banach theorem any nonzero $x \in L^p_+ := L^p(\mathbf{R}_+, \mathcal{F})$ can be separated from $\mathcal{C}$: there is a $z_x \in L^q$ such that $Ez_x x > 0$ and $Ez_x\xi \le 0$ for all $\xi \in \mathcal{C}$. Since $\mathcal{C} \supseteq -L^p_+$, the latter property yields that $z_x \ge 0$; we may assume that $\|z_x\|_q = 1$. Let us consider the family $\mathcal{G} := \{z_x > 0\}$. As any nonnull set Γ has a nonnull intersection with the set $\{z_x > 0\}$, $x = I_\Gamma$, the family $\mathcal{G}$ contains a countable subfamily of sets (say, corresponding to a sequence $\{x_i\}$) the union of which is of full measure. Thus, $z := \sum 2^{-i} z_{x_i} > 0$, and we can take $\tilde{P} := zP$. □

2.1.3 Proof of the DMW Theorem

The implications (b) ⇒ (a), (b) ⇒ (c), and (e) ⇒ (d) are trivial. The implication (d) ⇒ (a) is easy. Indeed, let $\xi \in A_T \cap L^0_+$, i.e., $0 \le \xi \le H \cdot S_T$. Since the conditional expectation with respect to the martingale measure $\tilde{E}(H_t \Delta S_t|\mathcal{F}_{t-1}) = 0$, we obtain by consecutive conditioning that $\tilde{E}H \cdot S_T = 0$. Thus, $\xi = 0$. To complete the proof, it remains to verify that (c) ⇒ (e) and (a) ⇒ (b).

(c) ⇒ (e). Notice that for any random variable η, there is an equivalent probability P' with bounded density such that $\eta \in L^1(P')$ (e.g., one can take $P' = Ce^{-|\eta|}P$). Property (c) (as well as (a) and (b)) is invariant under equivalent change of probability. This consideration allows us to assume that all S_t are integrable. The convex set $A^1_T := \bar{A}_T \cap L^1$ is closed in L^1. Since $A^1_T \cap L^1_+ = \{0\}$, Theorem 2.1.4 ensures the existence of $\tilde{P} \sim P$ with bounded density and such that $\tilde{E}\xi \le 0$ for all $\xi \in A^1_T$, in particular, for $\xi = \pm H_t \Delta S_t$ with bounded and $\mathcal{F}_{t-1}$-measurable H_t. Thus, $\tilde{E}(\Delta S_t|\mathcal{F}_{t-1}) = 0$.

(a) ⇒ (b). Lemma 2.1.2 allows us to establish the closedness of A_T by simple recursive arguments even without assuming that the σ-algebra $\mathcal{F}_0$ is trivial (of course, this does not add any generality but helps to start the induction in the time variable).

Let us consider the case $T = 1$. Let $H^n_1 \Delta S_1 - r^n \to \zeta$ a.s., where H^n_1 is $\mathcal{F}_0$-measurable, and $r^n \in L^0_+$. The closedness of A_1 means that $\zeta = H_1\Delta S_1 - r$ for some $\mathcal{F}_0$-measurable H_1 and $r \in L^0_+$. To show this, we represent each H^n_1 as a column vector and write the whole sequence of these column vectors as the infinite matrix

$$\mathbf{H}_1 := \begin{bmatrix} H_1^{11} & H_1^{21} & \dots & \dots & H_1^{n1} & \dots \\ H_1^{12} & H_1^{22} & \dots & \dots & H_1^{n2} & \dots \\ \dots & \dots & \dots & \dots & \dots & \dots \\ H_1^{1d} & H_1^{2d} & \dots & \dots & H_1^{nd} & \dots \end{bmatrix}.$$

If the matrix is zero, there is nothing to prove. Suppose that the assertion holds when this (random) matrix has, for each ω, at least m zero lines. We show that the claim holds true also when $\mathbf{H}_1$ has at least $m-1$ zero lines.

It is sufficient to find $\mathcal{F}_0$-measurable random variables $\tilde{H}^k_1$ convergent a.s. and $\tilde{r}^k \in L^0_+$ such that $\tilde{H}^k_1 \Delta S_1 - \tilde{r}^k \to \zeta$ a.s.

Let $\Omega_i \in \mathcal{F}_0$ form a finite partition of Ω. An important (though obvious) observation: we may argue on each Ω_i separately as on an autonomous measure space (considering the restrictions of random variables and traces of σ-algebras).

Let $\underline{H}_1 := \liminf |H_1^n|$. On $\Omega_1 := \{\underline{H}_1 < \infty\}$, we take, using Lemma 2.1.2, $\mathcal{F}_0$-measurable $\tilde{H}_1^k$ such that $\tilde{H}_1^k(\omega)$ is a convergent subsequence of $H_1^n(\omega)$ for every ω; $\tilde{r}^k$ are defined correspondingly. Thus, if Ω_1 is of full measure, the goal is achieved.

On $\Omega_2 := \{\underline{H}_1 = \infty\}$, we put $G_1^n := H_1^n/|H_1^n|$ and $h_1^n := r_1^n/|H_1^n|$. Clearly, $G_1^n \Delta S_1 - h_1^n \to 0$ a.s. By Lemma 2.1.2 we find $\mathcal{F}_0$-measurable $\tilde{G}_1^k$ such that $\tilde{G}_1^k(\omega)$ is a convergent subsequence of $G_1^n(\omega)$ for every ω. Denoting the limit by $\tilde{G}_1$, we obtain that $\tilde{G}_1 \Delta S_1 = \tilde{h}_1$ where $\tilde{h}_1$ is nonnegative; hence, in virtue of (a), $\tilde{G}_1 \Delta S_1 = 0$.

As $\tilde{G}_1(\omega) \neq 0$, there exists a partition of Ω_2 into d disjoint subsets $\Omega_2^i \in \mathcal{F}_0$ such that $\tilde{G}_1^i \neq 0$ on Ω_2^i. Define $\bar{H}_1^n := H_1^n - \beta^n \tilde{G}_1$, where $\beta^n := H_1^{ni}/\tilde{G}_1^i$ on Ω_2^i. Then $\bar{H}_1^n \Delta S_1 = H_1^n \Delta S_1$ on Ω_2. The matrix $\bar{\mathbf{H}}_1$ has, for each $\omega \in \Omega_2$, at least m zero lines: our operations did not affect the zero lines of $\mathbf{H}_1$, and a new one has appeared, namely, the ith one on Ω_2^i. We conclude by the induction hypothesis.

To establish the induction step in the time variable, we suppose that the claim is true for $(T-1)$-step models. Let $\sum_{t=1}^T H_t^n \Delta S_t - r^n \to \zeta$ a.s., where H_t^n are $\mathcal{F}_{t-1}$-measurable, and $r^n \in L_+^0$. As at the first step, we work with the matrix $\mathbf{H}_1$ using exactly the same reasoning.

On Ω_1 we take an increasing sequence of $\mathcal{F}_0$-random variables τ_k such that $H^k := H_1^{\tau_k}$ converges to H_1. Thus, $\sum_{t=2}^T H_t^{\tau_k} \Delta S_t - r^{\tau_k}$ converges as $k \to \infty$, and we have a reduction to a $(T-1)$-step model.

On Ω_2 we use again the same induction in m, the number of zero lines of $\mathbf{H}_1$. The only modification is that the identical operations (passage to subsequences, normalization by H_1^n, etc.) should be performed simultaneously over all other matrices $\mathbf{H}_2, \dots, \mathbf{H}_T$.

Remark 1. Exactly the same arguments as those used in the proof of the implication (a) $\Rightarrow$ (b) lead to the following assertion referred to as the Stricker lemma:

The set of results R_T is closed.

This property holds irrelevantly of the *NA*-condition. Indeed, the latter was used only to check that the nonnegative limit $\tilde{h}_1$ is, in fact, equal to zero. But this holds automatically if we start the arguments with $r_n = 0$.

Remark 2. The DMW theorem contains as a corollary the assertion that, in the discrete-time setting with finite horizon, any local martingale is a martingale with respect to a measure $\tilde{P} \sim P$ with bounded density. Moreover, this measure can be chosen in such a way that a given random variable ξ will be $\tilde{P}$-integrable. At the end of this chapter we show that, even in the model with infinite horizon, the local martingale is a martingale with respect to an equivalent probability measure.

2.1.4 Fast Proof of the DMW Theorem

Our detailed formulation of the DMW theorem, together with its proof, is intended to prepare the reader to the arguments developed for models with transaction costs. However, a short and elementary proof of the "main" equivalence (a) $\Leftrightarrow$ (e), a proof which can be used in introductory courses for mathematical students, is of separate interest. We give one here combining an optimization approach due to Chris Rogers with Lemma 2.1.2 on measurable subsequences. It is based on the one-step result the first condition of which is just an alternative reformulation of the NA-property.

Proposition 2.1.5 *Let $\xi \in L^0(\mathbf{R}^d)$, and let $\mathcal{G}$ be a sub-σ-algebra of $\mathcal{F}$. Then the following conditions are equivalent:*

(i) *for any $\alpha \in L^0(\mathbf{R}^d, \mathcal{G})$, the inequality $\alpha\xi \geq 0$ holds as the equality;*
(ii) *there exists a bounded random variable $\varrho > 0$ such that $E\varrho|\xi| < \infty$ and $E(\varrho\xi|\mathcal{G}) = 0$.*

Proof. One needs arguments only for the "difficult" implication (i) $\Rightarrow$ (ii).

First, examine the case where $\mathcal{G}$ is trivial. Let us consider the function $f(a) = Ee^{a\xi - |\xi|^2}$, $a \in \mathbf{R}^d$. If it attains its minimum at some point a_*, the problem is solved with $\rho = e^{a_*\xi - |\xi|^2}$, since at this point the derivative of f is zero: $E\xi e^{a_*\xi - |\xi|^2} = 0$. One can check that condition (i) excludes the possibility that the minimum is not attained—we do a verification below.

Let us turn to the general case. A dimension reduction argument allows us to work assuming that the relation $\alpha\xi = 0$ with $\alpha \in L^0(\mathbf{R}^d, \mathcal{G})$ holds only if $\alpha = 0$ (when $\mathcal{G}$ is trivial, this is just the linear independence of the components of ξ as elements of L^0). Let $Q(\omega, dx)$ be the regular conditional distribution of ξ with respect to $\mathcal{G}$. Define the function

$$f(\omega, a) := \int e^{ax - |x|^2} Q(\omega, dx)$$

continuous in a and $\mathcal{G}$-measurable in ω. Introduce the $\mathcal{G}$-measurable random variable $f_*(\omega) = \inf_a f(\omega, a)$ and consider, in the product space $\Omega \times \mathbf{R}^d$, the sets $\{(\omega, a) : f(\omega, a) < f_*(\omega) + 1/n\}$ with nonempty open ω-sections $\Gamma_n(\omega)$. Let α_n be a $\mathcal{G}$-measurable random variable with $\alpha_n(\omega) \in \Gamma_n(\omega)$. Such α_n can be constructed easily, without appealing to a measurable selection theorem, e.g., one can take $\alpha_n(\omega) := q_{\theta(n)}$, where

$$\theta(n) := \min\{k : \ f(\omega, q_k) < f_*(\omega) + 1/n\}$$

with an arbitrary countable dense subset $\{q_n\}$ in $\mathbf{R}^d$. Let us consider the set $\Omega_0 := \{\liminf |\alpha_n| < \infty\}$ with its complement Ω_1. Using Lemma 2.1.2, we may assume that on Ω_1 the sequence $\tilde{\alpha}_n := \alpha_n/|\alpha_n|$ converges to some β with $|\beta| = 1$ and, by the Fatou lemma,

$$\int e^{\lim |\alpha_n(\omega)|\beta(\omega)x-|x|^2} I_{\{\beta(\omega)x\neq 0\}} Q(\omega,dx)$$
$$\leq \liminf \int e^{\alpha_n(\omega)x-|x|^2} I_{\{\beta(\omega)x\neq 0\}} Q(\omega,dx) \leq f_*(\omega).$$

Necessarily, $Q(\omega,\{x: \ \beta(\omega)x > 0\}) = 0$, implying that $\beta\xi \leq 0$ (a.s.), and, therefore, in virtue of (i), we have that $\beta\xi = 0$. Due to our provision, this equality holds only if $\beta = 0$, and, hence, Ω_1 is a null set which does not matter. Again by Lemma 2.1.2 we may assume that on the set Ω_0 of full measure the sequence $\alpha_n(\omega)$ converges to some $\alpha_*(\omega)$. Clearly, $f(\omega,a)$ attains its minimum at $\alpha_*(\omega)$, and we conclude with $\varrho := e^{\alpha_*\xi-|\xi|^2}/c(\alpha_*)$, where the function $c(a) := \sup_x(1+|x|)e^{ax-|x|^2}$. □

The "difficult" implication (a) ⇒ (e) follows from the above proposition by backward induction. We claim that for each $t = 0,1,\dots,T-1$, there is a bounded random variable $\rho_t^T > 0$ such that $E\rho_t^T|\Delta S_u| < \infty$ and $E\rho_t^T\Delta S_u = 0$ for $u = t+1,\dots,T$. Since (a) implies the *NA*-property for each one-step model, the existence of ρ_{T-1}^T follows from the above proposition with $\xi = \Delta S_T$ and $\mathcal{G} = \mathcal{F}_{T-1}$. Suppose that we have already found ρ_t^T. Putting $\xi = E(\rho_t^T|\mathcal{F}_{t-1})\Delta S_{t-1}$ and $\mathcal{G} = \mathcal{F}_{t-2}$, we find bounded $\mathcal{F}_{t-1}$-measurable $\varrho_{t-1} > 0$ such that $E(\varrho_{t-1}E(\rho_t^T|\mathcal{F}_{t-1})|\Delta S_{t-1}|) < \infty$ and $E(\varrho_{t-1}E(\rho_t^T|\mathcal{F}_{t-1})\Delta S_{t-1}) = 0$. It is clear that ρ_{t-1}^T meets the requirements. Property (e) of the DMW theorem holds with $\rho_t := E(\rho_0^T|\mathcal{F}_t)$.

2.1.5 *NA* and Conditional Distributions of Price Increments

As shown by Jacod and Shiryaev, the long list of conditions equivalent to the *NA*-property can be completed by the following one involving the regular conditional distributions $Q_t(\omega,dx)$ of the price increments ΔS_t knowing $\mathcal{F}_{t-1}$:

(h) $0 \in \operatorname{ri}\operatorname{conv}\operatorname{supp} Q_t(\omega,dx)$ *a.s. for all* $t = 1,\dots,T$.

Recall that $Q_t(\omega,\Gamma)$ is an $\mathcal{F}_{t-1}$-measurable random variable in ω and a measure in Γ such that $P(\Delta S_t \in \Gamma|\mathcal{F}_{t-1}) = Q_t(\omega,\Gamma)$ (a.s.) for each Borel set Γ in $\mathbf{R}^d$. The topological support of the measure $Q_t(\omega,dx)$ is the intersection of all closed sets the complements of which are null sets for this measure. The abbreviation "ri" denotes the relative interior of a convex set, i.e., the interior in the relative topology of the smallest affine subspace containing it.

Comparing (h) and (g), we see that their equivalence follows from the next one-step result complementing Proposition 2.1.5.

Proposition 2.1.6 *Let $\xi \in L^0(\mathbf{R}^d)$, and let $Q(\omega, dx)$ be a regular conditional distribution of ξ with respect to a σ-algebra $\mathcal{G} \subseteq \mathcal{F}$. Then the NA-property (or the equivalent property* (ii)*) holds if and only if the following condition is satisfied:*

(iii) $0 \in \operatorname{ri}\operatorname{conv}\operatorname{supp} Q(\omega, dx)$ *a.s.*

Proof. (ii) $\Rightarrow$ (iii). Consider the case where $\mathcal{G}$ is trivial. If the origin does not belong to $A := \operatorname{ri}\operatorname{conv}\operatorname{supp} Q(dx)$, then there exists $a \in \mathbf{R}^d$ such that the latter set lies in the closed half-space $\{x : \ ax \geq 0\}$ but not in the subspace $\{x : \ ax = 0\}$ (to see this, apply the separation theorem in the linear subspace of minimal dimension containing A and extend the separating functional to a functional on the whole $\mathbf{R}^d$ vanishing on the orthogonal complement). So, $Q(x : \ ax > 0) > 0$, and for any strictly positive bounded random variable ρ measurable with respect to $\sigma\{\xi\}$ (i.e., of the form $\rho = r(\xi)$ with a Borel function r), we have

$$E\rho a\xi = \int r(x) ax I_{\{x:\ ax>0\}} Q(dx) > 0,$$

in contradiction with (ii).

In the general case we consider the set $\Gamma := \{(\omega, a) : \ Q(\omega, \{ax > 0\}) > 0\}$, which is measurable with respect to the product σ-algebra $\mathcal{G} \otimes \mathcal{B}^d$; let $\Gamma(\omega)$ be its ω-sections. If (iii) fails, then, as it was just shown, the projection $\mathrm{Pr}_\Omega \Gamma$ of Γ on Ω is nonnull. Due to the measurable selection theorem, there exists a $\mathcal{G}$-measurable $\mathbf{R}^d$-valued random variable α such that $\alpha(\omega) \in \Gamma(\omega)$ for almost all ω from $\mathrm{Pr}_\Omega \Gamma$. Now, take an arbitrary bounded strictly positive function $r(\omega, x)$ measurable with respect to $\sigma\{\mathcal{G}, \xi\} \otimes \mathcal{B}^d$ and put $\rho(\omega) := r(\omega, \xi(\omega))$. Then

$$E(r\alpha\xi|\mathcal{G}) = \int r(\omega, x)\alpha(\omega) x I_{\{x:\ \alpha(\omega)x>0\}} Q(\omega, dx) > 0 \quad \text{on } \mathrm{Pr}_\Omega \Gamma.$$

It is easy to see that this is a contradiction with (ii).

(iii) $\Rightarrow$ (ii). Again, let us first consider the case of trivial $\mathcal{G}$. Let L be the affine subspace of minimal dimension containing the set $A := \operatorname{ri}\operatorname{conv}\operatorname{supp} Q(dx)$.

The assumption $0 \in A$ implies that the function

$$f(a) := \int e^{ax-|x|^2} Q(dx) = \int e^{ax-|x|^2} I_L(x) Q(dx)$$

attains its minimum at some point a_*: otherwise, we could find, as in the proof of Proposition 2.1.5, a vector β such $\{x : \ \beta x > 0\} \cap L$ is a Q-null set. But this means that the origin is not in the relative interior of the convex hull of $\operatorname{supp} Q(dx)$. In the general case, we can find a $\mathcal{G}$-measurable random variable α_* such that $\alpha_*(\omega)$ is a minimizer of $f(\omega, a)$ and conclude in the same way as in Proposition 2.1.5. □

2.1.6 Comment on Absolute Continuous Martingale Measures

One may ask whether the existence of an absolute continuous martingale measure can be related with a certain no-arbitrage property. Indeed, in the case of finite number of states of nature, we have the following criterion:

Proposition 2.1.7 *Suppose that Ω is finite. Then the following conditions are equivalent:*

(a) $R_T \cap L^0(\mathbf{R}_+ \setminus \{0\}) = \emptyset$;
(b) *there is a probability measure* $\tilde{P} \ll P$ *such that* $S \in \mathcal{M}(\tilde{P})$.

Here the implication (b) $\Rightarrow$ (a) is obvious, while the converse follows easily from the finite-dimensional separation theorem applied to the disjoint convex sets $A_T \setminus \{0\}$ and $L^0(\mathbf{R}_+ \setminus \{0\})$: any separating functional after normalization is a density of probability measure with the needed property. The condition (a) means that there is no "universal" arbitrage strategy H, that is, such that $H \cdot S_T > 0$ (a.s.).

Unfortunately, the above proposition cannot be extended to the case of arbitrary Ω.

Example. Let us consider a one-period model with two risky assets whose price increments ΔS_1^1 and ΔS_1^2 are random variables defined on a countable probability space $\Omega = \{\omega_i\}_{i \geq 0}$ with all $P(\{\omega_i\}) > 0$. The initial σ-algebra is trivial. Let $\Delta S_1^1(\omega_0) = 1$, $\Delta S_1^1(\omega_i) = -i$, $i \geq 1$. Let $\Delta S_1^2(\omega_0) = 0$, $\Delta S_1^2(\omega_i) = 1$, $i \geq 1$. Apparently, the equalities $E_Q \Delta S_1^1 = 0$ and $E_Q \Delta S_1^2 = 0$ are incompatible, and, hence, there are no martingale measures. On the other hand, let $(H^1, H^2) \in \mathbf{R}^2$ be a "universal" arbitrage strategy. Then necessarily $H^1 > 0$, and we get a contradiction since in such a case the countable system of inequalities $-iH^1 + H^2 > 0$, $i \geq 1$, is incompatible whatever is H^2.

2.1.7 Complete Markets and Replicable Contingent Claims

As we observed, the set of results R_T is always closed in L^0. It is an easy exercise to deduce from this property that the set $\mathbf{R} + R_T$ is also closed. We use this remark in the proof of the following:

Proposition 2.1.8 *Suppose that the set $\mathcal{Q}^e$ of equivalent martingale measures is nonempty. Then the following conditions are equivalent:*

(a) $\mathcal{Q}^e$ *is a singleton;*
(b) $\mathbf{R} + R_T = L^0$.

Proof. (a) $\Rightarrow$ (b). We may assume without loss of generality that P is a martingale measure. Suppose that there is $\xi \in L^0$ which is not in the closed subspace $\mathbf{R} + R_T \subseteq L^0$. It follows that the random variables $\xi^n := \xi I_{\{|\xi| \leq n\}}$ are not in this subspace for all $n \geq N$. Applying the separation theorem, one can find η with $|\eta| \leq 1/2$ such that $E\eta\zeta = 0$ for the elements ζ from the closed subspace $(\mathbf{R}+R_T)\cap L^1$ of L^1 but $E\eta\xi^N>0$. Put $Q=(1+\eta)P$. Then $E_Q H\cdot S_T=0$ whatever is a bounded predictable process H. This means that Q is an equivalent martingale measure different from P, contradicting (a).

(b) $\Rightarrow$ (a). Take $\Gamma\in\mathcal{F}_T$. Then $I_\Gamma=c_\Gamma+H_\Gamma\cdot S_T$, where c_Γ is a constant. It follows that $Q(\Gamma)=c_\Gamma$ whatever is a martingale measure Q, i.e., the latter is unique.

The property (b), in financial literature referred to as the *market completeness*, means that any contingent claim can be replicated, that is, represented as the terminal value of a self-financing portfolio starting from a certain initial endowment. The above statement, asserting that an arbitrage-free market is complete if and only if there is only one equivalent martingale measure, sometimes is called the second fundamental theorem of asset pricing.

The closedness of the subspace $\mathbf{R}+R_T$ leads the next assertion concerning replicable claims on incomplete markets. In its formulation, Q_l and Q_l^e denote the sets of absolutely continuous and equivalent local martingale measures. □

Proposition 2.1.9 *Suppose that $\mathcal{Q}^e\neq\emptyset$. Let a random variable $\xi\geq 0$ be such that $a=\sup_{Q\in\mathcal{Q}_l^e}E_Q\xi<\infty$ and the supremum is attained on some measure Q^*. Then $\xi=a+H\cdot S_T$ for some predictable process (and, hence, the function $Q\mapsto E_Q\xi$ is constant on the set $\mathcal{Q}_l^e$).*

Proof. Supposing that the statement fails, we apply the Hahn–Banach theorem and separate ξ and the subspace $(\mathbf{R}+R_T)\cap L^1(Q^*)$ in $L^1(Q^*)$, that is, we find $\eta\in L^\infty$ such that $E_{Q^*}\xi\eta>0$ and $E_{Q^*}\eta\zeta=0$ for all ζ from the subspace. In particular, $E_{Q^*}\eta=0$ and $E_{Q^*}H\cdot S_T\eta=0$ whatever is a predictable process H such that $H\cdot S_T$ is integrable; in particular, the last equality holds for $H=I_{[0,\tau_n]}$, where τ_n is a localizing sequence for S. Normalizing, we may assume that $|\eta|\leq 1/2$. It follows that the measure $\tilde{Q}=(1+\eta)Q^*$ is an element of $\mathcal{Q}_l^e$ and $E_{\tilde{Q}}\xi=a+E_{Q^*}\xi\eta>a$ in an apparent contradiction with the definition of Q^*. □

2.1.8 DMW Theorem with Restricted Information

Let us consider the following setting, which is only slightly different from the classical one. Namely, assume that we are given a filtration $\mathbf{G}=(\mathcal{G}_t)_{t\leq T}$ with $\mathcal{G}_t\subseteq\mathcal{F}_t$. Suppose that the strategies are now predictable with respect to this smaller filtration (i.e., $H_t\in L^0(\mathcal{G}_{t-1})$), a situation which may happen when the portfolios are revised on the basis of restricted information, e.g., due to a delay. Again, we may define the sets R_T and A_T and give a definition of the arbitrage, which, in these symbols, looks exactly as (a) above, and we can list the corresponding necessary and sufficient conditions.

To this aim, we define the $\mathbf{G}$-optional projection X^o of an integrable process X by putting $X_t^o:=E(X_t|\mathcal{G}_t)$, $t\leq T$.

Theorem 2.1.10 *The following properties are equivalent:*

(a) $A_T \cap L^0_+ = \{0\}$ *(NA condition);*
(b) $A_T \cap L^0_+ = \{0\}$ *and* $A_T = \bar{A}_T$*;*
(c) $\bar{A}_T \cap L^0_+ = \{0\}$*;*
(d) *there is a strictly positive process* $\rho \in \mathcal{M}$ *with* $(\rho S)^o \in \mathcal{M}(\mathbf{G})$*;*
(e) *there is a bounded strictly positive process* $\rho \in \mathcal{M}$ *with* $(\rho S)^o \in \mathcal{M}(\mathbf{G})$*.*

The symbol $\mathcal{M}(\mathbf{G})$ stands here for the set of $\mathbf{G}$-martingales, and we presume tacitly in the last two conditions that $E\rho_t|S_t| < \infty$. Clearly, these conditions can be formulated in terms of existence of an equivalent probability $\tilde{P}$ such that $\tilde{E}(S_{t+1}|\mathcal{G}_t) = \tilde{E}(S_t|\mathcal{G}_t)$ for all $t \le T-1$.

We leave to the reader as an (easy) exercise to inspect that the arguments of the previous section go well for this theorem.

Remark. Curiously, this result, rather natural and important for practical applications, was established only recently. It happens that all numerous proofs, except one suggested in [131] and reproduced above in Sect. 2.1.3, in their most essential part concerning the construction of equivalent martingale measures given the *NA*-property, are based on the reduction to the one-step case with $T = 1$. Of course, (a) implies (g) (i.e., the *NA*-property for all one-step models). A clever argument in the Dalang–Morton–Willinger paper permits to assemble a required martingale density from martingale densities for one-step models. However, in the model with restricted information, the property (g) drops out from the list of equivalent conditions.

Example. Consider the model where $T = 2$, $\mathcal{G}_0 = \mathcal{G}_1 = \{\emptyset, \Omega\}$, but there is $A \in \mathcal{F}_2$ such that $0 < P(A) < 1$. Put

$$\Delta S_1 := I_A - \frac{1}{2} I_{A^c}, \qquad \Delta S_2 := -\frac{1}{2} I_A + I_{A^c}.$$

There is no arbitrage at each of two steps, but the constant process with $H_1 = H_2 = 1$ is an arbitrage strategy for the two-step model.

2.1.9 Hedging Theorem for European-Type Options

One of the most fundamental though simple ideas of mathematical finance is the arbitrage pricing of contingent claims.

A *contingent claim* or an *option* is a random variable ξ which can be interpreted as a pay-off of the option seller to the option buyer. For a European-type option, the payment is made at the terminal (maturity) date T and may depend on the whole history up to T. What is a "fair" price for such a contract payed at time zero? Apparently, and this is the basic principle, the option price should be such that neither of two parties has arbitrage opportunities, i.e., riskless profits.

Let us define the set

$$\Gamma := \Gamma(\xi) := \{x : \exists H \in \mathcal{P} \text{ such that } x + H \cdot S_T \geq \xi\}.$$

Clearly, if not empty, it is a semi-infinite interval (maybe, coinciding with the whole line). A priori, it can be either of the form $[\bar{x}, \infty[$ or $]\bar{x}, \infty[$.

The theorem below ensures, in particular, that $\bar{x} \in \Gamma$. If the contracted price of the option, say, x is strictly larger than $\bar{x}$, then the seller has a nonrisky profit by pocketing $x - \bar{x}$ and running a self-financing portfolio process in the underlying assets $\bar{x} + H \cdot S$, the terminal value of which dominates the terminal pay-off (so, selling the portfolio at the date T covers the liability).

Similarly, suppose that the right extremity $\underline{x}$ of the semi-infinite interval

$$-\Gamma(-\xi) = \{x : \exists H \in \mathcal{P} \text{ such that } -x + H \cdot S_T \geq -\xi\}$$

belongs to this interval. If x is strictly less than $\underline{x}$, then the option buyer will have an arbitrage opportunity. Indeed, in this case there exists a strategy H such that $-\underline{x} + H \cdot S_T \geq -\xi$. Thus, borrowing x at $t = 0$ to buy the option, the agent runs a portfolio $-x + H \cdot S$, which has a terminal value larger than $\underline{x} - x - \xi$. Therefore, after exercising the option, the agent will have a nonrisky profit $\underline{x} - x$.

These arguments show that "fair" prices lie in the interval $[\underline{x}, \bar{x}]$.

Remark 1. Note that it is tacitly assumed that the agent (option seller) may have a short position in option: for the discrete-time model, it is an innocent assumption, but it is questionable for continuous-time models, where the admissibility means that unbounded short positions even in the underlying are not allowed.

In the case where the contingent claim is redundant, that is, of the form $\xi = x + H^{\xi} \cdot S_T$, we necessarily have that $x = \underline{x} = \bar{x}$ is the no-arbitrage price of the option. Indeed, let us consider the hedging portfolio process $\bar{x} + H \cdot S$ for ξ. The absence of arbitrage implies that its terminal value must coincide with ξ and, in virtue of the "law of one price" (also due to NA, see the remark below), $x = \bar{x}$ and, by symmetry, $x = \underline{x}$. The same NA arguments show that if ξ is nonredundant, the hedging portfolio starting from $\bar{x}$ is an arbitrage opportunity. Thus, the range of no-arbitrage prices is either a singleton or an open interval $]\underline{x}, \bar{x}[$.

Remark 2. The law of one price (L1P) is the property asserting that the equality $x + H \cdot S_T = x' + H' \cdot S_T$ implies the equality $x = x'$. The NA-property is a sufficient condition for L1P that follows from the DMW theorem: the latter ensures that there is a measure under which the process $(H - H') \cdot S$ is a martingale. One may ask what is a necessary and sufficient condition for L1P. The answer is the following:

> *L1P holds if and only if there is a bounded martingale Z with $EZ_T = 1$ and $Z_0 > 0$ such that the process ZS is a martingale.*

In this formulation we do not suppose that the σ-algebra is trivial. Notice that L1P means that $R_T \cap L^0(\mathcal{F}_0) = \{0\}$, where, as already mentioned, the linear space R_T is closed. We hope that with this remark the proof of the nontrivial "only if" part will be an easy exercise for the reader.

Now we present the theorem giving a "dual" description of the set of initial capitals Γ, from which one can super-replicate (hedge) the contingent claim ξ.

Notation. Let $\mathcal{Q}$ (resp. $\mathcal{Q}^e$) be the set of all measures $Q \ll P$ (resp. $Q \sim P$) such that S is a martingale with respect to Q. We add to this notation the subscript l to denote larger sets of measures $\mathcal{Q}_l$ and $\mathcal{Q}_l^e$ for which S is only a local martingale. We shall denote by $\mathcal{Z}$, $\mathcal{Z}^e$, $\mathcal{Z}_l$, ... the density processes for measures from the corresponding sets.

Theorem 2.1.11 *Suppose that $\mathcal{Q}^e \neq \emptyset$. Let ξ be a bounded from below random variable such that $E_Q|\xi| < \infty$ for every $Q \in \mathcal{Q}^e$. Then*

$$\Gamma = \{x : \ x \geq E\rho_T\xi \ \textit{for all} \ \rho \in \mathcal{Z}^e\}. \tag{2.1.1}$$

In other words, $\bar{x} = \sup_{Q\in\mathcal{Q}^e} E_Q\xi$ and $\Gamma = [\bar{x}, \infty]$. An obvious corollary of this theorem (applied to the set $\Gamma(-\xi)$) is the assertion that $\underline{x} = \inf_{Q\in\mathcal{Q}^e} E_Q\xi$.

The direct proof of this result is not difficult, but we obtain it from two fundamental facts having their own interest. The first one usually is referred to as the *optional decomposition theorem*, which will be discussed in Sect. 2.1.12.

Theorem 2.1.12 *Suppose that $\mathcal{Q}^e \neq \emptyset$. Let $X = (X_t)$ be a bounded from below process which is a supermartingale with respect to each probability measure $Q \in \mathcal{Q}^e$. Then there exist a strategy H and an increasing process A such that $X = X_0 + H \cdot S - A$.*

Proposition 2.1.13 *Suppose that $Q^e \neq \emptyset$. Let ξ be a bounded from below random variable such that $\sup_{Q\in\mathcal{Q}^e} E_Q|\xi| < \infty$. Then the process X with*

$$X_t = \operatorname*{ess\,sup}_{Q\in\mathcal{Q}^e} E_Q(\xi|\mathcal{F}_t)$$

is a supermartingale with respect to every $Q \in \mathcal{Q}^e$.

For the proof of this result, we send the reader to Appendix (Proposition 5.3.7).

Proof of Theorem 2.1.11. The inclusion $\Gamma \subseteq [\bar{x}, \infty[$ is obvious: if $x + H \cdot S_T \geq \xi$, then $x \geq E_Q\xi$ for every $Q \in \mathcal{Q}^e$. To show the opposite inclusion, we may suppose that $\sup_{Q\in\mathcal{Q}} E_Q|\xi| < \infty$ (otherwise both sets are empty). Applying the optional decomposition theorem, we get that $X = \bar{x} + H \cdot S - A$. Since $\bar{x} + H \cdot S_T \geq X_T = \xi$, the result follows. □

2.1.10 Stochastic Discounting Factors

In this subsection we discuss financial aspects of the hedging theorem and give an interpretation of densities of martingale measures as stochastic discounting factors.

Let us consider a "practical example" where the option seller promised to deliver at the expiration date T a "basket" of d assets, namely, η^i units of the ith asset with positive price process S^i. Since the market is frictionless, this is same as to deliver ηS_T units of the numéraire, i.e., to make a payment $\xi = \eta S_T$. The hedging theorem asserts that the set of initial capitals allowing one to super-replicate ξ can be described in terms of prices. Namely, if the NA-property holds, one can hedge the pay-off from the initial capital x if and only if x dominates the expectation of "stochastically discounted" pay-off $\rho_T\xi = \xi S^\rho_T$ whatever is a martingale density ρ. In other words, the comparison should be done not by computing the "value" of the basket using the "true" price process but replacing the latter by a "consistent price system" $S^\rho = \rho S$ obtained by multiplying the "true" price process by the *stochastic discounting factor* ρ. The word "consistent" here reflects the fact that S^ρ_t is determined by S^ρ_T via the martingale property: $S^\rho_t = E(S^\rho_T|\mathcal{F}_T)$.

2.1.11 Hedging Theorem for American-Type Options

In the American-type option the buyer has the right to exercise at any date before T on the basis of the available information flow, so the exercise date τ is a stopping time; the buyer gets the amount Y_τ, the value of an adapted process Y at τ. The description of the pay-off process $Y = (Y_t)$ is a clause of the contract (as well as the final maturity date T).

By analogy with the case of European options, we define the set of initial capitals starting from which one can run a self-financing portfolio the values of which dominate the eventual pay-off on the considered time-interval:

$$\Gamma := \Gamma(Y) := \{x : \exists H \in \mathcal{P} \text{ such that } x + H \cdot S \geq Y\}.$$

Theorem 2.1.14 *Suppose that $\mathcal{Q}^e \neq \emptyset$. Let $Y = (Y_t)$ be an adapted process bounded from below and such that $E_Q|Y_t| < \infty$ for all $Q \in \mathcal{Q}^e$ and $t \leq T$. Then*

$$\Gamma = \big\{x : \; x \geq E\rho_\tau Y_\tau \textit{ for all } \rho \in \mathcal{Z}^e \textit{ and all stopping times } \tau \leq T\big\}. \quad (2.1.2)$$

The proof of this result based on application of the optional decomposition is exactly the same as of Theorem 2.1.11. The only difference is that now we take as X the process

$$X_t = \operatorname*{ess\,sup}_{Q\in\mathcal{Q}^e, \tau\in\mathcal{T}_t} E_Q(Y_\tau|\mathcal{F}_t),$$

where $\mathcal{T}_t$ is the set of stopping times with values in the set $\{t, t+1, \dots, T\}$. Under the assumption $\sup_{Q\in\mathcal{Q}^e} E_Q|Y_t| < \infty$ for each t, the process X is a supermartingale with respect to every $Q \in \mathcal{Q}^e$, see Proposition 5.3.8 in Appendix.

2.1.12 Optional Decomposition Theorem

We give here a slightly different formulation.

Theorem 2.1.15 *Suppose that $\mathcal{Q}_l^e \neq \emptyset$. Let $X = (X_t)$ be a process which is a generalized supermartingale with respect to each measure $Q \in \mathcal{Q}_l^e$. Then there are a strategy H and an increasing process A such that $X = X_0 + H \cdot S - A$.*

Proof. We start from a one-step version of the result. □

Lemma 2.1.16 *Let $\mathcal{G}$ be a sub-σ-algebra of $\mathcal{F}$, and let ξ and η be random variables with values in $\mathbf{R}$ and $\mathbf{R}^d$ and for which $E(|\xi| + |\eta||\mathcal{G}) < \infty$. Assume that $E(\alpha\xi|\mathcal{G}) \leq 0$ whatever is a random variable $\alpha > 0$ with $E(\alpha|\mathcal{G}) = 1$ such that $E(\alpha\eta|\mathcal{G}) = 0$ and $E(\alpha|\xi||\mathcal{G}) < \infty$, $E(\alpha|\eta||\mathcal{G}) < \infty$. Suppose that such α does exist. Then there is $\lambda \in L^0(\mathbf{R}^d, \mathcal{G})$ such that $\xi - \lambda\eta \leq 0$.*

Proof. First, we suppose without loss of generality that ξ and η are integrable (we may argue with $\tilde{\xi} := \xi/(1+E(|\xi|+|\eta||\mathcal{G}))$ and $\tilde{\eta} := \eta/(1+E(|\xi|+|\eta||\mathcal{G}))$). Define the set $A := \{\lambda\eta : \ \lambda \in L^0(\mathbf{R}^d, \mathcal{G})\} - L^0_+$. By the DMW theorem, it is closed in probability. Thus, the convex set $A^1 := A \cap L^1$ is closed in L^1. If the assertion of the lemma fails, $\xi \notin A^1$. Therefore, in virtue of the Hahn–Banach separation theorem, there is $\alpha \in L^\infty$ such that

$$E\alpha\xi > \sup_{\zeta \in A^1} E\alpha\zeta.$$

Necessarily, $\alpha \geq 0$: if not, the right-hand side of the above inequality would be infinite. By the same reason $E\alpha\lambda\eta = 0$ whatever is $\lambda \in L^\infty(\mathbf{R}^d, \mathcal{G})$. Hence, $E(\alpha\eta|\mathcal{G}) = 0$, and the supremum is equal to zero. That is, $E\alpha\xi > 0$. But this is incompatible with the inequality $E(\alpha\xi|\mathcal{G}) \leq 0$ we should have for such α. □

With the lemma, the proof of the theorem is easy. Indeed, let $\rho \in \mathcal{Z}_l^e$. Consider the obvious identity $\rho_t = \alpha_1 \dots \alpha_t$, where $\alpha_k := \rho_k/\rho_{k-1}$. The martingale property of ρ means that $E(\alpha_t|\mathcal{F}_{t-1}) = 1$. On the other hand, due to the coincidence of the classes of local and generalized martingales, $\rho \in \mathcal{Z}_l^e$ if and only if $E(\alpha_t|\Delta S_t||\mathcal{F}_{t-1}) < \infty$ and $E(\alpha_t \Delta S_t|\mathcal{F}_{t-1}) = 0$ for all $t \leq T$. Thus, by Lemma 2.1.16, there is $H_t \in L^0(\mathbf{R}^d|\mathcal{F}_{t-1})$ such that $\Delta X_t - H_t\Delta S_t \leq 0$. Denoting the right-hand side by $-\Delta A_t$ and putting $A_0 = 0$, we obtain the desired decomposition.

Remark. Let us return to the setting of Lemma 2.1.16, assuming that the σ-algebra $\mathcal{G}$ is trivial. Consider the maximization problem $E\alpha\xi \to \max$ under two equality constraints $E\alpha = 1$ and $E\alpha\eta = 0$, the constraint $\alpha > 0$ (a.s.), and "admissibility" assumptions on α to ensure the needed integrability. The hypothesis of the lemma says that the value of this problem does not exceed zero. It is not difficult to prove that there is a Lagrange multiplier λ "removing" the second equality constraint. For the new maximization problem, we also have that $E\alpha(\xi - \lambda\eta) \le 0$ for all α satisfying the remaining constraints. Clearly, this is possible only if $\xi - \lambda\eta \le 0$.

One may expect that these arguments can be extended for the general case, with conditional expectations. This is still easy for finite or countable Ω. This strategy of proof is feasible for arbitrary Ω, but one needs to look for a $\mathcal{G}$-measurable version of Lagrange multipliers by applying a delicate measurable selection result, requiring, in turn, specific preparations. However, this approach (inspired by the original proof of the DMW) works well also for continuous-time models, see [73]. We use it for an analysis of the structure of the set of equivalent martingale measures in the next subsection.

2.1.13 Martingale Measures with Bounded Densities

The following useful result gives, in particular, the positive answer to the question whether the set $\mathcal{Q}^e$ is norm-dense in $\mathcal{Q}^e_l$ (that is whether $\mathcal{Z}^e$ is dense in $\mathcal{Z}^e_l$ in the L^1-norm). Indeed, in virtue of the DMW theorem, $\mathcal{Q}^e_l \ne \emptyset$ if and only if $\mathcal{Q}^e \ne \emptyset$. It remains to take as the reference measure an arbitrary element of the latter set and apply the theorem below. This theorem happens to be useful to get a similar property for the discrete-time model with infinite horizon, which will be discussed in the next section.

Theorem 2.1.17 *Let $P \in \mathcal{Q}^e_l$. Then the set $\{Q \in \mathcal{Q}^e,\ dQ/dP \in L^\infty\}$ is norm-dense in $\mathcal{Q}^e_l$.*

Proof. It contains three steps. The first one is a simple lemma on the approximations of positive functions on the probability space $(\mathbf{R}^m, \mathcal{B}^n, \mu)$ by positive functions from $C(\bar{\mathbf{R}}^m)$, where $\bar{\mathbf{R}}^m$ is the one-point compactification of $\mathbf{R}^m$. □

Lemma 2.1.18 *Let $\phi : \mathbf{R}^m \to \mathbf{R}^l$ be a measurable mapping with $|\phi| \in L^1(\mu)$. Put $U := \{g \in L^1(\mu) :\ g > 0,\ g|\phi| \in L^1(\mu)\}$ and $U_C := U \cap C(\bar{\mathbf{R}}^m)$. Then for any $f \in U$ and $\varepsilon > 0$, there is $f^\varepsilon \in U_C$ such that $\|f - f^\varepsilon\|_{L^1(\mu)} < \varepsilon$ and*

$$E_\mu \phi f = E_\mu \phi f^\varepsilon. \tag{2.1.3}$$

Proof. Let $\mathcal{O}_\varepsilon(f)$ be an open ball in L^1_μ of radius ε with center at f. Define the convex sets $G := U \cap \mathcal{O}_\varepsilon(f)$ and $G_C := U_C \cap \mathcal{O}_\varepsilon(f)$ and consider the affine mapping $\Phi : G \to \mathbf{R}^l$ with $\Phi(g) = E_\mu(f - g)\phi$. We need to show that $0 \in \Phi(G_C)$. Notice that U_C is a dense subset of U, and, therefore, G_C is dense in G in L^1_μ. It follows that $\Phi(G_C)$ is dense in $\Phi(G)$. The convexity of these sets implies that $\operatorname{ri}\Phi(G_C) = \operatorname{ri}\Phi(G)$, and to complete the proof, it is sufficient to check that $0 \in \operatorname{ri}\Phi(G)$. To this aim we first observe that without loss of generality we may consider the case where $f\phi^i$, $i = 1, \dots, l$, are linearly independent elements of L^1_μ. Suppose that $0 \notin \operatorname{ri}\Phi(G)$. Let us consider the smallest hyperplane H containing $\Phi(G)$. Since $0 \in \Phi(G)$, it is a subspace. By the separation theorem, there is a nontrivial linear functional y on H such that

$yx \geq 0$ for all $x \in \Phi(G)$. Extending y to a linear functional on the whole $\mathbf{R}^l$, we may rewrite this as $E_\mu(f-g)y\phi \geq 0$ whatever is $g \in G$. Using functions of the form $g = f \pm \delta f I_\Gamma$ where Γ is a measurable set and $\delta \in]0,1[$ is such that $g \in \mathcal{O}_\varepsilon(f)$, we get from here that $E_\mu I_\Gamma f y\phi = 0$ for any Γ. Hence, $yf\phi = 0$, in contradiction with the assumed linear independence of components. □

With this preparatory result, we can easily prove the claim for the one-period model.

Lemma 2.1.19 *Let $\mathcal{G}$ be a (complete) sub-σ-algebra of $\mathcal{F}$, and let α and η be random variables taking values, respectively, in $\mathbf{R}_+ \setminus \{0\}$ and $\mathbf{R}^d$ such that $E((1+\alpha)|\eta||\mathcal{G}) < \infty$. Assume that $E(\alpha|\mathcal{G}) = 1$, $E(\eta|\mathcal{G}) = 0$, and $E(\alpha\eta|\mathcal{G}) = 0$. Then there are bounded random variables $\alpha^n > 0$ converging to α a.s. and such that $E(\alpha^n|\mathcal{G}) = 1$, $E(\alpha^n\eta|\mathcal{G}) = 0$.*

Proof. Let $\mu(dx,\omega)$ be a regular conditional distribution of the random vector (α,η) knowing $\mathcal{G}$. Define on $\mathbf{R}^{d+1}$ the functions $f(x) := x^1$ and $\phi(x) := (1, x^2, \dots, x^{d+1})$. Writing the conditional expectations as the integrals with respect to conditional distribution, we express properties of α as follows: $E_{\mu(.,\omega)} f\phi = e_1$ (the first orth in $\mathbf{R}^{d+1}$) for all ω except a null-set. The set

$$\Gamma^n := \big\{(\omega, g) \in \Omega \times C(\bar{\mathbf{R}}^{d+1}) : \ g > 0, \ E_{\mu(.,\omega)} g\phi = e_1, \\ E_{\mu(.,\omega)}|f-g| < 1/n\big\}$$

is $\mathcal{G} \otimes \mathcal{B}(C(\bar{\mathbf{R}}^{d+1}))$-measurable and, according to the previous lemma, has the projection on Ω of full measure. By the classical measurable selection theorem Γ^n admits a $\mathcal{G}$-measurable selector $f^n : \Omega \to C(\bar{\mathbf{R}}^{d+1})$. The function of two variables $f^n(\omega, x)$, being $\mathcal{G}$-measurable in ω and continuous in x, is $\mathcal{G} \otimes \mathcal{B}^{d+1}$-measurable. The random variables $\tilde{\alpha}^n = f^n(\omega, (\alpha(\omega), \eta(\omega)))$ converge to α in L^1 and, hence, in probability. Let us define the bounded random variables $\tilde{\alpha}^{n,k}(\omega) := \tilde{f}^{n,k}(\omega, (\alpha(\omega), \eta(\omega)))$, where

$$\tilde{f}^{n,k}(\omega, x) = f^n(\omega, x) I_{\{\|f^n(\omega,.)\| \leq k\}} + I_{\{\|f^n(\omega,.)\| > k\}},$$

and $\|.\|$ is a uniform norm in x. Since $E_{\mu(.,\omega)} g\phi = e_1$, we have the equalities $E(\tilde{\alpha}^{n,k}|\mathcal{G}) = 1$, $E(\tilde{\alpha}^{n,k}\eta|\mathcal{G}) = 0$.

Obviously, $\tilde{\alpha}^{n,k}$ converge to $\tilde{\alpha}^n$ in probability. The convergence in probability is a convergence in a metric space, and, therefore, one can take a subsequence k_n such that $\alpha^n := \tilde{\alpha}^{n,k_n}$ converge to α in probability. But then there is a subsequence of α^n convergent to α a.s.

The third, concluding step, is also simple. Note first that we may replace the reference measure by any other from Q_l^e with bounded density. According to the DMW theorem, between such measures, there are measures from Q^e, and so we may assume without loss of generality that already $P \in Q^e$.

We again use the multiplicative representation of the density $\rho_T = dQ/dP$, namely, $\rho_T = \alpha_1 \dots \alpha_T$ with $\alpha_t := \rho_t/\rho_{t-1}$. The property $\rho \in \mathcal{Z}_l^e$ holds if and only if $E(\alpha_t|\mathcal{F}_{t-1}) = 1$, $E(\alpha_t|\Delta S_t||\mathcal{F}_{t-1}) < \infty$ and $E(\alpha_t \Delta S_t|\mathcal{F}_{t-1}) = 0$ for all $t \le T$. Applying the preceding lemma, we define the measure $P^n := \rho_T^n P \in Q^e$ with bounded density $\rho_T^n := \alpha_1^n \dots \alpha_T^n$ convergent to ρ_T a.s. But by the Scheffe theorem, here we also have the convergence in L^1. □

Remark. Theorem 2.1.17 has several obvious corollaries. For example, if $P \in Q_l^e$, then the set of $Q \in \mathcal{Q}_l^e$ with bounded densities $\rho_T = dQ/dP$ and $\rho_T^{-1} = dP/dQ$ is dense in $\mathcal{Q}_l^e$. This fact is easily seen by considering the convex combinations $Q^n = (1-1/n)Q + (1/n)P$ and letting n tend to infinity. Noticing that $\mathcal{Q}_l^e$ is dense in the set $\mathcal{Q}_l$ (by the similar consideration), one can further strengthen the claim in another direction, etc.

It is not difficult to check that the set of local martingale measures with finite entropy (i.e., with $E\rho_T \ln \rho_T < \infty$), if nonempty, is also dense in Q_l^e. We explain the idea by establishing a more general result, which has applications in portfolio optimization problems.

Let $\varphi :]0,\infty[\to \mathbf{R}$ be a measurable function bounded from below, and let

$$\mathcal{Q}_\varphi^e := \{Q \in \mathcal{Q}^e : \ E\varphi(dQ/dP) < \infty\}.$$

Proposition 2.1.20 *If the set $\mathcal{Q}_\varphi^e \ne \emptyset$, then it is dense in $\mathcal{Q}^e$ in the following two cases:*

(a) *for every $c \ge 1$, there exist constants $r_1(c), r_2(c) \ge 0$ such that*

$$\varphi(\lambda y) \le r_1(c)\varphi(y) + r_2(c)(y+1), \quad y \in]0,\infty[, \ \lambda \in [c^{-1}, c]; \tag{2.1.4}$$

(b) *φ is convex, and $\mathcal{Q}_\varphi^e = \mathcal{Q}_{\varphi_\lambda}^e$ for any $\lambda > 0$, where $\varphi_\lambda(y) := \varphi(\lambda y)$.*

Proof. (a) Let $\tilde{P} \in \mathcal{Q}_\varphi^e$. Take an arbitrary measure $Q \in \mathcal{Q}^e$. By the above theorem and the accompanying remark, there exists a sequence $Q^n \in \mathcal{Q}^e$ convergent to Q with the densities $dQ^n/d\tilde{P}$ taking values in intervals $[c_n^{-1}, c_n]$. We have

$$E\varphi\left(\frac{dQ^n}{dP}\right) = E\varphi\left(\frac{dQ^n}{d\tilde{P}}\frac{d\tilde{P}}{dP}\right) \le r_1(c_n)E\varphi\left(\frac{d\tilde{P}}{dP}\right) + 2r_2(c_n) < \infty.$$

Hence, $Q^n \in \mathcal{Q}_\varphi^e$, and the result follows.

(b) We may assume without loss of generality that $\varphi \ge 0$ (by adding a constant) and repeat the same arguments modifying only the last step. Clearly, $dQ^n/d\tilde{P} = \alpha_n c_n^{-1} + (1-\alpha_n)c_n$, where α_n is a random variable taking values in $[0,1]$. By convexity,

$$\begin{aligned} E\varphi\left(\frac{dQ^n}{d\tilde{P}}\frac{d\tilde{P}}{dP}\right) &\le E\left[\alpha_n\varphi\left(c_n^{-1}\frac{d\tilde{P}}{dP}\right) + (1-\alpha_n)\varphi\left(c_n\frac{d\tilde{P}}{dP}\right)\right] \\ &\le E\varphi\left(c_n^{-1}\frac{d\tilde{P}}{dP}\right) + E\varphi\left(c_n\frac{d\tilde{P}}{dP}\right) < \infty \end{aligned}$$

in virtue of assumption, and we conclude as before.

Note that for convex φ, condition (a) implies (b). The latter hypothesis entangles properties of φ and $\mathcal{Q}^e$.

In financial applications, typically, $\varphi(y) = y^p$, $p > 0$, or $\varphi(y) = y \ln y$. In particular, if nonempty, the set $\mathcal{Q}^e_{y \ln y}$ of martingale measures with finite entropy is dense in the set of equivalent martingale measures $\mathcal{Q}^e$.

More generally, let $u : \mathbf{R} \to \mathbf{R}$ be an increasing *concave* differentiable function, and let u^* be its Fenchel dual (which is, by definition, the Fenchel dual of the *convex* function $-u(-.)$), i.e., the convex function

$$u^*(y) = \sup_x \big(u(x) - xy\big).$$

For example, the dual of the exponential utility function $u(x) = 1 - e^{-x}$ is the function $u^*(y) = y \ln y - y + 1$, $y \geq 0$, and $u^*(y) = \infty$, $y < 0$.

Suppose that u has a "reasonable" asymptotic elasticity, i.e.,

$$AE_+(u) := \limsup_{x \to \infty} \frac{xu'(x)}{u(x)} < 1, \qquad AE_-(u) := \liminf_{x \to -\infty} \frac{xu'(x)}{u(x)} > 1.$$

It can be shown that the function $\varphi = u^*$ satisfies the growth condition (a) of Proposition 2.1.20. □

2.1.14 Utility Maximization and Convex Duality

In this subsection we explain the importance of the set of equivalent martingale measures in the problem of portfolio optimization. Namely, we consider the simplest model with finite number of states of nature where the investor maximizes the mean value of an exponential utility function of the terminal value of his portfolio. Applying the classical Fenchel theorem, we show that the dual problem involves martingale measures.

So, Ω is finite, and hence, the space L^0 can be identified with a finite-dimensional Euclidean space. As usual, R_T is the set of random variables $H \cdot S_T$, and $\mathcal{Z}^e_T$ (respectively, $\mathcal{Z}^e$) is the set of densities (respectively, density processes) of equivalent martingale measures.

It is *supposed* in the following discussion that $\mathcal{Z}^e_T \neq \emptyset$.

We are interested in the portfolio optimization problem the value of which is

$$J^o := \sup_{\eta \in R_T} E\big(1 - e^{-\eta}\big). \qquad (2.1.5)$$

It will be studied jointly with the minimization problem $EZ_T \ln Z_T \to \min$ over the set of equivalent martingale densities $\mathcal{Z}^e_T$; let its value be

$$\underline{J} := \inf_{\xi \in \mathcal{Z}^e_T} E\xi \ln \xi. \qquad (2.1.6)$$

The latter problem, by abuse of language, sometimes is referred to as the "dual" one. As we shall see below, this terminology deviates from the standard one of the convex analysis.

The continuous function $\xi \to E\xi \ln \xi$ attains its minimum $\underline{J}$ on the compact set $\mathcal{Z}_T^a$ which is a closure of $\mathcal{Z}_T^e$. Due to strict convexity, the minimizer ξ^o is unique. The derivative of the function $\varphi(x) = x \ln x$ (with $\varphi(0) = 0$) at zero is $-\infty$, and this property implies that ξ^o is strictly positive. Indeed, let us take an arbitrary point ξ of the set $\mathcal{Z}_T^e$ (assumed nonempty) and consider on $[0, T]$ the function $F_t = Ef_t$, where $f_t := \varphi(t\xi + (1-t)\xi^o)$. As F attains its minimum at $t = 0$, the derivative $F_0' \geq 0$. But $F_0' = Ef_0'$. Since $f_0' = \varphi'(0)\xi = -\infty$ on the set $\{\xi^o = 0\}$, the probability of the latter is zero.

The measure $P^o = \xi^o$ is called the *entropy minimal* equivalent martingale measure.

Proposition 2.1.21 $J^o = 1 - e^{-\underline{J}}$.

This result is a direct consequence of the fundamental Fenchel theorem. We recall the simplest version of its formulation (in its traditional form, for convex functions).

Let X be a Hilbert space, and let $f : X \to \mathbf{R} \cup \{\infty\}$, $g : X \to \mathbf{R} \cup \{\infty\}$ be two convex lower semicontinuous functions not identically equal to infinity, i.e., $\operatorname{dom} f \neq \emptyset$ and $\operatorname{dom} g \neq \emptyset$.

Let us consider two minimization problems, the primal

$$f(\eta) + g(\eta) \to \min \quad \text{on } X \tag{2.1.7}$$

and the dual

$$f^*(-\xi) + g^*(\xi) \to \min \quad \text{on } X^*(= X). \tag{2.1.8}$$

We denote their values $v := \inf_x [f(x) + g(x)]$ and $v_* := \inf_y [f^*(-y) + g^*(y)]$.

Suppose that

$$\operatorname{dom} f \cap \operatorname{dom} g \neq \emptyset, \qquad (-\operatorname{dom} f^*) \cap \operatorname{dom} g^* \neq \emptyset;$$

we rewrite these conditions, to relate them with those in the formulation of theorem below, as

$$0 \in \operatorname{dom} f - \operatorname{dom} g, \qquad 0 \in \operatorname{dom} f^* + \operatorname{dom} g^*.$$

They ensure that $v < \infty$ and $v_* < \infty$.

Note that always $v + v_* \geq 0$ because by the Fenchel inequality

$$f(\eta) + g(\eta) + f^*(-\xi) + g^*(\xi) \geq (-\eta, \xi) + (\eta, \xi) = 0.$$

The following result is a particular case of the Fenchel theorem.

Theorem 2.1.22 (a) *Let* $0 \in \operatorname{int}(\operatorname{dom} f - \operatorname{dom} g)$. *Then the dual problem (2.1.8) has a solution, and* $v + v_* = 0$.

(b) *Let* $0 \in \operatorname{int}(\operatorname{dom} g^* + \operatorname{dom} f^*)$. *Then the primal problem (2.1.7) has a solution, and* $v + v_* = 0$.

Let us consider the minimization problem

$$f(\eta) + g(\eta) \to \min \quad \text{on } L^0, \tag{2.1.9}$$

where $f(\eta) := E(e^\eta - 1)$, and $g = \delta_{R_T}$, the indicator function (in the sense of convex analysis), which is equal to zero on R_T and infinity on the complement. Clearly, f^* is calculated via the dual to the convex function $e^x - 1$, namely, $f^*(-\xi) = E(\xi \ln \xi - \xi + 1)\delta_{[0,\infty[}(\xi)$, and $g^* = \delta_{R_T^\circ}$. In our case the polar R_T° is just $R_T^\perp$, the subspace orthogonal to R_T. The conditions of the Fenchel theorem, part (a), are obviously fulfilled. Thus, J^o coincides with the (attained) value of the dual problem

$$f^*(-\xi) + g^*(\xi) \to \min \quad \text{on } L^0,$$

i.e., J^o is equal to the minimum of $f^*(-\xi)$ on the set $R_T^\perp \cap L_+^0 = \mathbf{R}_+ \mathcal{Z}_T^a$. Since

$$\inf_{\xi \in \mathcal{Z}} \inf_{t \geq 0} E(t\xi \ln t\xi - t\xi + 1) = \inf_{\xi \in \mathcal{Z}_T^a} \left(1 - e^{-E\xi \ln \xi}\right) = 1 - e^{-\underline{J}},$$

we get the result.

Remark. If $\mathcal{Z}^e \neq \emptyset$, the hypothesis of the Fenchel theorem, part (*b*), holds, ensuring the existence in the primal problem. In the case where $\mathcal{Z}^e = \emptyset$, there is an arbitrage strategy H^a with $\eta^a := H^a \cdot S_T \geq 0$ and $\eta^a \neq 0$. Clearly, for any $\eta \in R_T$, the value of the functional in (2.1.5) at $\eta^a + \eta$ is strictly larger than at η.

Proposition 2.1.23 *Let H^o be the optimal strategy for the problem of portfolio optimization, Then the random variable*

$$\xi^o := e^{-H^o \cdot S_T} / E e^{-H^o \cdot S_T} \tag{2.1.10}$$

is the density of the minimal entropy equivalent martingale measure P^o.

Proof. The right-hand side of (2.1.10) is the density of a martingale measure. Indeed, for any strategy H, the function

$$f_H(\lambda) = 1 - E e^{-H^o \cdot S_T + \lambda H \cdot S_T}$$

attains its maximum at $\lambda = 0$, and, therefore, $f'_H(0) = 0$, i.e.,

$$E(H \cdot S_T) e^{-H^o \cdot S_T} = 0,$$

implying the claimed property. Using it, we can easily verify that

$$1 - e^{-E\xi^o \ln \xi^o} = 1 - E e^{-H^o \cdot S_T}.$$

Accordingly to Proposition 2.1.21, this equality may hold only if ξ^o is the solution of the problem of the entropy minimization. □

2.2 Discrete-Time Infinite-Horizon Model

The aim of this section is to present relations between the absence of arbitrage and the existence of an equivalent martingale measure for the model with an $\mathbf{R}^d$-valued price process $S = (S_t)_{t=0,1,\dots}$ defined on some filtered space $(\Omega, \mathcal{F}, \mathbf{F} = (\mathcal{F})_{t=0,1,\dots})$. We assume that the initial σ-algebra is trivial. In the first subsection we discuss some purely probabilistic questions. In particular, we show that if S admits an equivalent local martingale measure, then it admits an equivalent martingale measure. Moreover, the latter can be chosen to ensure the integrability of an arbitrary adapted process fixed in advance. Afterwards we introduce some substitutes for the no-arbitrage property and prove necessary and sufficient conditions for them.

2.2.1 Martingale Measures in Infinite-Horizon Model

We consider the discrete-time infinite-horizon model with an $\mathbf{R}^d$-valued process $S = (S_t)_{t\ge 0}$ and introduce, for $p \ge 1$, the set $\mathcal{Q}^{e,p}$ of probability measures $Q \sim P$ such that S is a Q-martingale and $S_t \in L^p(Q)$ for all $t \ge 0$. We also use the standard notation $S_t^* := \sup_{s\le t} |S_s|$.

Theorem 2.2.1 *Let $S \in \mathcal{M}_{\mathrm{loc}}(P)$. Then there exists a probability measure $\tilde{P} \sim P$ such that $S \in \mathcal{M}(\tilde{P})$.*

In the case of finite time-horizon this assertion is a direct corollary of the DMW theorem (and the measure $\tilde{P}_T \sim P$ even can be chosen with the bounded density $d\tilde{P}_T/dP$). For the infinite-time horizon, we get it from the following much more general assertion.

Theorem 2.2.2 *Let S be a local martingale, and $Y = (Y_t)$ be an adapted process dominating S^*. Let $\varepsilon > 0$. Then there exists a measure $P' \sim P$ such that S is a local P'-martingale, $Y_t \in L^1(P')$ for every t, and $\|P' - P\| \le \varepsilon$.*

As an obvious corollary, we have:

Theorem 2.2.3 *The set $\mathcal{Q}^{e,p}$ is dense in the set $\mathcal{Q}_l^e$.*

Theorem 2.2.2 is a generalization of Theorem 2.1.17. It is interesting that the reference to the latter constitutes the essential ingredients of the proof.

Lemma 2.2.4 *Let $S = (S_t)_{t\le T}$ be a local martingale in $\mathbf{R}^d$, and let $\xi \in L^0_+$. Then for any $\varepsilon > 0$, there is a probability measure $P^\varepsilon \sim P$ with density Z_T^ε such that $S = (S_t)_{t\le T}$ is a local martingale with respect to P^ε, $E|Z_T^\varepsilon - 1| < \varepsilon$, and $Z_T^\varepsilon(1+\xi)$ is bounded.*

Proof. We introduce the probability measure $P^1 = ce^{-\xi}P$. Since the NA-property holds for P, it holds for P^1. By the DMW theorem there is $P^2 \sim P^1$ with $dP^2/dP^1 \in L^\infty$ such that $S \in \mathcal{M}(P^2)$. Applying Theorem 2.1.17 with P^2 as the reference measure, we obtain that there exists a measure $P^\varepsilon \sim P^2$ with $dP^\varepsilon/dP^2 \in L^\infty$ such that $\|P^\varepsilon - P\| < \varepsilon$ and $S \in \mathcal{M}(P^\varepsilon)$. The measure P^ε meets the requirements. □

Proof of Theorem 2.2.2. We may suppose that $\varepsilon < 1$. Take a sequence $\varepsilon_n > 0$ such that $\sum \varepsilon_n < \varepsilon/3$. We define recursively an auxiliary sequence of probability measures $P^n \sim P$ with bounded $\mathcal{F}_n$-measurable densities dP^n/dP. Let ζ^k denote the density process of P^k with respect to P, that is, $\zeta_t^k = E(dP^k/dP|\mathcal{F}_t)$. Put

$$Z_t^n := \zeta_t^1 \dots \zeta_t^n, \qquad \bar{Z}_t^n := E(Z_\infty^n|\mathcal{F}_t).$$

The construction will ensure the following properties:

(a) $\zeta_n^n(1+Y_n) \le c_n$ for some constant c_n;
(b) $E(\bar{Z}_t^n S_t|\mathcal{F}_{t-1}) = \bar{Z}_{t-1}^n S_{t-1}$ for all t, i.e., $\bar{Z}^n S$ is a local martingale;
(c) $\|P^n - P\| \le \tilde{\varepsilon}_n := \varepsilon_n/(1+c_0 \dots c_{n-1})$.

Using Lemma 2.2.4, we define a probability measure $P^1 \sim P$ with $\mathcal{F}_1$-measurable density dP^1/dP such that $\zeta_1^1(1+Y_1) \le c_1$ for some constant c_1, the process $(S_t)_{t\le 1}$ is a P^1-martingale, and $\|P^1 - P\| \le \varepsilon_1$. Note that the whole process S remains a local martingale with respect to P^1.

Suppose that the measures P^k for $k \le n-1$ are already constructed. Applying Lemma 2.2.4 to the $(d+1)$-dimensional local martingale $(\bar{Z}_t^{n-1}S_t, \bar{Z}_t^{n-1})_{t\le n}$, we find a measure $P^n \sim P$ with $\mathcal{F}_n$-measurable density such that the properties (a) and (c) hold and $(\bar{Z}_t^{n-1}S_t, \bar{Z}_t^{n-1})_{t\le n}$ is a local P^n-martingale. The latter property means that $(\bar{Z}_t^{n-1}\zeta_t^n S_t, \bar{Z}_t^{n-1}\zeta_t^n)_{t\le n}$ is a local martingale. The martingale $(\bar{Z}_t^{n-1}\zeta_t^n)$, having at the date $t=n$ the value $Z_n^{n-1}\zeta_n^n = Z_n^n = Z_\infty^n$, coincides with $(\bar{Z}_t^n)$. Thus, $\bar{Z}^n S$ is a local martingale, and the property (b) holds.

By virtue of our construction,

$$E\sum|\zeta_\infty^n - 1| \le \sum \varepsilon_n < \infty,$$

and hence $\sum|\zeta_\infty^n - 1| < \infty$ a.s. Thus, Z_∞^n converges almost surely to some finite random variable $Z_\infty > 0$. Moreover, the convergence holds also in L^1 because

$$E\big|Z_\infty^n - Z_\infty^{n-1}\big| = EZ_\infty^{n-1}\big|\zeta_\infty^n - 1\big| \le c_0 \dots c_{n-1} E\big|\zeta_\infty^n - 1\big| \le \varepsilon_n$$

and $\sum \varepsilon_n$ is finite. Also,

$$E|Z_\infty - 1| \le \sum E\big|Z_\infty^n - Z_\infty^{n-1}\big| \le \sum \varepsilon_n < \varepsilon/3.$$

It remains to check that the probability measure $P' = (Z_\infty/EZ_\infty)P$ meets the requirements. Since $EZ_\infty \ge 1 - \varepsilon/3 \ge 2/3$, we have that

$$\|P' - P\| = E|Z_\infty/EZ_\infty - 1| \le \frac{2E|Z_\infty - 1|}{EZ_\infty} \le \varepsilon.$$

It is easy to check that, for each fixed t, the sequence $Z_\infty^n Y_t$, $n = t, t+1, \dots$, is fundamental in L^1. Indeed, we again use the properties (a) and (c):

$$E\left|Z_\infty^n Y_t - Z_\infty^{n-1} Y_t\right| \le E\left|\zeta_n^n - 1\right|\zeta_1^1 \ldots \zeta_t^t Y_t \zeta_{t+1}^{t+1} \ldots \zeta_{n-1}^{n-1}$$
$$\le c_0 \ldots c_{n-1} E\left|\zeta_n^n - 1\right| \le \varepsilon_n.$$

It follows that $Z_\infty^n Y_t$ and $Z_\infty^n S_t$ converges in L^1 to integrable random variables. Thus, $Z_\infty Y_t \in L^1$, and, in virtue of (b),

$$E(Z_\infty \Delta S_t | \mathcal{F}_{t-1}) = 0,$$

i.e., P' is a martingale measure. □

2.2.2 No Free Lunch for Models with Infinite Time Horizon

Infinite-horizon discrete-time market models based on the price process $(S_t)_{t=0,1,\ldots}$ pose new interesting mathematical problems related with the so-called doubling strategies or the St.-Petersburg game. It is well known that if S is a symmetric random walk on integers and, hence, a martingale, the strategy $H_t = 2^t I_{\{t \le \tau\}}$ where $\tau := \inf\{t \ge 1 : \ \Delta S_t = 1\}$ looks as an arbitrage opportunity: $H \cdot S_\infty = 1$. This strategy vanishes after the stopping time τ, which is finite but not bounded. So, certain restrictions on strategies are needed to exclude such a one. A satisfactory criterion relating the existence of an equivalent martingale measure with a strengthened no-arbitrage property can be obtained by assuming that there is no trading after some bounded stopping time where the bound depend on the strategy. Using the concepts and notation developed above, we can formalize this easily.

Let R_∞ be the union of all sets R_T, $T \in \mathbf{N}$, and let $A_\infty := R_\infty - L^0_+$.

The infinite-horizon model has the *NA-property* if $R_\infty \cap L^0_+ = \{0\}$ (or, equivalently, $A_\infty \cap L^0_+ = \{0\}$). In general, *NA* is weaker than the *EMM*-property claiming the existence of a probability measure $\tilde{P} \sim P$ such that S is a $\tilde{P}$-martingale. The simplest reinforcing of *NA* is the *NFL*-property ("no-free-lunch") suggested by Kreps: $\bar{C}^w_\infty \cap L^\infty_+ = \{0\}$, where $\bar{C}^w_\infty$ is the closure of the set $C_\infty := A_\infty \cap L^\infty$ in the topology $\sigma(L^\infty, L^1)$ (i.e., the weak* closure).

Theorem 2.2.5 *The following properties are equivalent:*

(a) $\bar{C}^w_\infty \cap L^\infty_+ = \{0\}$ *(NFL);*
(b) *there exists* $\tilde{P} \sim P$ *such that* $S \in \mathcal{M}_{\mathrm{loc}}(\tilde{P})$*;*
(c) *there exists* $\tilde{P} \sim P$ *such that* $S \in \mathcal{M}(\tilde{P})$.

Proof. The Kreps–Yan theorem says that the NFL-property holds if and only if there exists $P' \sim P$ such that $E'\xi \le 0$ for all $\xi \in \bar{C}^w_\infty$. This P' can be called a *separating measure* since its density is a functional from L^1 which separates $\bar{C}^w_\infty$ and L^∞_+. Of course, a local martingale measure $\tilde{P}$ is a separating one. Indeed, if $H \cdot S_T$ is bounded from below, then the process $(H \cdot S_t)_{t \le T}$ is a $\tilde{P}$-martingale. Hence, for any bounded from below random variable $\xi = H \cdot S_T - h$ where $h \in L^0_+$, we have the inequality $\tilde{E}\xi \le 0$. It follows that this inequality holds for any $\xi \in \bar{C}^w_\infty$. This gives us the implication (b) ⇒ (a). The more difficult implication (a) $\Rightarrow$ (b) follows from Theorem 2.2.6 below ensuring that, amongst the equivalent separating measures, there is a local martingale measure. The equivalence (b) $\Leftrightarrow$ (c) follows from Theorem 2.2.2. □

Theorem 2.2.6 *Any neighborhood of a separating measure contains an equivalent probability measure P' under which S is a local martingale.*

Proof. We assume without loss of generality that the reference measure P is separating. Fix $\varepsilon > 0$ and a sequence of numbers $\varepsilon_s > 0$ such that $\sum_{s\geq 1} \varepsilon_s < \varepsilon$. The theorem will be proven if, for each $s \geq 1$, we find an $\mathcal{F}_s$-random variable $\alpha_s > 0$ with the following properties:

(i) $E(\alpha_s|\mathcal{F}_{s-1}) = 1$;
(ii) $E(|1-\alpha_s||\mathcal{F}_{s-1}) \leq \varepsilon_s$;
(iii) $E(\alpha_s|\Delta S_s||\mathcal{F}_{s-1}) < \infty$ and $E(\alpha_s \Delta S_s|\mathcal{F}_{s-1}) = 0$.

Indeed, let us consider the process $Z_t := \alpha_1 \dots \alpha_t$, $t \geq 1$, $Z_0 = 1$, which is a martingale in virtue of (i). In virtue of (ii),

$$E|\Delta Z_s| = EZ_{s-1}E\big(|\alpha_s - 1||\mathcal{F}_{s-1}\big) \leq \varepsilon_s.$$

The martingale Z, being dominated by an integrable random variable, namely, by $1 + \sum |\Delta Z_s|$, is uniformly integrable. Also, $E\sum|\alpha_s - 1| < \varepsilon$. Therefore, $\sum|\alpha_s - 1| < \infty$ a.s., and the infinite product $Z_\infty > 0$ a.s. Thus, the probability measure $\tilde{P} = Z_\infty P$ is equivalent to P. In virtue of (iii), the process S is a generalized martingale under $\tilde{P}$, i.e., belongs to the class coinciding with $\mathcal{M}_{\text{loc}}(\tilde{P})$. Moreover,

$$E|Z_\infty - 1| \leq E\sum_{s\geq 1}|\Delta Z_s| < \varepsilon.$$

Let $H_s \in L^0(\mathbf{R}^d, \mathcal{F}_{s-1})$ be such that the random variable $H_s \Delta S_s$ is bounded from below. Then $(H_s \Delta S_s) \wedge n$, being an element of C_∞, has a negative expectation—we assumed that P is a separating measure. By the Fatou lemma $EH_s\Delta S_s \leq 0$. In the proposition below we show that this ensures the existence of α_s with the required properties. □

So, we need the following one-step result.

Proposition 2.2.7 *Let $\mathcal{G}$ be a sub-σ-algebra of $\mathcal{F}$. Suppose that $\eta \in L^0(\mathbf{R}^d)$ is such that $E\gamma\eta \leq 0$ for any $\gamma \in L^0(\mathbf{R}^d, \mathcal{G})$ for which $\gamma\eta$ is bounded from below. Let $\varepsilon > 0$. Then there is a strictly positive random variable α such that $E(\alpha|\mathcal{G}) = 1$, $E(|1-\alpha||\mathcal{G}) \leq \varepsilon$, $E(\alpha|\eta||\mathcal{G}) < \infty$, and $E(\alpha\eta|\mathcal{G}) = 0$.*

Proof. Let $\mu(dx, \omega)$ be a regular conditional distribution of η with respect to $\mathcal{G}$. In the space $\Omega \times C(\bar{\mathbf{R}}^d)$ we consider the $\mathcal{G}\otimes\mathcal{B}(C(\bar{\mathbf{R}}^d))$-measurable set Γ defined as the intersection of the sets

$$\big\{(\omega, g) :\ g > 0,\ E_{\mu(.,\omega)}g = 1,\ E_{\mu(.,\omega)}|1-g| \leq \varepsilon\big\}$$

and

$$\{(\omega,g):\ E_{\mu(.,\omega)}g|x|<\infty,\ E_{\mu(.,\omega)}gx=0\}.$$

If the projection of Γ on Ω is of full measure, we apply the measurable selection theorem, take an arbitrary $\mathcal{G}$-measurable selector $f:\Omega\to C(\bar{\mathbf{R}}^d)$, and conclude by putting $\alpha(\omega)=f(\omega,\eta(\omega))$.

Let Δ_ω be the image of the convex set

$$\{g\in C(\bar{\mathbf{R}}^d):\ g>0,\ E_{\mu(.,\omega)}g=1,\ E_{\mu(.,\omega)}g|x|<\infty,\ E_{\mu(.,\omega)}|1-g|\le\varepsilon\}$$

under the linear mapping $\Psi_\omega:=g\mapsto E_{\mu(.,\omega)}gx$. The full projection property means that, for almost all ω, the set Δ_ω contains the origin.

Let us first consider the case $d=1$, where Δ_ω is just an interval. Define the $\mathcal{G}$-measurable random variables

$$\zeta'(\omega)=\inf\{t:\ \mu(]-\infty,t],\omega)>0\},\qquad \zeta''(\omega)=\sup\{t:\ \mu(]-\infty,t],\omega)<1\}.$$

The random variables $I_AI_{\{-n\le\xi'\}}\eta$, where $A\in\mathcal{G}$, being bounded from below, have negative expectations. Hence, $I_{\{-\infty<\xi'\}}E(\eta|\mathcal{G})\le 0$. This implies that

$$I_{\{-\infty<\xi'\}}E(\eta^+|\mathcal{G})\le I_{\{-\infty<\xi'\}}E(\eta^-|\mathcal{G})<\infty.$$

Therefore, $\Psi(1)\le 0$ on the set $\{-\infty<\xi'\}$ and, by symmetry, $\Psi(1)\ge 0$ on the set $\{\xi''<\infty\}$ (a.s.). Thus, on the intersection of these sets, $\Psi(1)=0$. It follows from the elementary lemma below that the interval $\Delta_\omega\supseteq[0,\infty[$ for almost all $\omega\in\{-\infty<\xi',\ \xi''=\infty\}$. By symmetry, the interval $\Delta_\omega\supseteq]-\infty,0]$ for almost all $\omega\in\{-\infty=\xi',\ \xi''<\infty\}$. □

In the following assertion, ω is fixed and omitted in notation.

Lemma 2.2.8 *If $\xi''=\infty$, then Δ is unbounded from above.*

Proof. Fix $\varepsilon\in]0,1]$ and $a>0$ such that $\mu(\{a\})=0$. Consider the subset $W_{\gamma,a}$ formed by the continuous functions g such that $g(a)=1$, $xg(x)\to 0$ as $x\to\pm\infty$, and $E_\mu gI_{[a,\infty[}=\varepsilon$. Note that $\sup_{g\in W_{\varepsilon,a}}E_\mu xgI_{[a,\infty[}=\infty$. Indeed, as the support of μ is unbounded, we can find a continuous function g_0 with a compact support contained in the interval $]a,\infty[$ such that $E\mu g_0<\varepsilon$ while the value $E_\mu xg$ is arbitrarily large. Adding to g_0 the function $e^{-\lambda|x-a|}$ with an appropriately chosen parameter λ, we obtain a function $g\in W_{\varepsilon,a}$ with $E_\mu xgI_{[a,\infty[}\ge E_\mu g_0$.

Take $a>0$ such that $\mu(\{a\})=0$ and $\mu(\{x:\ |x|\ge a\})\le\delta/2$. Take $f=e^{-\lambda|x+a|}$ and choose the parameter λ to ensure that

$$\varepsilon:=\mu(\{x:\ |x|\ge a\})-E\mu fI_{]-\infty,a]}>0.$$

By the above, for any $N>0$, we can find $f_N\in W_{\varepsilon,a}$ such that $E_\mu xf_NI_{[a,\infty[}\ge N$. The assertion became obvious since, for the function

$$g_N:=fI_{]-\infty,-a[}+I_{]-a,a[}+g_NI_{]a,a]},$$

we have $\Psi(g_N)\ge N$. The lemma is proven. □

For the case of $d > 1$ and ω for which $0 \notin \Delta_\omega$, there is, in virtue of the Hahn–Banach theorem, $l(\omega) \in \mathbf{R}^d$ such that $|l(\omega)| = 1$ and $l(\omega)x > 0$ for all $x \in \Delta_\omega$. Put $l(\omega) = 0$ if $0 \in \Delta_\omega$. Using a measurable version of the Hahn–Banach theorem, one can choose the separating functionals in such a way that the function $\omega \mapsto l(\omega)$ is $\mathcal{G}$-measurable. Applying the above reasoning to the scalar random variable $\eta^l := l\eta$, we find a function $f^l(\omega, y)$ on $\Omega \times \mathbf{R}$ which is $\mathcal{G}$-measurable in ω and continuous in x. Denoting by $\mu^l(dy,.)$ the regular conditional distribution of η^l with respect to $\mathcal{G}$, we get, by the change of variable, that

$$l(\omega)\int_{\mathbf{R}^d} xf^l(\omega, l(\omega)x)\mu(dx,\omega) = \int_{\mathbf{R}} yf^l(\omega,y)\mu^l(dy,\omega) = 0.$$

Thus, $l = 0$ (a.s.), and the required property holds.

Remark. Let P be a probability measure under which S is a local martingale, and let H be a strategy such that the process $H \cdot S$ is bounded from below. Then this process is a true martingale converging at infinity to a random variable $H \cdot S_\infty$ almost surely. By the Fatou lemma, $H \cdot S_t \geq E(H \cdot S_\infty|\mathcal{F}_t)$. Therefore, for this strategy, $H \cdot S \geq 0$ if and only if $H \cdot S_\infty \geq 0$. These considerations show that there is hope to get conditions for the existence of an equivalent local martingale measure based on strategies of such a type. This is done in the next section.

In the following model the *NA*-property is fulfilled, but there is no equivalent separating measure. Namely, $R_\infty \subset L^\infty$, $R_\infty \cap L^\infty_+ = \{0\}$, but $\bar{C}^w_\infty = L^\infty$!

Example. Let $\Omega = \mathbf{N}$, $P(\{2k-1\}) = P(\{2k\}) = 2^{-k-1}$, $k \geq 1$, and let $\mathcal{F}_t := \sigma\{\{1\},\dots,\{2t\}\}$. Put $S_0 := 0$,

$$\Delta S_k = 2^{5k}I_{\{2k-1\}} + 2^{2k}I_{\{2k\}} - 2^{-k}I_{\{2k+1,\dots\}}.$$

Since $\mathcal{F}_T$ is finite, the random variables $H \cdot S_T$ are bounded, and $R_\infty \subseteq L^\infty$.

Let $0 \leq \xi \leq 1$. Then $S_T \wedge \xi \in C_\infty$, and $S_T \wedge \xi \to \xi$ in probability as $T \to \infty$. Hence, $\xi \in \bar{C}^w_\infty$. It follows that $\bar{C}^w_\infty = L^\infty$.

Let $\eta \neq 0$ be a random variable from $R_\infty \cap L^\infty_+$, i.e., of the form $H \cdot S_T$. Let k be the first integer for which at least one of the values $\eta(2k-1)$ or $\eta(2k)$ is strictly positive. Inspecting sequentially the increments $H_t\Delta S_t$, we deduce that $H_t = 0$ for $t < k$, while the $\mathcal{F}_{k-1}$-measurable random variable $H_k(j)$ is equal to $a > 0$ for $j \geq k-1$. It follows that $H_k(j)\Delta S_k(j) \leq -ae^{-k}$ for $j \geq 2k+1$. The negative values at elementary events $2k+1$ and $2k+2$ can be compensated only if $H_{k+1}(j) \geq a2^{-k}$ for $j \geq 2k+1$. Continuing this inspection, we arrive at the last increment $H_T\Delta S_T$ the negative values of which on elementary events $2T+1,\dots$ cannot be compensated.

More surprisingly, in this example the closure of R_∞ in the L^1-norm intersects L^∞_+ only at zero. This can be shown by a similar sequential inspection of $\lim_n H^n_t\Delta S_t$. To ensure the positivity of η, these random variables should take such large positive values at the elementary events with odd numbers larger than k that the L^1-norm of η would be infinite in an apparent contradiction.

2.2.3 No Free Lunch with Vanishing Risk

The *NFL*-condition can be criticized because the weak* closure has no good financial interpretation.[1] Fortunately, it can be replaced by the more attractive *NFLVR*-property.

To describe the latter we introduce the class of *admissible* strategies H whose value processes $H \cdot S$ are bounded from below (by constants depending on H) and converge a.s. to finite limits. Denoting by $R_{\rm ad}$ the set of random variables $H \cdot S_\infty$, we define the sets $A_{\rm ad} := R_{\rm ad} - L^0_+$ and $C_{\rm ad} := A_{\rm ad} \cap L^\infty$.

We say that the process S has the *NFLVR-property* (no free lunch with vanishing risk) if $\bar{C}_{\rm ad} \cap L^\infty = \{0\}$, where $\bar{C}_{\rm ad}$ is the norm-closure of $C_{\rm ad}$. "Financial" motivation of the terminology is based on the alternative description: *NFLVR-property* holds if and only if $P\text{-}\lim \xi_n = 0$ for every sequence $\xi_n \in C_{\rm ad}$ such that $\|\xi_n^-\|_{L^\infty} \to 0$, see Lemma 2.2.11.

Though the sets A and $A_{\rm ad}$ may be not related by an inclusion, the property $A_{\rm ad} \cap L^0_+ = \{0\}$ ensures the property $A_\infty \cap L^0_+ = \{0\}$. Indeed, the former implies that, for any finite T, there is no arbitrage in the class of strategies with the value processes $(H \cdot S_t)_{t \le T}$ bounded from below. As we know, this is equivalent to the absence of arbitrage in the class of all strategies and, hence, to the existence of an equivalent martingale measure on $\mathcal{F}_T$. It follows that the property $A_{\rm ad} \cap L^0_+ = \{0\}$ implies that the bound $H \cdot S_T \le c$ propagates backwards and $C_\infty \subseteq C_{\rm ad}$.

Theorem 2.2.9 *NFLVR holds if and only if there is $P' \sim P$ such that $S \in \mathcal{M}_{\rm loc}(P')$.*

Proof. It is easy to see (using the Fatou lemma) that a local martingale measure (and even a separating measure for R_∞) separates $\bar{C}_{\rm ad}$ and L^∞_+. So, the implication "if" is obvious. On the other hand, the condition $\bar{C}^w_{\rm ad} \cap L^\infty_+ = \{0\}$, ensuring that $C_\infty \subseteq C_{\rm ad}$, implies the *NFL*-property, and the needed measure P' does exist in virtue of Theorem 2.2.5. But according to Theorem 2.2.10 below, such a condition holds because under *NFLVR* the set $\bar{C}^w_{\rm ad}$ coincides with $\bar{C}_{\rm ad}$. □

Theorem 2.2.10 *Suppose that $\bar{C}_{\rm ad} \cap L^\infty = \{0\}$. Then $C_{\rm ad} = \bar{C}^w_{\rm ad}$.*

Before the proof we establish some simple facts from functional analysis. Let $\lfloor \eta, \infty \lfloor$ be the set of $\xi \in L^0$ such that $\xi \ge \eta$.

Lemma 2.2.11 *Let C be a convex cone in L^∞ containing $-L^\infty_+$. Then the following properties are equivalent:*

[1] Of course, the definition of weak* closure involving only halfspaces is even simpler than that of the norm closure. The intuition, though, appeals to the "interior" description, in terms of limits. In general, the weak* sequential closure lies strictly between the norm-closure and weak* closure. To get all points of the latter as limits, one needs to consider convergence along the nets, which is, indeed, not intuitive.

(a) $\bar{C} \cap L^\infty_+ = \{0\}$;
(b) P-$\lim \xi_n = 0$ *for every sequence* $\xi_n \in C$ *such that* $\|\xi_n^-\|_{L^\infty} \to 0$;
(c) *the set* $C \cap \lfloor -1, \infty \lfloor$ *is bounded in probability.*

Proof. (a) $\Rightarrow$ (b). If the assertion fails, one can find a sequence $\xi_n \in C$ such that $\xi_n \geq -1/n$ and $P(\xi_n > \varepsilon) \geq \varepsilon$ for some $\varepsilon > 0$. Since $\xi_n \wedge 1 \in C$, we may assume that $\xi_n \leq 1$. By the von Weizsäcker theorem there are random variables of the form $\bar{\xi}_k = k^{-1} \sum_{i=1}^k \xi_{n_i}$ (thus, elements of C) convergent to a certain random variable ξ a.s. Note that the negative parts of $\bar{\xi}_k$ converge to zero in L^∞. On the other hand, ξ is not zero. Indeed, ξ is also the limit of $k^{-1} \sum_{i=1}^k \tilde{\xi}_{n_i}$, where $\tilde{\xi}_n := \xi_n + 1/n \geq 0$ and $P(\tilde{\xi}_n > \varepsilon) \geq \varepsilon$. It is easy to see that

$$Ee^{-\tilde{\xi}_n} \leq P(\tilde{\xi}_n \leq \varepsilon) + e^{-\varepsilon} P(\tilde{\xi}_n > \varepsilon) \leq 1 - \varepsilon + e^{-\varepsilon}\varepsilon < 1.$$

Due to convexity of the exponential, the same bound holds for the convex combinations of $\tilde{\xi}_n$ and, thus, for the limit ξ. So, $\beta := P(\xi > 0) > 0$. By the Egorov theorem, there is a measurable set Γ with $P(\Gamma) > 1 - \beta/2$ on which the convergence $\bar{\xi}_n \to \xi$ is uniform. But then the sequence $\bar{\xi}_n^+ I_\Gamma - \bar{\xi}_n^-$ of elements of C converges in L^∞ to a nonzero random variable $\xi I_\Gamma \geq 0$, in contradiction with (a).

(b) $\Rightarrow$ (c). If the set $C \cap \lfloor 1, \infty \lfloor$ is unbounded in probability, then it contains a sequence of random variables $\xi_n^0 \geq 1$ such that $\lim P(\xi_n^0 \geq n) > 0$. But then the sequence $\xi_n := \xi_n^0/n$ violates condition (b).

(c) $\Rightarrow$ (a). If (a) fails to be true, there exist a sequence $\xi_n \in C$ and a nonzero $\xi \in L^\infty_+$ such that $\|\xi - \xi_n\|_{L^\infty} \leq 1/n$. It follows that $\|\xi_n^-\|_{L^\infty} \leq 1/n$. Then the random variables $n\xi_n$ belong to $C \cap \lfloor -1, \infty \lfloor$ and form a sequence divergent to infinity on the set $\{\xi > 0\}$ and, therefore, not bounded in probability. □

The next lemma, comparatively with the previous one, requires a specific structure of the cone C. We use the notation $\bar{K}^P$ for the closure of K in L^0.

Lemma 2.2.12 *Let* $C = (K - L^0_+) \cap L^\infty$ *where* K *is a cone,* $K \subseteq \lfloor -1, \infty \lfloor$. *Suppose that* K *is bounded in probability. Let* ξ_n *be a sequence in* $C \cap \lfloor -1, \infty \lfloor$ *convergent to* ξ *a.s. Then the set* $\bar{K}^P \cap \lfloor \xi, \infty \lfloor$ *is nonempty and contains a maximal element* η_0.

Proof. In virtue of the assumed structure of the set C, there are $\eta_n \in K$ such that $\eta_n \geq \xi_n$. Applying the von Weizsäcker theorem, we find a subsequence such that $\bar{\eta}_k := k^{-1} \sum_{i=1}^k \eta_{n_i}$ converge a.s. to some $\bar{\eta} \geq \xi$. Since K is bounded in probability, so is the set $\bar{K}^P$. Thus, $\bar{\eta}$ is finite and belongs to $\bar{K}^P \cap \lfloor \xi, \infty \lfloor \neq \emptyset$. It remains to recall that any nonempty closed bounded subset of L^0 has a maximal element with respect to the natural partial ordering (each linearly ordered subset $\{\zeta_\alpha\}$ has as a majorant $\operatorname{ess\,sup}_\alpha \zeta_\alpha < \infty$, and the existence of the maximal element holds by the Zorn lemma). □

Lemma 2.2.13 *Let* $C_{\rm ad} \cap L_+ = \{0\}$. *If* H *is an admissible integrand, then* $H \cdot S_\infty \geq -1$ *if and only if the process* $H \cdot S \geq -1$.

Proof. Suppose that H is admissible and $H \cdot S_\infty \geq -1$ but there is u such that $P(\Gamma_u) > 0$, where $\Gamma_u := \{H \cdot S_u < -1\}$. Then the strategy $HI_{[u,\infty[}I_{\Gamma_u}$ is admissible, and the random variable $HI_{]u,\infty[}I_{\Gamma_u} \cdot S_\infty \geq 0$ is strictly positive on Γ_u. This is a contradiction with the assumption of the lemma. □

Proof of Theorem 2.2.10. According to the Krein–Šmulian theorem, a convex set is closed in $\sigma\{L^\infty, L^1\}$ if an only if its intersection with every ball of L^∞ is closed in probability. Obviously, the last condition follows if the set is *Fatou-closed*, that is, if it contains the limit of any bounded from below sequence of its elements convergent almost surely. So, let ξ_n be a sequence in C_{ad} convergent to ξ a.s. and such that all $\xi_n \geq -c$. It is sufficient to argue with $c = 1$. We apply Lemma 2.2.12 with $K = R_{\text{ad}} \cap \lfloor -1, \infty \lfloor$, which is bounded in probability by virtue of Lemma 2.2.11. The theorem will be proven if we show that a maximal element η_0 in $\bar{K}^P \cap \lfloor \xi, \infty \lfloor \neq \emptyset$ belongs to K. So, we have a sequence $V^n := H^n \cdot S \geq -1$ with $V^n_\infty \to \eta$ a.s. We claim that $\sup_t |V^n_t - V^m_t| \to 0$ in probability as $n, m \to \infty$. If this not true, then $P((\sup_t (V^{i_k}_t - V^{j_k}_t)^+ > \varepsilon) \geq \varepsilon$ with some $\varepsilon > 0$ and $i_k, j_k \to \infty$. For $T_k := \inf\{t : V^{i_k}_t - V^{j_k}_t > \alpha\}$, we have $P(T_k < \infty) \geq \varepsilon$. Let us consider the process

$$\tilde{V}^k := \left(I_{[0,T_k]} H^{i_k} + I_{]T_k,T]} H^{j_k}\right) \cdot S,$$

which is an element of $K = R_{\text{ad}} \cap \lfloor -1, \infty \lfloor$. Note that

$$\tilde{V}^k_\infty = V^{i_k}_\infty I_{\{T_k = \infty\}} + V^{j_k}_\infty I_{\{T_k < \infty\}} + \xi_k,$$

where $\xi_k := (V^{i_k}_{T_k} - V^{j_k}_{T_k}) I_{\{T_k < \infty\}} \geq 0$, and $P(\xi_k \geq \varepsilon) \geq \varepsilon$. Using the von Weizsäcker theorem in the same way as in the proof of Lemma 2.2.11, we find a sequence $\bar{V}^k \in K$ such that $\bar{V}^k_\infty \to \eta_0 + \xi$, where $\xi \in L^0$ and $\xi \neq 0$. This contradicts the maximality of η_0.

Taking a subsequence, we may assume that $\sup_t |V^n_t - V^m_t| \to 0$ a.s. Thus, there is a process V which is a uniform limit of V^n (a.s.). Obviously, $V \geq -1$, and the limit V_∞ exists and is finite. Since $\Delta V^n_t = H^n_t \Delta S_t$ converges to ΔV_t and R_T is closed, we have $\Delta V_t = H_t \Delta S_t$. □

2.2.4 Example: "Retiring" Process

Here we present an example where a martingale measure can be constructed in a rather straightforward way. We shall use the result later, in the study of models with transaction costs.

Let $S = (S_t)_{t \geq 0}$ be an $\mathbf{R}^d$-valued discrete-time adapted process. Put $\xi_t = \Delta S_t$, $\Gamma_t := \{\xi_t = 0\}$.

Proposition 2.2.14 *Suppose that the following conditions hold:*

(i) *for each finite T, the process $(S_t)_{t \leq T}$ has the NA-property;*
(ii) *$I_{\Gamma_t} \uparrow 1$ a.s.;*
(iii) *$E(I_{\Gamma_t} | \mathcal{F}_{t-1}) > 0$ a.s. on Γ^c_{t-1} for each $t \geq 1$.*

Then there exists a probability $Q \sim P$ such that S is a Q-martingale bounded in $L^2(Q)$ (hence, uniformly integrable with respect to Q).

Proof. By the DMW theorem condition (i) is equivalent to the NA-property for each one-step model: the relation $\gamma\xi_t \geq 0$ with $\gamma \in L^0(\mathbf{R}^d, \mathcal{F}_{t-1})$ may hold only if $\gamma\xi_t = 0$. The same theorem asserts that each ξ_t admits an equivalent martingale measure which can be chosen to ensure the integrability of any fixed finite random variable, e.g., $|\xi_t|^2$. In terms of densities this means that there are $\mathcal{F}_t$-measurable random variables $\bar{\alpha}_t > 0$ such that $E(\bar{\alpha}_t\xi_t|\mathcal{F}_{t-1}) = 0$ and $c_t := E(\bar{\alpha}_t|\xi_t|^2|\mathcal{F}_{t-1}) < \infty$. Normalizing, to this we can also add the property $E(\bar{\alpha}_t|\mathcal{F}_{t-1}) = 1$.

We define the $\mathcal{F}_t$-measurable random variable $\alpha_t > 0$ by the formula

$$\alpha_t = I_{\Gamma_{t-1}} + \left[\frac{(1-\delta_t)I_{\Gamma_t}}{E(I_{\Gamma_t}|\mathcal{F}_{t-1})} + \frac{\delta_t\bar{\alpha}_t I_{\Gamma_t^c}}{E(\bar{\alpha}_t I_{\Gamma_t^c}|\mathcal{F}_{t-1})}\right] I_{\Gamma_{t-1}^c \cap A_t} + I_{\Gamma_{t-1}^c \cap A_t^c},$$

where $A_t := \{E(\bar{\alpha}_t I_{\Gamma_t^c}|\mathcal{F}_{t-1}) > 0\}$ and $\delta_t := 2^{-t}E(\bar{\alpha}_t I_{\Gamma_t^c}|\mathcal{F}_{t-1})/(1+c_t)$. Clearly, $E(\alpha_t|\mathcal{F}_{t-1}) = 1$.

Noting that $\bar{\alpha}_t I_{\Gamma_t^c} I_{A_t^c} = 0$ (a.s.), we obtain that $E(\alpha_t\xi_t^2|\mathcal{F}_{t-1}) \leq 2^{-t}$ and $E(\alpha_t\xi_t|\mathcal{F}_{t-1}) = 0$.

The process $Z_t := \alpha_1 \dots \alpha_t$ is a martingale which converges (stationarily) a.s. to a random variable $Z_\infty > 0$ with $EZ_\infty \leq 1$. Recalling that $I_{\Gamma_t} \uparrow 1$ (a.s.) and using the identity $Z_\infty I_{\Gamma_t} = Z_t I_{\Gamma_t}$, we obtain that

$$EZ_\infty = E\lim_t Z_\infty I_{\Gamma_t} = \lim_t EZ_\infty I_{\Gamma_t} = \lim_t EZ_t I_{\Gamma_t} = 1 - \lim_t EZ_t I_{\Gamma_t^c}.$$

It follows that $EZ_\infty = 1$ (i.e., (Z_t) is a uniformly integrable martingale). Indeed, $E(\alpha_k I_{\Gamma_k^c}|\mathcal{F}_{k-1} \leq 2^{-k}$, and, hence,

$$EI_{\Gamma_t^c} Z_t = E\prod_{k\leq t} \alpha_k I_{\Gamma_k^c} \leq \prod_{k\leq t} 2^{-k} \to 0.$$

Thus, $Q := Z_\infty P$ is a probability measure under which S is a martingale. At last,

$$E_Q S_t^2 = \sum_{k\leq t} EZ_k\xi_k^2 \leq \sum_{k\leq t} 2^{-k} \leq 1,$$

i.e., S_t belongs to the unit ball of $L^2(Q)$. □

Remark 1. Condition (iii) cannot be omitted. Indeed, let S be the symmetric random walk starting from zero and stopped at the moment when it hits unit. It is a martingale, and condition (ii) holds. Since $S_\infty = 1$ a.s., the process S cannot be a uniformly integrable martingale with respect to a measure Q equivalent to P.

Remark 2. Fix $f : \mathbf{R} \to \mathbf{R}_+$. A minor modification of the arguments leads to a martingale measure Q for which $E_Q \sup_t f(S_t) < \infty$. Indeed, let (η_t) be an adapted process with $\eta_t = \eta_t I_{\Gamma_t^c} \geq 0$. As above, we can find α_t with the extra property $E(\alpha_t f(S_t)|\mathcal{F}_{t-1}) \leq 2^{-t}$ implying that $E\sum_t \eta_t < \infty$. It remains to take $\eta_t = f(S_t)I_{\Gamma_t^c}$ and note that $\sup_t f(S_t) \leq \sum_t \eta_t$.

2.2.5 The Delbaen–Schachemayer Theory in Continuous Time

This book is addressed to the reader from whom we do not expect the knowledge of stochastic calculus beyond standard textbooks. Luckily, the theory of markets with transaction costs, in current state of art, does not require such a knowledge, in a surprising contrast to the classical continuous-time *NA*-theory initiated by Kreps and largely developed in a series of papers by Delbaen and Schachermayer collected in [57]. However, it seems to be useful to provide a short abstract of the main results of the latter, which will serve as a background for a discussion explaining this difference.

In the classical continuous-time theory we are given a set $\mathcal{X}$ of scalar semimartingales X on a compact interval $[0,T]$ interpreted as value processes; the elements of $R_T := \{X_T : X \in \mathcal{X}\}$ are the investor's "results"; the *NA*-property means that $R_T \cap L^0_+ = \{0\}$. Typically, $\mathcal{X}$ is the set of stochastic integrals $H \cdot S$, where S is a fixed d-dimensional semimartingale (interpreted as the price processes of risky assets), and H is a d-dimensional predictable process for which the integral is defined and is bounded from below by a constant depending on H. The condition on H ("admissibility") rules out the doubling strategies. The experience with discrete-time models gives a hint that martingale densities can be obtained by a suitable separation theorem. Put $C_T := (R_T - L^0_+) \cap L^\infty$ (the set of bounded contingent claims hedgeable from zero initial endowment) and introduce the "no-free-lunch condition" (*NFL*): $\bar{C}^w_T \cap L^\infty_+ = \{0\}$, where $\bar{C}^w_T$ is a closure of C_T in the weak* topology, i.e., $\sigma\{L^\infty, L^1\}$. The Kreps–Yan theorem (Theorem 2.1.4) says that *NFL* holds if and only if there exists an equivalent "separating" measure $P' \sim P$ such that $E'\xi \le 0$ for all ξ from $\bar{C}^w_T$ (or R_T). It is easy to see that in the model with a bounded (resp., locally bounded) price process S, the latter is a martingale (resp. local martingale).

The above result established by Kreps in the context of financial modelling ("FTAP") was completed by Delbaen and Schachermayer by a number of important observations for the model based on the price process S. We indicate here only a few.

First, they observed that in the Kreps theorem the condition *NFL* can be replaced by a visibly weaker (but, in fact, equivalent) condition "no-free-lunch condition with vanishing risk" (*NFLVR*): $\bar{C}_T \cap L^\infty_+ = \{0\}$, where $\bar{C}_T$ is the norm-closure of C_T in L^∞. The reason for this is in the following simply formulated (but difficult to prove) result from stochastic calculus:

Theorem 2.2.15 *Let the NFLVR-condition be fulfilled. Then* $C_T = \bar{C}^w_T$.

This result, which is a generalization of Theorem 2.2.9, can be formulated in a more abstract way for a convex set $\mathcal{X}$ of bounded from below semimartingales which satisfies some closedness and concatenation properties.

Second, they establish that in any neighborhood of a separating measure P', there exists an equivalent probability measure $\tilde{P}$ (also a separating one) such that the semimartingale S with respect to $\tilde{P}$ is a σ-martingale (i.e., for some predictable integrants G^i with values in $]0,1]$, the processes $G^i \cdot S^i$, $i = 1, \ldots, d$, are $\tilde{P}$-martingales). The situation for the continuous time is rather different even with respect to infinite-horizon discrete-time models: one cannot claim the existence of an equivalent local martingale measure! The reason for this is clear: in discrete time there is no difference between local martingales and σ-martingales (which are just generalized martingales).

As we shall see further, for the model with transaction costs, the portfolio processes are vector-valued, and their dynamics can be described using only the Lebesgue integrals. In the case of zero transaction cost, one can make a reduction to scalar wealth processes $H \cdot S$, but the resulting H are (vector-valued) processes of *bounded variation* and not arbitrary integrands, which is, apparently, an additional complication. In the general case the problem of no-arbitrage criteria has also other particularities arising even in the discrete-time framework.

3

Arbitrage Theory under Transaction Costs

3.1 Models with Transaction Costs

3.1.1 Basic Model

We describe here a financial framework leading to a "standard" discrete-time model with proportional transaction costs with complete information.

Suppose that the agent portfolio contains d assets which we prefer to interpret as currencies. Their quotes are given in units of a certain *numéraire* which may not be a traded security. At time t, the quotes are expressed by the vector of prices $S_t = (S_t^1, \dots, S_t^d)$ with strictly positive components.

The agent's positions can be described either by the vector of "physical" quantities $\widehat{V}_t = (\widehat{V}_t^1, \dots, \widehat{V}_t^d)$ or by the vector $V = (V_t^1, \dots, V_t^d)$ of values invested in each asset; they are related as follows:

$$\widehat{V}_t^i = V_t^i / S_t^i, \quad i \le d.$$

This formula suggests the notation $\widehat{V}_t = V_t / S_t$. More formally, introducing the diagonal operator

$$\phi_t : (x^1, \dots, x^d) \mapsto (x^1/S_t^1, \dots, x^d/S_t^d), \tag{3.1.1}$$

we may write that

$$\widehat{V}_t = \phi_t V_t.$$

In the considered market any asset can be exchanged to any other. At time t, the increase of the value of ith position in one unit of the numéraire by changing the value of jth position requires diminishing the value of the latter in $1 + \lambda_t^{ji}$ units of the numéraire. The matrix of transaction cost coefficients $\Lambda_t = (\lambda_t^{ij})$ has nonnegative entries and the zero diagonal.

In the dynamical multiperiod setting, $S = (S_t)$ and $\Lambda = (\Lambda_t)$ are adapted processes; it is convenient to choose the scales to have $S_0^i = 1$ for all i and assume as a convention that $S_{0-}^i = 1$.

Yu. Kabanov, M. Safarian, *Markets with Transaction Costs*,
Springer Finance, DOI 10.1007/978-3-540-68121-2_3,

The portfolio evolution can be described by the initial condition $V_{-0} = v$ (the endowments of the agent when entering the market) and the increments at dates $t \geq 0$:

$$\Delta V_t^i = \widehat{V}_{t-1}^i \Delta S_t^i + \Delta B_t^i \tag{3.1.2}$$

with

$$\Delta B_t^i := \sum_{j=1}^{d} \Delta L_t^{ji} - \sum_{j=1}^{d} (1 + \lambda_t^{ij}) \Delta L_t^{ij}, \tag{3.1.3}$$

where $\Delta L_t^{ji} \in L^0(\mathbf{R}_+, \mathcal{F}_t)$ represents the net amount transferred from the position j to the position i at the date t. The first term in the right-hand side of (3.1.2) is due to the price increments, while the second corresponds to the agent's own actions at the date t (made after the instant when the new prices were announced). These actions are charged by the amount

$$-\sum_{i=1}^{d} \Delta B_t^i = \sum_{i=1}^{d} \sum_{j=1}^{d} \lambda_t^{ij} \Delta L_t^{ij}$$

(broker's fees, taxes, etc.) diminishing the nominal portfolio value.

One can interpret the matrix (ΔL_t^{ij}) as the investor orders immediately executed by the trader: the entry (i, j) means "increase the jth position by ΔL_t^{ij} units of the numéraire in exchange with the ith position; the transaction costs indicate that for this, the trader has to decrease the value of the latter by $(1 + \lambda_t^{ij}) \Delta L_t^{ij}$ units of the numéraire. The latter quantities also can be interpreted as orders.

Note that in the present setting the orders "to increase" ("to get") and "to decrease" ("to send") are related by a simple $\mathcal{F}_t$-measurable bijection $(\Delta L_t^{ij}) \mapsto (1 + \lambda_t^{ij}) \Delta L_t^{ij}$. So, only one of them is needed to describe the portfolio evolution. In the setting where the information available to the investor is given by a smaller filtration, the control of the portfolio can be done using orders of both types and not only in units of the numéraire but also in physical units. We shall discuss this issue in Sect. 3.5.

With every $\mathbf{M}_+^d$-valued process $L = (L_t)$ (in our notation $\mathbf{M}_+^d$ stands for the set of matrices with positive elements) and any initial endowment $v \in \mathbf{R}^d$, we associate, by the formula (3.1.2), a value process $V = (V_t)$, $t = 0, \dots, T$. The terminal values of such processes form the set of "results" R_T^v.

Notice that, in "reasonable" orders, we can expect that $\Delta L_t^{ij} \Delta L_t^{ji} = 0$, i.e., there are no bidirectional fund transfers. However, it is convenient not to exclude "unreasonable orders": their inclusion has no effect on the results we are interested in, namely, on no-arbitrage criteria and hedging theorems. Similarly, one may assume the "free disposal" of assets, that is, enlarge the class of strategies by extracting from B^i increasing adapted processes (in the notation we develop below, this means that $\Delta B_t \in -L^0(K_t, \mathcal{F}_t)$ rather than $\Delta B_t \in -L^0(M_t, \mathcal{F}_t)$).

Finishing with the modeling issues, we look now for an appropriate mathematical setting.

Observe that relation (3.1.2) is, in fact, a linear controlled difference equation (a vector one) of a very simple structure with the components connected only via controls:

$$\Delta V_t^i = V_{t-1}^i \Delta Y_t^i + \Delta B_t^i, \qquad V_{-1}^i = v^i, \tag{3.1.4}$$

where

$$\Delta Y_t^i = \frac{\Delta S_t^i}{S_{t-1}^i}, \qquad Y_0^i = 1, \tag{3.1.5}$$

with B given by (3.1.3). Since the dynamics are driven by the d-dimensional process B, we can diminish the dimension of the phase space of controls and choose B as the control strategy. Indeed, any $\Delta L_t \in L^0(\mathbf{M}_+^d, \mathcal{F}_t)$ defines the $\mathcal{F}_t$-measurable random variable ΔB_t with values in the set $-M_t$ where

$$M_t := \left\{ x \in \mathbf{R}^d : \; \exists a \in \mathbf{M}_+^d \text{ such that } x^i = \sum_{j=1}^d \left[(1+\lambda_t^{ij})a^{ij} - a^{ji}\right], \; i \le d \right\}.$$

Vice versa, a simple measurable selection argument shows that any portfolio increment $\Delta B_t \in L^0(-M_t, \mathcal{F}_t)$ is generated by a certain (in general, not unique) order $\Delta L_t \in L^0(\mathbf{M}_+^d, \mathcal{F}_t)$. Note that this does work only in the case of full information: if the investor's actions are measurable with respect to a smaller filtration, such a reduction is impossible.

We shall denote by $\mathcal{B}$ the set of control strategies, i.e., of the processes $B = (B_t)$ with $\Delta B_t \in -M_t$ (i.e., more formally, with $\Delta B_t \in L^0(-M_t, \mathcal{F}_t)$).

It is useful to look at the dynamics of the portfolio in "physical units." It is given by the simpler formula

$$\Delta \widehat{V}_t^i = \frac{\Delta B_t^i}{S_t^i}, \quad V_{-1}^i = v^i, \quad i \le d, \tag{3.1.6}$$

which can be written also as

$$\Delta \widehat{V}_t = \widehat{\Delta B}_t, \qquad -\widehat{\Delta B}_t \in \widehat{M}_t := \phi_t M_t.$$

Financially, it is absolutely obvious (the increments of positions now are due to fund transfers only and do not depend on price movements), but this can be also checked formally. A closer look reveals that the formula is a half-step to solve the linear nonhomogeneous equation: the second half-step yields the solution via the discrete analog of the Cauchy formula (with S, the solution of the linear equation (3.1.5) playing the role of the "exponential" of Y):

$$V_t^i = S_t^i \widehat{V}_t^i = S_t^i \left(v^i + \sum_{s=0}^t \frac{\Delta B_s^i}{S_s^i} \right).$$

These trivial observations underlie the whole development of the discrete-time theory.

An important concept in the above setting is the *solvency cone* K_t (depending, in general, on ω). It is defined as the set of vectors $x \in \mathbf{R}^d$ for which one can find a matrix $a \in \mathbf{M}^d_+$ such that

$$x^i + \sum_{j=1}^{d} \left[a^{ji} - (1+\lambda_t^{ij})a^{ij}\right] \geq 0, \quad i \leq d.$$

In other words, K_t is the set of portfolios (denominated in units of the numéraire) which can be converted at time t, paying the transactions costs, to portfolios without short positions (i.e., without debts in any asset). Clearly, $K_t = M_t + \mathbf{R}^d_+$, and $\widehat{K}_t = \widehat{M}_t + \mathbf{R}^d_+$, the solvency cone when the accounting of assets (e.g., currencies) is done in terms of physical units.

In this model the *contingent claim* ξ is just a d-dimensional random variable. To *hedge* this contingent claim means to find a portfolio process V such that $V_T - \xi \in K_T$ (a.s.). Denoting by the symbol $\geq_T$ the partial ordering in $\mathbf{R}^d$ associated with the cone K_T (as usual, $x \geq_T 0$ means that $x \in K_T$), we may write this as the "inequality" $V_T \geq_T \xi$.

It is easy to see that the claim ξ is hedgeable if and only if there exists a portfolio process such that $V_T \geq \xi$ componentwise, i.e., in the sense of the partial ordering generated by the (smaller) cone $\mathbf{R}^d_+$. Indeed, if $V_T \geq_T \xi$, then $V_T - \xi \in K_T$ and, therefore, $V_T - \xi = \eta + \rho$, where $\eta \in L^0(M_T, \mathcal{F}_T)$ and $\rho \in L^0(\mathbf{R}^d_+, \mathcal{F}_T)$. Replacing the last transaction ΔB_T by $\Delta B_T - \eta$, we obtain a new value process the terminal value of which is $\xi + \rho$ and, hence, dominates ξ componentwise.

Introducing the model, we formulate two fundamental questions:

Problem 1. What is the analog of FTAP?

Problem 2. What is the analog of the hedging theorem?

Of course, the answers to these questions necessitates not only to define appropriate concepts of absence of arbitrage but also to find an analog to the notion of equivalent martingale measure or martingale density. As we shall see further, this can be easily done by placing the model in an adequate mathematical framework.

Turning back to the description of the cone M_t, we observe that it is an image of the polyhedral cone $\mathbf{M}^d_+$ (or even of the smaller one $\tilde{\mathbf{M}}^d_+$ of matrices with positive entries and e zero diagonal) in the space $\mathbf{M}^d$ of $d \times d$ matrices under the linear mapping $\Psi : \mathbf{M}^d \to \mathbf{R}^d$ with

$$\left[\Psi((a^{ij}))\right]^i := \sum_{j=1}^{d} \left[(1+\lambda_t^{ij})a^{ij} - a^{ji}\right].$$

Thus, M_t is also a *polyhedral cone*. The cone $\tilde{\mathbf{M}}^d_+$ (which can be identified with $\mathbf{R}^{d(d-1)}_+$) is generated by $d \times (d-1)$ elements which are matrices with all zero entries except a single one equal to unit. The image of this set of generators is a generating set for M_t. Therefore, M_t is a (random) polyhedral cone, namely,

$$M_t = \operatorname{cone}\big\{(1+\lambda_t^{ij})e_i - e_j,\ 1 \le i,j \le d\big\}.$$

Its dual positive cone

$$\begin{aligned} M_t^* &:= \{w:\ wx \ge 0\ \forall x \in M_t\} \\ &= \big\{w:\ (1+\lambda_t^{ij})w^i - w^j \ge 0,\ 1 \le i,j \le d\big\}. \end{aligned}$$

Similarly, the cone K_t is the image of $\tilde{\mathbf{M}}_+^d \otimes \mathbf{R}_+^d$ under a linear mapping,

$$K_t = \operatorname{cone}\big\{(1+\lambda_t^{ij})e_i - e_j,\ e_i,\ 1 \le i,j \le d\big\},$$

and its positive dual is

$$K_t^* = M_t^* \cap \mathbf{R}_+^d = \big\{w \in \mathbf{R}_+^d:\ (1+\lambda_t^{ij})w^i - w^j \ge 0,\ 1 \le i,j \le d\big\}.$$

It is an easy exercise to check that

$$\widehat{K}_t = \phi_t K_t = \operatorname{cone}\big\{\pi_t^{ij} e_i - e_j,\ e_i,\ 1 \le i,j \le d\big\},$$

where

$$\pi_t^{ij} := (1+\lambda_t^{ij}) S_t^j / S_t^i.$$

Note also that if there is a nonzero transaction cost coefficient λ_t^{ij}, then all basis orths e_i belong to $M_t = K_t$.

The above consideration shows that our "basic" model is nothing but a linear difference equation with additive control subjected to polyhedral cone constraints. The posed questions are related with properties of attainability sets of such equations, and, of course, they can be addressed to more general ones. Linear equations can be solved, and this makes the analysis relatively easy, especially, in the case of finite Ω.

The solvency cone K_t can be generated by many matrices Λ_t. In theoretical analysis it is convenient to consider the matrix with minimal absolute norm $\sum_{ij} \lambda_t^{ij}$. Note that, for this matrix,

$$1 + \lambda_t^{ij} \le (1+\lambda_t^{ik})(1+\lambda_t^{kj}) \quad \forall i,j,k. \tag{3.1.7}$$

Indeed, if we have an opposite inequality, then both vectors $(1+\lambda_t^{ij})e_i - e_j$ and $(1+\tilde{\lambda}_t^{ij})e_i - e_j$ with $\tilde{\lambda}_t^{ij} := (1+\lambda_t^{ik})(1+\lambda_t^{kj}) - 1$ are conic combinations of the other generating vectors of K_t. Thus, replacing λ^{ij} by $\tilde{\lambda}^{ij}$, we obtain the same cone K_t diminishing the norm of Λ_t.

The financial interpretation of (3.1.7) is obvious: an "intelligent" investor will first try all possible chains of transfers from the ith position to the position j and act accordingly to a cheapest one, i.e., replacing effectively a given matrix of transaction costs by that with the minimal norm. However, in practice it is not always done (the real situation is more complicated than the model considered: the proportional transaction costs are already a simplification).

The linear space $K_t^0 := K_t \cap (-K_t)$ also has a simple financial interpretation. It is composed by the positions which can be converted to zero without paying transaction costs and vice versa. Indeed, let $x \in K_t \cap (-K_t)$. According to the definition,

$$x^i = \sum_{j=1}^{d} [(1+\lambda_t^{ij})a^{ij} - a^{ji}] + h^i,$$
$$-x^i = \sum_{j=1}^{d} [(1+\lambda_t^{ij})\tilde{a}^{ij} - \tilde{a}^{ji}] + \tilde{h}^i.$$

Summing up, we get that

$$\sum_{i=1}^{d}\sum_{j=1}^{d} \lambda_t^{ij}(a^{ij} + \tilde{a}^{ij}) + \sum_{i=1}^{d}(h^i + \tilde{h}^i) = 0.$$

It follows that all summands here are zero, and this leads to the claimed property.

Before going further, we give in the next subsections a survey of other approaches or parameterizations which lead, essentially, to the same model with proportional transaction costs or its particular case. However, all this variety can be studied in a framework of the geometrical formalism we develop in this chapter.

3.1.2 Variants

1. Alternative parameterizations. In the literature one can find various specifications for transaction cost coefficients. To explain the situation, let us put $\Delta\tilde{L}_t^{ij} := (1+\lambda_t^{ij})\Delta L^{ij}$. The increment of the value of the ith position due to the agent's action can be written as

$$\Delta B_t^i = \sum_{j=1}^{d} \mu_t^{ji} \Delta\tilde{L}_t^{ji} - \sum_{j=1}^{d} \Delta\tilde{L}_t^{ij},$$

where $\mu_t^{ji} := 1/(1+\lambda_t^{ji})$ take value in the interval $]0,1]$. The matrix (μ_t^{ij}) can be specified as the matrix of the transaction cost coefficients.

Historically, the theory of models with transaction costs was initiated having in mind the interpretation of the stock market with an obvious *numéraire*: cash or bank account. In models with a traded numéraire, i.e., a nonrisky asset, a mixture of both specifications is frequent. For example, for the two-asset model, the dynamics quite often is written as

$$\Delta V_t^1 = (1-\mu_t)\Delta M_t - (1+\lambda_t)\Delta L_t,$$
$$\Delta V_t^2 = V_{t-1}^2 \Delta Y_t^2 + \Delta L_t - \Delta M_t,$$

where $\Delta L_t \geq 0$ and $M_t \geq 0$ are $\mathcal{F}_t$-measurable random variables.

2. A model of stock market. In this model it is assumed that all transactions pass through the money: so the orders are either "buy a stock" or "sell a stock." At time t, they can be represented by $\mathcal{F}_t$-measurable random vectors $(\Delta L_t^2, \dots, \Delta L_t^d)$ and $(\Delta M_t^2, \dots, \Delta M_t^d)$.

The corresponding d-asset dynamics is given by the system

$$\Delta V_t^1 = \sum_{j=2}^d (1-\mu_t^j)\Delta M_t^j - \sum_{j=2}^d (1+\lambda_t^j)\Delta L_t^j,$$
$$\Delta V_t^i = V_{t-1}^i \Delta Y_t^i + \Delta L_t^i - \Delta M_t^i, \quad i = 2, \dots, d.$$

The generators of the cone M_t in $\mathbf{R}^d$ are the vectors $-(1+\lambda_t^j)e_1 + e_j$, $(1-\mu_t^j)e_1 - e_j$, $j = 2, \dots, d$. The solvency cone K_t is generated by this set augmented by the vector e_1. It is not difficult to see that it can be described as follows:

$$K_t = \left\{ x \in \mathbf{R}^d : \; x^1 + \sum_{j=2}^d \left[(1-\mu_t^j)x^j I_{\{x^j>0\}} - (1+\lambda_t^j)x^j I_{\{x^j<0\}}\right] \geq 0 \right\}.$$

Comparing this with the model of currency market (given by a matrix of transaction cost coefficients), we notice that it can be imbedded into the former by choosing sufficiently large transaction cost coefficients. Of course, this leads to a larger set of value processes, but such a procedure has no effect on the arbitrage properties.

3. Modeling in physical units. Let us consider a kind of a barter market where we are given not quotes of assets in terms of a numéraire but only the matrix $\Pi = (\pi^{ij})$ (depending on t and ω) with generic entry $\pi^{ij} > 0$ representing a number of units of the ith asset needed to get in exchange one unit of the jth asset (of course, $\pi^{ii} = 1$). In the literature it is usually assumed that $\pi^{ij} \leq \pi^{ik}\pi^{kj}$, i.e., the direct exchange is better than two consecutive ones. As explained above, this is not a loss of generality: an "intelligent" agent will act not according to a given matrix of exchange rates $\tilde{\Pi}$ but to a corrected one with

$$\pi^{ij} = \min\{\tilde{\pi}^{ii_1} \dots \tilde{\pi}^{i_n j}\}.$$

Now the solvency region, i.e., the set of $y \in \mathbf{R}^d$ for which one can find $c \in \mathbf{M}_+^d$ such that

$$y^i \geq \sum_{j=1}^d \left[\pi_t^{ij}(\omega)c^{ij} - c^{ji}\right], \quad i \leq d,$$

is the cone $\{e_i, \; \pi^{ij}e_i - e_j, \; 1 \leq i,j \leq d\}$.

In dynamics, $\Pi = (\Pi_t)$ is an adapted process, sometimes called in the literature the bid-ask process.

Obviously, the market model with the price quotes S_t and the transaction cost coefficients Λ_t can be reformulated in terms of Π: in this case we have

$$\pi_t^{ij} = (1+\lambda_t^{ij})S_t^j/S_t^i, \quad 1 \le i,j \le d.$$

Reciprocally, one can introduce in the barter market "money" and generate a price process S and a matrix Λ of transaction cost coefficients. Indeed, take an arbitrary $S_t \in L^0(\widehat{K}_t^* \setminus \{0\}, \mathcal{F}_t)$ and note that the components of this process are strictly positive. Put

$$\lambda_t^{ij} := \pi_t^{ij} S_t^i/S_t^j - 1.$$

Clearly, $\lambda_t^{ij} \ge 0$ because by duality $S_t(\pi_t^{ij} e_i - e_j) \ge 0$. It remains to recall that the solvency cone in terms of physical units for the model described by S and $\Lambda_t = (\lambda_t^{ij})$ coincides with $\widehat{K}_t$.

Thus, modeling via Π is nothing but just another parameterization of the model considered in the first subsection. The former has certain advantages: first, it follows the tradition already established in the financial literature; second, suits better to study portfolio optimization problems; and, third, allows for a rather straightforward generalization for the continuous-time setting.

4. Models where the transactions charge the bank account. In this case the dynamics is given as follows:

$$\Delta V_t^1 = \sum_{j=2}^{d}(\Delta L_t^{j1} - \Delta L_t^{1j}) - \sum_{i,j=1}^{d} \gamma_t^{ij} \Delta L_t^{ij},$$

$$\Delta V_t^i = \widehat{V}_{t-1}^i \Delta S_t^i + \sum_{j=1}^{d} \Delta L_t^{ji} - \sum_{j=1}^{d} \Delta L_t^{ij}, \quad i = 2,\dots,d,$$

where $\gamma_t^{ij} \in [0,1[$, $\gamma^{ii} = 0$. For this model, again linear and with polyhedral cone constraints on the controls, the solvency cone is always polyhedral:

$$K_t = \text{cone}\{\gamma^{ij}e_1 + e_i,\ (1+\gamma^{1i})e_1 - e_i,\ (-1+\gamma^{j1})e_1 + e_j,\ e_i,\ 1 \le i,j \le d\}.$$

5. Models with a price spread. Usually such a model is designed for stock markets, i.e., transactions are only buying or selling shares according to two price processes $\bar{S}$ and $\underline{S}$, where $\bar{S}^j \ge \underline{S}^j > 0$, $j = 2,\dots,d$. Clearly, it can be given in terms of a single price or quote process and transaction cost coefficients. For example, one can put $S_t := (\bar{S}_t + \underline{S}_t)$ and define $\lambda_t^j := \bar{S}_t^j/S_t^j - 1$, $\mu_t^j := 1 - \underline{S}_t^j/S_t^j$. The absence of arbitrage opportunities means that $R_T \cap L_+^0 = \{0\}$ where the "results" here are terminal values of the money component of the portfolio processes (in our terminology this will correspond to the NA^w-property).

Historically, the first criterion of absence of arbitrage was obtained for a model described in terms of bid and ask prices. The Jouini–Kallal theorem claims (under some conditions) that there is no-arbitrage if and only if there exist a probability measure $\tilde{P} \sim P$ and an $\mathbf{R}^{d-1}$-valued $\tilde{P}$-martingale $\tilde{S}$ such that $\underline{S}_t^j \le \tilde{S}_t^i \le \bar{S}_t^i$, $i = 2, \dots, d$. In the case where $\underline{S} = \bar{S}$, this assertion coincides with the DMW theorem.

As we shall see further, for the model with finite Ω (or in the case where $d = 2$), the NA^w-property is equivalent to the existence of the martingale Z with strictly positive components which evolves in the duals to the solvency cones in physical units, i.e., such that $Z_t \in L^0(\widehat{K}_t^* \setminus \{0\}, \mathcal{F}_t)$. The latter property is more general and, for specific models, can be rewritten in the formulation due to Jouini and Kallal.

Let us consider the basic model assuming in addition that $S^1 \equiv 1$, i.e., the first asset is the numéraire ("money"), and for all i and j,

$$(1 + \lambda^{i1})(1 + \lambda^{1j}) \le 1 + \lambda^{ij}.$$

This means that the direct exchanges are more expensive than those via money; they can be excluded at all (as it is usually done in stock market models). The cone K^* consists of all $w \in \mathbf{R}_+^d$ satisfying the inequalities

$$\frac{1}{1 + \lambda^{i1}} w^1 \le w^i \le (1 + \lambda^{1i}) w^1, \quad i > 1.$$

Indeed, other inequalities defining K^* hold automatically: for any pair i, j, we have

$$w^j \le (1 + \lambda^{1j}) w^1 \le (1 + \lambda^{i1})(1 + \lambda^{1j}) w^i \le (1 + \lambda^{ij}) w^i.$$

Let $Z_t \in L^0(\widehat{K}_t^* \setminus \{0\}, \mathcal{F}_t)$ be a martingale. Normalizing, we can assume that $EZ_T^1 = 1$ and define the probability measure $\tilde{P} = Z_T^1 P$. The condition that Z evolves in $\widehat{K}^*$ reads as

$$\frac{1}{1 + \lambda^{i1}} Z^1 \le \frac{Z^i}{S^i} \le (1 + \lambda^{1i}) Z^1, \quad i > 1.$$

Introducing the selling and buying prices

$$\underline{S}^i := \frac{1}{1 + \lambda^{i1}} S^i, \qquad \overline{S}^i := (1 + \lambda^{1i}) S^i,$$

we obtain that the process $\tilde{S} := Z/Z^1$ is a martingale with respect to $\tilde{P}$ and

$$\underline{S}^i \le \tilde{S}^i \le \overline{S}^i, \quad i > 1.$$

Thus, for models with this particular structure of the solvency cones, the NA-criteria can be written in the formulation suggested by Jouini and Kallal.

3.1.3 No-arbitrage problem: NA^w for finite Ω

Considering our basic model described above, we define the *strict arbitrage opportunity* as a strategy $B \in \mathcal{B}$ such that the terminal value V_T of the portfolio process $V = V^B$ given by (3.1.2) with $V_{0-} = 0$ belongs to $L^0(\mathbf{R}^d_+)$ but is not equal to zero. We shall say that a model has the *weak no-arbitrage property* (in symbols: NA^w) if it does not admit strict arbitrage opportunities. Denoting by R^0_T the set of terminal values of portfolios with zero initial endowments, we may rewrite the definition of NA^w in a conformity with the previous section: $R^0_T \cap L^0(\mathbf{R}^d_+) = \{0\}$ or, equivalently, $\widehat{R}^0_T \cap L^0(\mathbf{R}^d_+) = \{0\}$, where $\widehat{R}^0_T = \phi_T R^0_T$ is the set of attainable results in physical units.

We define also the set A^0_T of hedgeable claims $A^0_T := R^0_T - L^0(K_T, \mathcal{F}_T)$.

Let denote by $\mathcal{M}^T_0(\widehat{K}^* \setminus \{0\})$ the set of martingales $Z = (Z_t)_{t \le T}$ such that $Z_t \in L^0(\widehat{K}^*_t \setminus \{0\})$ for all t.

In the literature the elements of $\mathcal{M}(\widehat{K}^* \setminus \{0\})$ are sometimes referred to as *consistent price systems.* In the model with finite number of states of the nature, the existence of the latter is equivalent to the absence of strict arbitrage opportunities.

Theorem 3.1.1 *Suppose that Ω is finite. Then the following conditions are equivalent:*

(a) $R^0_T \cap L^0(\mathbf{R}^d_+) = \{0\}$ *(i.e., NA^w);*

(b) $A^0_T \cap L^0(\mathbf{R}^d_+) = \{0\}$;

(c) $\mathcal{M}^T_0(\widehat{K}^* \setminus \{0\}) \ne \emptyset$.

Proof. Conditions (a) and (b) are obviously equivalent. Let us check the equivalence of (a) and (c). Without loss of generality we may assume that all elementary events of $\Omega = \{\omega_1, \dots, \omega_N\}$ have strictly positive probabilities. Thus, the space of d-dimensional random variables can be identified with the Euclidean space of dimension $d \times N$ with the scalar product $(\xi, \eta) := E\xi\eta$. Note that if $G = \text{cone}\{\xi_1, \dots, \xi_m\}$ is a random cone generated by $\mathcal{F}_t$-measurable $\mathbf{R}^d$-valued random variables, then $L^0(G, \mathcal{F}_t)$ is the polyhedral cone generated by the random variables $\xi_i I_{\Gamma_j}$ where Γ_j are the atoms of the σ-algebra $\mathcal{F}_t$. Thus, being a sum of polyhedral cones, $\widehat{R}^0_T$ is also a polyhedral cone, and so is $L^0(\mathbf{R}^d_+)$; the positive dual of the latter coincides with the primal one. Applying the Stiemke lemma as it is formulated in Lemma 5.1.1 of the Appendix, we obtain that condition (a) holds if and only if there exists a d-dimensional random variable η in the intersection of $(-\widehat{R}^{0*}_T)$ with the interior of $L^0(\mathbf{R}^d_+)$. This means that $E\xi\eta \le 0$ for all $\xi \in \widehat{R}^0_T$ and that all components of η are strictly positive. It remains to notice that the martingale $Z_t = E(\eta|\mathcal{F}_t)$ belongs to $\mathcal{M}^T_0(\widehat{K}^* \setminus \{0\})$. Indeed, for any $\zeta \in L^0(\widehat{K}_t, \mathcal{F}_t) \subseteq -\widehat{R}^0_T + L^0(\mathbf{R}^d_+)$, we have that

$$EZ_t\zeta = E\eta\zeta \ge 0.$$

This property obviously means that $Z_t \in L^0(\widehat{K}^*_t, \mathcal{F}_t)$. $\square$

The above statement contains as a particular case the Harrison–Pliska theorem. To see this, suppose that $\Lambda = 0$ and the first asset is the numéraire, i.e., $\Delta S_t^1 = 0$. Let $\bar{V}_t = \sum_{i\le d} V_t^i$. Summing up relations (3.1.2), we get that the dynamics of $\bar{V}$ is given by

$$\Delta \bar{V}_t = \sum_{i=1}^{d} \widehat{V}_{t-1}^i \Delta S_t^i = H_t \Delta S_t,$$

where $H_t \in L^0(\mathbf{R}^d, \mathcal{F}_{t-1})$. There is a single linear relation for the components $\widehat{V}_{t-1}^i$, namely, $\bar{V}_{t-1} = \bar{V}_t$, but it is of no importance since $\Delta S_t^1 = 0$. Thus, the set of attainable random variables $\bar{V}_T$ is exactly the same as R_T in our model of frictionless market, and the classical *NA*-condition $R_T \cap L_+^0 = \{0\}$ is equivalent to the NA^w-condition.

On the other hand, in the case $\Lambda = 0$ the cone $\widehat{K}_t^* = \mathbf{R}_+ S_t$, and, hence, the property $Z_t \in L^0(\widehat{K}_t^*, \mathcal{F}_t)$ means simply that $Z_t = \rho_t S_t$ for some $\rho_t \ge 0$. Thus, $Z \in \mathcal{M}_0^T(\widehat{K}^* \setminus \{0\})$ if and only if there exists a strictly positive martingale $\rho = (\rho_t)$ such that ρS is a martingale; normalizing, we may always assume that $E\rho_t = 1$.

It is worth noticing that the NA^w-condition can be reformulated in many various ways. In particular, below one can replace R_T^0 and K_T by $\widehat{R}_T^0$ and $\widehat{K}_T$.

Proposition 3.1.2 *The following conditions are equivalent:*

(a) $R_T^0 \cap L^0(K_T, \mathcal{F}_T) \subseteq L^0(\partial K_T, \mathcal{F}_T)$;
(b) $R_T^0 \cap L^0(\mathbf{R}_+^d, \mathcal{F}_T) = \{0\}$.

Proof. The implication (a) $\Rightarrow$ (b) holds because $\mathbf{R}_+^d \cap \partial K_T = \{0\}$. To prove the implication (b) $\Rightarrow$ (a), we notice that if $V_T^B \in L^0(K_T, \mathcal{F}_T)$ where $B \in \mathcal{B}$, then there exists $B' \in \mathcal{B}$ such that $V_T^{B'} \in L^0(\mathbf{R}_+^d, \mathcal{F}_T)$ and $V_T^{B'}(\omega) \ne 0$ on the set $V_T^B(\omega) \notin \partial K_T(\omega)$. To construct such B', it is sufficient to modify only ΔB_T by combining the last transfer with the liquidation of the negative positions. □

As we shall see later, a straightforward generalization of Theorem 3.1.1 for arbitrary Ω, unlike the cases of Harrison–Pliska and DWW theorems, fails to be true. Thus, one can try to find other definitions of arbitrage opportunities permitting to extend NA-criteria beyond models with finite number of states of the nature. The alternative definitions of the NA^w excluding the final liquidation gives a hint to alternative approaches.

3.1.4 No-arbitrage Problem: NA^s for Finite Ω

Suppose that, at time t, investor's portfolio is a subject of an audit. Auditors are not interested in real transaction costs needed to liquidate negative positions. Summing up the black and red figures (of course, the latter with minus), they may obtain a strictly positive value. From their point of view, the investor has an arbitrage. Judging this as an exaggeration, we consider now a minor modification of the NA^w-property more acceptable from the point of view of the investor.

We say that a strategy B realized a *weak arbitrage opportunity* at time $t \le T$ if $V_t^B \in K_t$ but $P(V_t^B \notin K_t^0) > 0$, where $K_t^0 := K_t \cap (-K_t)$. Respectively, the absence of such a one at time t is referred to as a *strict no-arbitrage* property NA_t^s:

$$R_t^0 \cap L^0(K_t, \mathcal{F}_t) \subseteq L^0(K_t^0, \mathcal{F}_t),$$

or, equivalently, in the realm of physical values,

$$\widehat{R}_t^0 \cap L^0(\widehat{K}_t, \mathcal{F}_t) \subseteq L^0(\widehat{K}_t^0, \mathcal{F}_t).$$

We use the notation NA^s when NA_t^s holds for every $t \le T$. Clearly, in the no-friction case this definition coincides with the classical one.

Theorem 3.1.3 *The following conditions are equivalent:*

(a) $R_T^0 \cap L^0(K_T, \mathcal{F}_T) \subseteq L^0(K_T^0, \mathcal{F}_T)$ *(i.e., NA_T^s holds);*
(b) $A_T^0 \cap L^0(K_T, \mathcal{F}_T) \subseteq L^0(K_T^0, \mathcal{F}_T)$*;*
(c) *there exists $Z^{(T)} \in \mathcal{M}_0^T(\widehat{K}^* \setminus \{0\})$ such that $Z_T^{(T)} \in L^1(\mathrm{ri}\, \widehat{K}_T^*, \mathcal{F}_T)$.*

Proof. The equivalence of (a) and (b) is obvious. The proof of the equivalence (a) $\Leftrightarrow$ (c) differs from that of Theorem 3.1.1 only in that we now use Theorem 5.1.3 instead of Lemma 5.1.1. □

As a direct corollary, we obtain the following:

Theorem 3.1.4 *The following conditions are equivalent:*

(a) $R_t^0 \cap L^0(K_t, \mathcal{F}_t) \subseteq L^0(K_t^0, \mathcal{F}_t)$ *for all t (i.e., NA^s holds);*
(b) $A_t^0 \cap L^0(K_t, \mathcal{F}_t) \subseteq L^0(K_t^0, \mathcal{F}_t)$ *for all t;*
(c) *for each $t \le T$, there exists a process $Z^{(t)} \in \mathcal{M}_0^t(\widehat{K}^* \setminus \{0\})$ such that $Z_t^{(t)} \in L^1(\mathrm{ri}\, \widehat{K}_t^*, \mathcal{F}_t)$.*

Note that NA_T^s does not imply NA_t^s for $t < T$. In other words, weak arbitrage opportunities may disappear next day.

Example. Let us consider the (deterministic) one-period two-asset model with $(S_0^1, S_0^2) = (1,1)$ and $(S_1^1, S_2^2) = (1,2)$. Assume that the entries of Λ are equal to zero except $\lambda^{12} = \lambda$. The vectors $(1,1)$ and $(1, 1+\lambda)$ are generators of K^*. Clearly, transfers at $T = 1$ cannot increase the value, so the only strategy to be inspected is with $\Delta B_0 = (-(1+\lambda), 1)$ (the transfers are $\Delta L_0^{12} = 1$, $\Delta L_0^{21} = 0$) and $\Delta B_1 = (0,0)$. So, $V_1^B = (-(1+\lambda), 2)$. For $\lambda \in [0, 1[$, we have $V_1^B \in \mathrm{int}\, K$, i.e., B is a strict arbitrage opportunity; for $\lambda = 1$, the model satisfies NA_1^w condition, but the strategy B is a weak arbitrage opportunity; and if $\lambda > 1$, the model enjoys the NA_1^s property.

We can extend in time this model by assuming that at the second period S_2^2 takes values ε and $1/\varepsilon$, say, with probabilities $1/2$. For $\lambda = 1$, this model satisfies NA_2^s when the parameter $\varepsilon > 2$ (i.e., the price increment ΔS_2^2 takes a negative and a positive value).

A specific feature of the NA^s-property is that the above criterion holds true in the general case (i.e., without finiteness assumption on Ω) for important classes of models, namely, for models with constant coefficients and also for models with *efficient friction* with $K_t^0 = \{0\}$. In both cases (c) is equivalent to the condition $\mathcal{M}_0^T(\operatorname{ri}\widehat{K}^*) \neq \emptyset$, see Theorem 3.2.1 and the accompanying discussion. However, without further assumptions, this is not true. It happened that, for a general Ω, the condition $\mathcal{M}_0^T(\operatorname{ri}\widehat{K}^*) \neq \emptyset$ is equivalent to the so-called robust no-arbitrage property NA^r which excludes arbitrage even under better investment opportunities, i.e., when the NA^w-property holds for a certain process $\tilde{\Lambda}_t = (\tilde{\lambda}_t^{ij})$, where $\tilde{\lambda}_t^{ij}(\omega) < \lambda_t^{ij}(\omega)$ if the latter coefficient is not zero. We shall analyze this in a more general geometric context.

Remark. The following example shows that even in the case of Ω consisting of two elementary events ω_1, ω_2 having equal probabilities, condition (c) does not imply that $\mathcal{M}_0^T(\operatorname{ri}\widehat{K}^*) \neq \emptyset$, and, hence, NA^s is weaker than NA^r. Indeed, let $\mathcal{F}_0$ be trivial, $\mathcal{F}_1 = \mathcal{F}_2 = \mathcal{F}$. Take now $\widehat{K}_0^* = \widehat{K}_2^* = \operatorname{cone}\{e_1 + e_2\}$ and let $\widehat{K}_1^*(\omega_1) = \operatorname{cone}\{e_1 + e_2, e_1 + 2e_2\}$ and $\widehat{K}_1^*(\omega_2) = \operatorname{cone}\{e_1 + e_2, 2e_1 + e_2\}$. Clearly, one can find martingales $Z^{(t)}$, $t \leq 2$, satisfying (c), but there is no one with values in $\operatorname{ri}\widehat{K}_t^*$, $t = 0, 1, 2$.

3.2 No-arbitrage Problem: Abstract Approach

3.2.1 NA^r- and NA^s-Properties: Formulations

We consider now the case of arbitrary Ω in an abstract framework which covers various discrete-time models of financial markets with proportional transaction costs (eventually, with additional linear inequality constraints). The development below is purely mathematical. To make the idea more transparent, we prefer not to insist here on financial aspects. We use the notation G_t instead of our traditional $\widehat{K}_t$ hoping that the reader understands well that we have in mind the corresponding structure in the world of physical units.

In the present framework we are given a sequence of set-valued mappings $G = (G_t)$, called C*-valued process*, specified by a countable sequence of adapted $\mathbf{R}^d$-valued processes $X^n = (X_t^n)$ such that, for all t and ω, only a finite but nonzero number of $X_t^n(\omega)$ is different from zero and

$$G_t(\omega) := \operatorname{cone}\{X_t^n(\omega),\ n \in \mathbf{N}\},$$

i.e., $G_t(\omega)$ is a polyhedral cone generated by the finite set $\{X_t^n(\omega),\ n \in \mathbf{N}\}$.

Let G and $\tilde{G}$ be closed cones. We shall say that G is *dominated* by $\tilde{G}$ if $G \setminus G^0 \subseteq \operatorname{ri}\tilde{G}$, where G^0 is the linear space $G \cap (-G)$. We extend this notion in the obvious way to $\mathcal{C}$-valued or cone-valued processes. It can be formulated in terms of the dual cones because

$$G \setminus G^0 \subseteq \operatorname{ri} \tilde{G} \quad \Longleftrightarrow \quad \tilde{G}^* \setminus \tilde{G}^{*0} \subseteq \operatorname{ri} G^*.$$

In particular, if G has an interior (which is always the case for financial models),

$$G \setminus G^0 \subseteq \operatorname{int} \tilde{G} \quad \Longleftrightarrow \quad \tilde{G}^* \setminus \{0\} \subseteq \operatorname{ri} G^*.$$

For a $\mathcal{C}$-valued process G, let $A_t(G) := -\sum_{s=0}^{t} L^0(G_s, \mathcal{F}_s)$.

Extending to the abstract setting the concepts introduced in a financial context, we say that G satisfies:

- *weak no-arbitrage property* NA^w if
$$A_t(G) \cap L^0(G_t, \mathcal{F}_t) \subseteq L^0(\partial G_t, \mathcal{F}_t) \quad \forall t \le T;$$
- *strict no-arbitrage property* NA^s if
$$A_t(G) \cap L^0(G_t, \mathcal{F}_t) \subseteq L^0(G_t^0, \mathcal{F}_t) \quad \forall t \le T.$$
- *robust no-arbitrage property* NA^r if G is dominated by $\tilde{G}$ satisfying NA^w.

It is an easy exercise to check that if G dominates the constant process $\mathbf{R}_+^d$, then NA^w holds if and only if $A_T(G) \cap L^0(\mathbf{R}_+^d) = \{0\}$.

The main results can be formulated as follows:

Theorem 3.2.1 *Assume that G dominates $\mathbf{R}_+^d$. Then*

$$NA^r \quad \Longleftrightarrow \quad \mathcal{M}_0^T(\operatorname{ri} G^*) \ne \emptyset.$$

Theorem 3.2.2 *Assume that $L^0(G_s^0, \mathcal{F}_{s-1}) \subseteq L^0(G_{s-1}^0, \mathcal{F}_{s-1})$ $\forall s \le T$. Then*

$$NA^s \quad \Longleftrightarrow \quad \mathcal{M}_0^T(\operatorname{ri} G^*) \ne \emptyset.$$

We postpone the proof of extended versions of these theorems to the next subsections. The "easy" implication $\Leftarrow$ will be established without extra assumptions.

The hypothesis of the second theorem holds trivially when $G^0 = \{0\}$ (an efficient friction condition in financial context). More interesting, it is fulfilled also for the setting corresponding to the market model for which the subspace $K_t^0 = K_t \cap (-K_t)$ is constant over time (e.g., the transaction costs are constant) and NA^s holds. Therefore, in such a case the properties NA^r and NA^s coincide.

Indeed, let us suppose that $G_t = \phi_t K_t$ where K is a $\mathcal{C}$-valued process dominating $\mathbf{R}_+^d$,

$$\phi_t(\omega) : (x^1, \dots, x^d) \mapsto (x^1/S_t^1(\omega), \dots, x^d/S_t^d(\omega)),$$

and S_t^i are strictly positive $\mathcal{F}_t$-measurable random variables.

For $J \subseteq \{1, \dots, d\}$, we put $\mathbf{1}_J := \sum_{i \in J} e_i$, where $\{e_i\}$ is the canonical basis in $\mathbf{R}^d$.

Proposition 3.2.3 *Suppose that there is a partition $J_1, \dots, J_l$ of $\{1, \dots, d\}$ such that $K_t^{0\perp} = \operatorname{lin}\{\mathbf{1}_{J_1}, \dots, \mathbf{1}_{J_l}\}$. If G satisfies the NA^w-property, then the assumption of Theorem 3.2.2 is fulfilled.*

Proof. If the claim fails, there is $\xi \in L^0(G_s^0, \mathcal{F}_{s-1})$ such that $\{\xi \notin G_{s-1}^0\}$ is a nonnull set. Without loss of generality we may assume that ξ is equal to zero outside it. Necessarily, some set $\{\phi_{s-1}^{-1}\xi\mathbf{1}_{J_k} \neq 0\}$ is nonnull. We may assume that $\{\phi_{s-1}^{-1}\xi\mathbf{1}_{J_k} > 0\}$, the random variable ξ is zero outside this set and, moreover, all components of ξ vanish except those corresponding to J_k. Notice that J_k is not a singleton because ϕ_s and ϕ_{s-1} are diagonal operators. Take $i_0 \in J_k$ such that $\xi^{i_0} > 0$ and consider ξ' different of ξ only by the i_0th component

$$\xi'^{i_0} := -\frac{1}{S_{s-1}^{i_0}} \sum_{i \in J_k \setminus \{i_0\}} \xi^i S_{s-1}^i.$$

Clearly, $\xi' \in L^0(G_{s-1}^0, \mathcal{F}_{s-1})$ and $\xi - \xi' = h$, where h is equal to zero except the nontrivial component $h^{i_0} \geq 0$. This violates the NA^w property. □

In a specific financial context, where K_t is the solvency cone in values (generated by the matrix of transaction cost coefficients Λ_t), and S is the price process, the linear space $K_t^{0\perp}$ is always the linear span of the random vectors $\mathbf{1}_{J_1(t)}, \dots, \mathbf{1}_{J_l(t)}$, where $J_i(t)$ are the classes of "equivalent" assets (i.e., the assets which can be converted one into another without transaction costs). Of course, in the case of constant transaction costs these vectors do not evolve in time. Since NA^w is weaker than NA^s, the latter implies that $\mathcal{M}_0^T(\operatorname{ri} G^*) \neq \emptyset$.

Remark. The hypothesis of Theorem 3.2.1 can be slightly relaxed by demanding that G dominates an increasing $\mathcal{C}$-valued process H such that all H_t have nonempty interiors.

As in the frictionless case, one can modify the right-hand of the equivalence in the suggested theorems by taking into account the next result which is an easy corollary of the DMW theorem.

Lemma 3.2.4 *Let $\mathcal{M}_0^T(\operatorname{ri} G^*)$ be nonempty, and let $\tilde{P} \sim P$. Then $\mathcal{M}_0^T(\operatorname{ri} G^*, \tilde{P})$ is also nonempty and even contains a bounded martingale.*

Proof. Let $Z \in \mathcal{M}_0^T(\operatorname{ri} G^*)$, and let $\zeta := 1 + \sup_{t \leq T} |Z_t|$. By the DMW theorem ("easy" part) the P-martingale $(1, Z)$ has the NA-property. The latter, being invariant under equivalent change of probability measure, holds also with respect to $P' := c\zeta^{-1}P$. Again by the same theorem but this time its "difficult" part, there is a bounded density process $\rho > 0$ such that ρZ is a P'-martingale, or, equivalently, the process $\tilde{Z}_t := E(\zeta^{-1}|\mathcal{F}_t)\rho_t Z_t$ is a P-martingale. It is bounded and, since $\operatorname{ri} G_t^*(\omega)$ are cones, belongs to $\mathcal{M}_0^T(\operatorname{ri} G^*)$. Using the same idea, we can obtain from $\tilde{Z}$ a bounded element of $\mathcal{M}_0^T(\operatorname{ri} G^*, \tilde{P})$. □

In the proof we used only that the values of $\operatorname{ri} G^*$ are cones. The extension of the lemma (as well as its version below) to arbitrary cone-valued processes is obvious.

Lemma 3.2.5 *Let* $\mathcal{M}_0^T(\operatorname{ri} G^*) \neq \emptyset$, $\tilde{P} \sim P$, *and* $\alpha \geq 0$ *be a (finite) random variable. Then* $\mathcal{M}_0^T(\operatorname{ri} G^*, \tilde{P})$ *contains* Z *such that* $(1+\alpha)|Z|$ *is bounded.*

Of course, the previous proof still works with ζ replaced by $\zeta + \alpha$.

Remark. The above result is extremely useful. In many cases it removes technical difficulties related with integrability and provides a great flexibility in arguments. Note also that we can fabricate a "convenient" vector-valued martingale just by multiplying the given one by a strictly positive scalar process.

Recall that for any sequence of random variables, there exists an equivalent probability measure $\tilde{P}$ with bounded density such that all these random variables are integrable, and, moreover, if, initially, the sequence was convergent almost surely, it will be convergent in $L^1(\tilde{P})$. Lemma 3.2.4 (more precisely, its variant with $\mathcal{M}_0^T(G^* \setminus \{0\})$), combined with this elementary fact, makes almost obvious the following assertion:

Proposition 3.2.6 *Let* $\mathcal{M}_0^T(G^* \setminus \{0\}) \neq \emptyset$. *Then*

$$\bar{A}_0^T \cap L^0(G_T, \mathcal{F}_T) \subseteq L^0(\partial G_T, \mathcal{F}_T).$$

Proof. Suppose that the sequence $\zeta^n = \sum_{t=0}^T \xi_t^n$, where $\xi_t^n \in -L^0(G_t, \mathcal{F}_t)$, converge a.s. to some $\zeta \in L^0(G_T, \mathcal{F}_T)$. According to the above remark, there is no loss of generality to assume, by a suitable choice of the reference measure, that $\xi_t^n \in -L^1(G_t, \mathcal{F}_t)$ and $\zeta^n \to \zeta$ in L^1. Take a bounded $Z_t \in \mathcal{M}_0^T(G^* \setminus \{0\})$. Then

$$EZ_T\zeta^n = \sum_{t=0}^T EZ_t\xi_t^n \leq 0.$$

On the other hand, $EZ_T\zeta \geq 0$, and this inequality is strict if $\zeta \in \operatorname{int} G_T$ with positive probability. This implies that $\zeta \in L^0(\partial G_T, \mathcal{F}_T)$. □

3.2.2 *NA^r*- and *NA^s*-Properties: Proofs

We start the proof with certain useful properties. It is convenient to formulate them in a more general framework.

Let N_s, $s = 0, 1, \ldots, T$, be closed convex cones in $L^0(\mathbf{R}^d, \mathcal{F}_s)$ stable under multiplication by the elements of $L^0(\mathbf{R}_+, \mathcal{F}_s)$. The last property implies that N_s are decomposable and, hence, can be represented as $L^0(G_s, \mathcal{F}_s)$ for some cone-valued process (G_s), see Appendix 5.4.

Let $N_s^0 := N_s \cap (-N_s)$, $A_t := -\sum_{s=0}^t N_s$.

We introduce the following conditions:

(i) $A_T \cap N_t \subseteq N_t^0$ for every $t = 0, \ldots, T$;
(ii) $A_{t-1} \cap N_t \subseteq N_t^0$ for every $t = 1, \ldots, T$;
(iii) the relation $\sum_{s=0}^T \xi_s = 0$ with $\xi_s \in N_s$ implies that all $\xi_s \in N_s^0$.

In the case where $N_s = L^0(G_s, \mathcal{F}_s)$, condition (i) coincides with the NA^s-property.

Lemma 3.2.7 (iii) $\Rightarrow$ (i).

Proof. Suppose that $\sum_{s=0}^{T} \zeta_s = -\eta$, where $\zeta_s \in N_s$ and $\eta \in N_t$. In virtue of (iii), we have that $\xi_t := \zeta_t + \eta$ is an element of N_t^0. Thus, $\eta = \xi_t - \zeta_t$ is in $-N_t$, i.e., $\eta \in N_t^0$. □

Remark. Trivially, (i) $\Rightarrow$ (ii). In general, the implication (ii) $\Rightarrow$ (iii) may fail. However, it is easily seen that it holds if all $N_s^0 = \{0\}$, and in this case these three properties are equivalent.

The following lemma shows that condition (iii) ensures the closedness of A_T in probability. Though, its proof is based on the same arguments as for the "difficult" implication in the DMW theorem, and we spell out it here.

Lemma 3.2.8 *If* (iii) *holds, then* $A_T = \bar{A}_T$.

Proof. We proceed by induction. For $T = 0$, there is nothing to prove. Suppose that the claim holds up to $T-1$ periods. Let $\sum_{s=0}^{T} \xi_s^n \to \xi$ a.s., where $\xi_s^n \in N_s$. The question is whether $\xi = \sum_{s=0}^{T} \xi_s$ with $\xi_s \in N_s$. If $\Omega_i \in \mathcal{F}_0$ form a partition of Ω, we may argue separately with each part as it were the whole Ω, find appropriate representations, and "assemble" ξ_s from separate pieces.

The case $\Omega = \{\liminf |\xi_0^n| < \infty\}$ is simple: by Lemmas 2.1.2 we may assume that ξ_0^n converge to $\xi_0 \in N_0$ and, hence, $\sum_{s=1}^{T} \xi_s^n$ converge a.s. to a random variable ζ which is in $\sum_{s=1}^{T} N_s$ by the induction hypothesis.

In the case $\Omega = \{\liminf |\xi_0^n| = \infty\}$, we put $\tilde{\xi}_s^n := \xi_s^n/|\xi_0^n|$ (with the convention $0/0 = 0$). As $|\tilde{\xi}_0^n| \le 1$, we again may assume that $\tilde{\xi}_0^n$ converge to some $\tilde{\xi}_0 \in N_0$. Then $\sum_{s=1}^{T} \tilde{\xi}_s^n$ converge a.s. to a random variable which can be represented by the induction hypothesis as $\sum_{s=1}^{T} \tilde{\xi}_s$, where $\tilde{\xi}_s \in N_s$. Since $\xi/|\xi_0^n| \to 0$ a.s., the limit of the whole normalized sum is zero, i.e., $\sum_{s=0}^{T} \tilde{\xi}_s = 0$. By the assumption all $\tilde{\xi}_s \in N_s^0$. Since $|\tilde{\xi}_0| = 1$, there are disjoint sets $\Gamma_i \in \mathcal{F}_0$ such that $\Omega = \bigcup_{i=1}^{d} \Gamma_i$ and $\Gamma_i \subseteq \{\tilde{\xi}_0^i \neq 0\}$.

Put $\bar{\xi}_s^n = \sum_{i=1}^{d} (\xi_s^n + \beta^{ni} \tilde{\xi}_s) I_{\Gamma_i}$, where $\beta^n = -\xi_0^{ni}/\tilde{\xi}_0^i$. Clearly, $\bar{\xi}_s^n \in N_s$, and $\sum_{s=0}^{T} \bar{\xi}_s^n$ converge to ξ a.s. The situation is reproduced. It is instructive to represent sequences ξ_0^n and $\bar{\xi}_0^n$ as infinite-dimensional matrices with d-dimensional columns. Of course, every zero line of the first matrix remains zero line of the second one. But the second matrix contains one more zero line (namely, the ith for $\omega \in \Gamma_i$). Thus, if the first matrix contains one nonzero line a.s., the proof is accomplished (all $\bar{\xi}_0^n = 0$, and we can use the induction hypothesis). If not, we repeat the whole procedure with the sequence of processes $(\bar{\xi}_s^n)$, etc. □

Lemma 3.2.9 *Assume that* (iii) *holds. Then for any* $\zeta \in N_t$, $t \le T$, *there is a bounded* $\mathbf{R}^d$*-valued martingale* Z^ζ *such that:*

(1) $Z_s^\zeta \xi \ge 0$ *for any* $\xi \in N_s$, $s \le T$;
(2) $\zeta I_{\{Z_t^\zeta \zeta = 0\}} \in N_t^0$.

Proof. Let $A_T^1 := A_T \cap L^1$ and $\mathcal{Z}_T := \{Z_T \in L^\infty(\mathbf{R}^d) : E\eta\xi \le 0 \ \forall \xi \in A_T^1\}$. With each $Z_T \in \mathcal{Z}_T$, we associate the martingale Z with $Z_s := E(Z_T|\mathcal{F}_s)$. It satisfies condition (1), since otherwise we could find $\xi \in N_s \cap L^1$ such that the set $\Gamma := \{Z_s\xi < 0\}$ is nonnull and, therefore, $EZ_T(\xi I_\Gamma) = EZ_s(\xi I_\Gamma) < 0$, contradicting the definition of $\mathcal{Z}_T$. Let $c_t := \sup_{Z_T \in \mathcal{Z}_T} P(Z_t\zeta > 0)$. There exists $Z_T^\zeta \in \mathcal{Z}_T$ such that, for the corresponding martingale Z^ζ, the supremum is attained, i.e., we have $c_t = P(Z_t^\zeta \zeta > 0)$. To see this, take the martingales Z^n generated by $Z_T^n \in \mathcal{Z}_T$ with $P(Z_t^n\zeta > 0) \to c_t$ and put $Z_T^\zeta := \sum 2^{-n} Z_T^n/\|Z_T^n\|_\infty$.

If (2) fails, then, for a sufficiently large, $\zeta^a := \zeta I_{\{Z_t^\zeta \zeta = 0,\ |\zeta| \le a\}}$ does not belong to N_t^0 and, being in $N_t \cap L^1$, cannot not belong, in virtue of condition (i) implied by (iii), to the convex cone A_T^1, which is closed in L^1 accordingly to Lemma 3.2.8. By the Hahn–Banach theorem one can separate ζ^a and A_T^1, that is, to find $\eta \in L^\infty(\mathbf{R}^d)$ such that

$$\sup_{\xi \in A_T^1} E\eta\xi < E\eta\zeta^a.$$

Since A_T^1 is a cone, the supremum above is equal to zero, ensuring that $\eta \in \mathcal{Z}_T$ and $E\eta\zeta^a > 0$. The latter inequality implies that, for $Z_t = E(\eta|\mathcal{F}_t)$, the product $Z_t\zeta$ (always ≥ 0) is strictly positive on a nonnull set. Thus, for the martingale $\tilde{Z} := Z + Z^\zeta$ with the terminal value $\tilde{Z}_T := \eta + Z_T\zeta$, we have

$$P(\tilde{Z}_t\zeta > 0) > P\big(Z_t^\zeta\zeta > 0\big) = c_t.$$

This contradiction shows that (2) holds. □

Lemma 3.2.10 *Assume that* (iii) *holds. Let Γ be a countable subset of $\bigcup_{s \le T} N_s$. Then there is a bounded $\mathbf{R}^d$-valued martingale Z such that, for all $s \le T$, we have:*

(1) $Z_s\xi \ge 0$ *for any* $\xi \in N_s$;
(2′) $\zeta I_{\{Z_s\zeta=0\}} \in N_s^0$ *whatever is* $\zeta = \sum_n \alpha_n \xi^n$ *with* $\xi^n \in \Gamma \cap N_s$ *and* $\alpha_n \in L_+^0(\mathcal{F}_s)$.

Proof. One can take as Z any (countable) convex combination with strictly positive coefficients of all elements of the family $\{Z^\zeta\}_{\zeta\in\Gamma}$ with $|Z_T^\zeta| \le 1$. Taking into account that N_s is stable under multiplication on the elements of $L_+^0(\mathcal{F}_s)$, we verify consecutively that property 2′) holds for ξ^n, $\alpha_n\xi^n$, and, at last, for ζ which are represented as series of ξ_n with $\mathcal{F}_s$-measurable positive coefficients. □

In the next lemma we argue in the case where $N_t = L^0(G_t, \mathcal{F}_t)$, where (G_t) is a $\mathcal{C}$-valued process and, hence, $N_t^0 = L^0(G_t^0, \mathcal{F}_t)$ and $A_t = A_t(G)$.

Lemma 3.2.11 *If* (iii) *holds, then* $\mathcal{M}_0^T(\operatorname{ri} G^*)$ *is nonempty.*

Proof. Consider the process Z from the previous lemma corresponding to the set Γ which is the union of all X_t^n, $t \le T$, $n \in \mathbf{N}$. Property (1) means that $Z \in \mathcal{M}_0^T(G^*)$. Recall also that $Z_s(\omega) \notin \operatorname{ri} G_s^*(\omega)$ if and only if there exists $x \in G_s(\omega) \setminus G_s^0(\omega)$ such that $Z_s(\omega)x = 0$. Thus, if the property $Z_s \in L^0(\operatorname{ri} G_s^*, \mathcal{F}_s)$ is violated, the $\mathcal{F}_s \otimes \mathcal{B}^d$-measurable set

$$\Delta := \{(\omega, x) : \ Z_s(\omega)x = 0, \ x \in G_s(\omega) \setminus G_s^0(\omega)\}$$

has a nonnull projection onto Ω. Take its $\mathcal{F}_s$-measurable selector ζ putting it zero outside the projection of Δ. Then $\zeta = \sum_n \alpha_n X_s^n$ with $\alpha_n \in L_+^0(\mathcal{F}_s)$ (for each ω, this sum has only a finite number of nonzero summands). Thus, according to property 2′), we have $\zeta I_{\{Z_s\zeta=0\}} \in G_s^0$, in an apparent contradiction with our assumption. □

Lemma 3.2.12 *Suppose that a* C*-valued process* G *satisfies the* NA^r*-property. If* G *dominates* $\mathbf{R}_+^d$*, then* (iii) *holds.*

Proof. Let $\tilde G$ dominate G, and let $A_T(\tilde G) \cap L^0(\mathbf{R}_+^d, \mathcal{F}_T) = \{0\}$. Assume that in the identity $\sum_{s=0}^T \xi_s = 0$ where $\xi_s \in L^0(G_s, \mathcal{F}_s)$, a random variable ξ_t does not belong to $L^0(G_t^0, \mathcal{F}_t)$. This means that $\xi_t(\omega) \in \operatorname{int} \tilde G_t(\omega)$ on a set B of positive probability. It follows that there is a random variable $\epsilon \in L^0(\mathbf{R}_+^d, \mathcal{F}_t)$ strictly positive on B such that $\xi_t - \epsilon$ is still in $L^0(\tilde G_t, \mathcal{F}_t)$. The nontrivial random variable $\epsilon = -\sum_{s=0}^T \xi_s'$ where $\xi_s' := \xi_s$, $s \ne t$, $\xi_t' := \xi_t - \epsilon$, being in $A_T(\tilde G) \cap L^0(\mathbf{R}_+^d, \mathcal{F}_T)$, violates the NA^w-property of $\tilde G$. □

Lemma 3.2.13 *Suppose that a cone-valued process* G *has the* NA^s*-property. If, in addition,*

$$L^0(G_s^0, \mathcal{F}_{s-1}) \subseteq L^0(G_{s-1}^0, \mathcal{F}_{s-1}) \quad \forall s \le T, \tag{3.2.8}$$

then condition (iii) *holds.*

Proof. This can be shown by induction starting trivially and with an easy step. The equality $\sum_{s=0}^{T-1} \xi_s = -\xi_T$ implies that ξ_T is $\mathcal{F}_{T-1}$-measurable and, in virtue of the NA^s-property, belongs to $L^0(G_T^0, \mathcal{F}_T)$. By the assumed inclusion ξ_T belongs also to $L^0(G_{T-1}^0, \mathcal{F}_{T-1})$ and can be combined with ξ_{T-1}, reducing the sum to $T-1$ terms which are elements of $L^0(G_s^0, \mathcal{F}_s)$ (the induction hypothesis). In particular, $\xi_{T-1} + \xi_T$ belongs to $L^0(G_{T-1}^0, \mathcal{F}_{T-1})$ as well as both summands. □

The "difficult" implications (⇒) in Theorems 3.2.1 and 3.2.2 follow from the last three lemmas. Now we establish the "easy" implications (⇐).

Lemma 3.2.14 *If* $\mathcal{M}_0^T(\operatorname{ri} G^*) \ne \emptyset$*, then* NA^s *and* NA^r *hold.*

Proof. Let us check first the NA^s-property. To this aim we apply Lemma 3.2.7 with $N_s = L^0(G_s, \mathcal{F}_s)$, reducing the problem to a verification of (iii). So, let

$\sum_{s=0}^{T} \xi_s = 0$ with $\xi_s \in N_s$. Choose a reference measure such that all ξ_s are integrable. By Lemma 3.2.4 there is a bounded martingale $Z \in \mathcal{M}_0^T(\operatorname{ri} G^*)$. Multiplying the identity by Z_T and taking the expectation, we obtain that $\sum_{s=0}^{T} EZ_s\xi_s = 0$. Since $Z_s\xi_s \geq 0$, this is possible only when all $Z_s\xi_s = 0$, i.e., when $\xi_s \in L^0(G_s^0, \mathcal{F}_s)$.

To verify the NA^r-property, we take an arbitrary $Z \in \mathcal{M}_0^T(\operatorname{ri} G^*)$ and define the $\mathcal{C}$-valued process $\tilde{G}$ whose values are half-spaces $\tilde{G}_s(\omega) = (\mathbf{R}_+ Z_s(\omega))^*$. Applying the above arguments with $N_s = L^0(\tilde{G}_s, \mathcal{F}_s)$, we conclude that $\tilde{G}$ has the NA^s-property coinciding with the NA^w-property because $\partial\tilde{G}_s = \tilde{G}_s^0$. Since $\tilde{G}$ dominates G, the latter has the NA^r-property. □

Remark 1. Inspecting the first part of the above proof, it is not difficult to see that condition (c) in Theorem 3.1.4 implies (iii) and, hence, the NA^s-property. Unlike the situation of finite Ω, the converse is not true. However, if all $G_t^0 = \{0\}$, then (3.2.8) holds trivially, and NA^s implies that $\mathcal{M}_0^T(\operatorname{ri} G^*) \neq \emptyset$, i.e., in this particular case the latter condition, condition (c), and NA^s are equivalent.

Remark 2. Recently, T. Pennanen and I. Penner, [177], suggested a simple argument showing that for the cone-valued processes the NA^r-property holds if and only if $\mathcal{M}_0^T(\operatorname{ri} G^*) \neq \emptyset$. Indeed, it is not difficult to check that if (G_t) has the NA^r-property, then there is a cone-valued process (G_t') dominating (G_t) still having the NA^r-property. From the above general lemmas we infer that the cone $A_T(G')$ is closed in L^0, and, using the Kreps–Yan theorem in L^1, we obtain that $\mathcal{M}_0^T(G'^* \setminus \{0\})$ is non-empty. But $G'^* \setminus \{0\} \subseteq \operatorname{ri} G^*$.

The condition $\mathcal{M}_0^T(\operatorname{ri} G^*) \neq \emptyset$ guarantees that $A_T(G)$ is closed. This topological property looks indispensable in the theory of no-arbitrage criteria. A simple example shows that NA^w does implies the closedness of $A_T(G)$, and, therefore, the extension of NA^w-criteria given in Theorem 3.1.1 are impossible. More surprising is the result by Grigoriev asserting that, for two-asset model, the necessary and sufficient conditions of this theorem, except (b), hold for arbitrary Ω.

3.2.3 The Grigoriev Theorem

Throughout this subsection we shall work with a $\mathcal{C}$-valued process G dominating the constant process $\mathbf{R}_+^2$. The initial σ-algebra $\mathcal{F}_0$ is not assumed trivial.

The proof of the theorem involves models with initial dates different from zero. By this reason we shall use the notation A_0^T, A_1^T, etc., giving an idea about the considered time range.

We formulate the statement in the same way as the Dalang–Morton–Willinger theorem, omitting, of course, condition (B) (which looks exactly as (b) of the latter): being stronger than (A), it is sufficient but not necessary for the NA^w-property.

Theorem 3.2.15 *Let $d = 2$. Then the following conditions are equivalent:*

(A) $A_0^T \cap L^0(\mathbf{R}_+^d) = \{0\}$;
(C) $\bar{A}_0^T \cap L^0(\mathbf{R}_+^d) = \{0\}$;
(D) $\mathcal{M}_0^T(G^* \setminus \{0\}) \neq \emptyset$;
(E) $\mathcal{M}_0^T(G^* \setminus \{0\}, \tilde{P})$ *contains a bounded process whatever is* $\tilde{P} \sim P$.

Proof. The equivalence of (D) and (E) was already discussed for the general setting (see Lemma 3.2.14 and the accompanying remark). The implication (E) $\Rightarrow$ (A) is obvious because a freedom in the choice of the reference measure allows us to avoid the problems with integrability, and we can argue in the same way as for the case of finite Ω.

Thus, the problem poses only the implication (A) $\Rightarrow$ (C) the proof of which uses essentially the specificity of the plane, where the cones $G_t(\omega)$ are just sectors containing the first orthant. □

Let us define the set

$$\Gamma_1^T := \{\xi \in L^\infty(\mathbf{R}^2, \mathcal{F}_0) : \; E\xi Z_1 \leq 0 \; \forall Z \in \mathcal{M}_1^T(G^* \setminus \{0\})\}.$$

Proposition 3.2.16 *Suppose that $(G_t)_{0 \leq t \leq T}$ satisfies the NA^w-property. Then:*

(a) *we have the inclusion $\Gamma_1^T \subseteq A_1^T$;*
(b) *if $\xi \in \Gamma_1^T$ is such that $E\xi Z_1 < 0$ for all $Z \in \mathcal{M}_1^T(G^* \setminus \{0\})$, then there is $\epsilon \in L^0(\mathbf{R}_+^2, \mathcal{F}_T)$, $\epsilon \neq 0$, such that $\xi + \epsilon \in A_1^T$.*

The above proposition plays a crucial role. It makes easy the induction step on the length of the time interval for the implication (A) $\Rightarrow$ (C). Reciprocally, the NA^w-criterion permits us to prove the induction step for this proposition. Let T_N and P_N denote the assertions of Theorem 3.2.15 and Proposition 3.2.16 for the N-step model. Symbolically, the arguments can be chained in the following way:

(1) $(T_N, P_{N+1}) \Rightarrow T_{N+1}$,
(2) $(T_{N+1}, P_{N+1}) \Rightarrow P_{N+2}$.

Proof of the implication $(T_N, P_{N+1}) \Rightarrow T_{N+1}$.

So, let NA^w hold for the $(N+1)$-step model. We are looking for a martingale $Z \in \mathcal{M}_0^{N+1}(G^* \setminus \{0\})$. By the induction hypothesis we know that there is $Z^0 \in \mathcal{M}_1^{N+1}(G^* \setminus \{0\})$. We want to show that there is an element of the latter set which can be extended one step backward as a martingale with initial value in $G_0^* \setminus \{0\}$.

Case 1. For any nonnull set $\Gamma \in \mathcal{F}_0$, there is $Z^\Gamma \in \mathcal{M}_1^{N+1}(G^* \setminus \{0\})$ such that the intersection $\Gamma \cap \{E(Z_1^\Gamma|\mathcal{F}_0) \in G_0^*\}$ is nonnull. By Lemma 2.1.3 we can find at most countable family of sets $\Delta_i := \{E(Z_1^{\Gamma_i}|\mathcal{F}_0) \in G_0^*\}$ the union of which is of full measure. Normalizing, we may assume that

$E|Z^{\Gamma_i}_{N+1}| = 1$. It is easily seen that $Z_t := \sum_i 2^{-i} I_{\Delta_i} E(Z^{\Gamma_i}_{N+1}|\mathcal{F}_t)$ is an element of $\mathcal{M}^{N+1}_0(G^* \setminus \{0\})$.

Case 2. There is a nonnull set $\Gamma \in \mathcal{F}_0$ such that $\Gamma \cap \{E(Z_1|\mathcal{F}_0) \in G^*_0\}$ is a null set whatever is $Z \in \mathcal{M}^{N+1}_1(G^* \setminus \{0\})$. The arguments below exclude this case.

Let g_i, $i = 1, 2$, be the random vectors of unit length generating the boundary rays of the sector G_0, and let

$$g := g_1 I_{H_1} + g_2 I_{H_2 \setminus H_1}$$

where $H_i := \{g_i E(Z^0_1|\mathcal{F}_0) < 0\}$. Since

$$\{E(Z^0_1|\mathcal{F}_0) \notin G^*_0\} = H_1 \cup H_2,$$

we have $gE(Z^0_1|\mathcal{F}_0) < 0$ on Γ (a.s.). Moreover, $gE(Z_1|\mathcal{F}_0) < 0$ on Γ (a.s.) whatever is $Z \in \mathcal{M}^{N+1}_1(G^*\setminus\{0\})$. The latter assertion holds because, in the opposite case, we could find in Γ a nonnull subset $\Gamma' \in \mathcal{F}_0$ and $Y \in \mathcal{M}^{N+1}_1(G^* \setminus \{0\})$ such that $gE(Y_1|\mathcal{F}_0) \geq 0$ on Γ'. It is easy to see that there is a scalar $\mathcal{F}_0$-measurable random variable $\alpha \geq 0$ such that, on Γ',

$$\alpha g E(Z^0_1|\mathcal{F}_0) + gE(Y_1|\mathcal{F}_0) = 0.$$

It follows that, for the process $Z := \alpha Z^0 + Y$ from $\mathcal{M}^{N+1}_1(G^* \setminus \{0\})$, we have on Γ' the equality $gp = 0$ where $p := E(Z_1|\mathcal{F}_0)$. But for the considered two-dimensional model, this means that a.s. on Γ' the vector $p(\omega)$ with strictly positive components belongs to one of two boundary rays of $G^*_0(\omega)$. This is an apparent contradiction with the defining property of Γ.

Applying P_{N+1} with $\xi := gI_\Gamma$, we infer that there exists a nontrivial $\epsilon \in L^0(\mathbf{R}^2_+, \mathcal{F}_{N+1})$ such that $gI_\Gamma + \epsilon \in \widehat{A}^{N+1}_1$. Since $gI_{\Gamma'} \in L^0(G_0, \mathcal{F}_0)$, this contradicts to the NA^w-property of G^{N+1}_0.

Proof of the implication $(T_{N+1}, P_{N+1}) \Rightarrow P_{N+2}$.

(a) We shall prove that

$$\Gamma^{N+2}_1 := \{\xi \in L^\infty(\mathbf{R}^2, \mathcal{F}_0) : \ E\xi Z_1 \leq 0 \ \forall Z \in \mathcal{M}^{N+2}_1(G^* \setminus \{0\})\} \subseteq A^{N+2}_1,$$

assuming NA^w for G^{N+2}_1 and knowing already that the latter property is equivalent to the existence of $Z^0 \in \mathcal{M}^{N+2}_1(G^* \setminus \{0\})$ with $|Z^0_1| = 1$; moreover, the claim holds for all shorter time ranges. Note that the existence of Z^0 implies that in the definition of Γ^{N+2}_1 we can replace the set $\mathcal{M}^{N+2}_1(G^* \setminus \{0\})$ by the larger set $\mathcal{M}^{N+2}_1(G^*)$ (due to the "regularization" $Z + \varepsilon Z^0$ with $\varepsilon \downarrow 0$).

If subsets $\Omega_i \in \mathcal{F}_1$ form a finite partition of Ω, it is sufficient to establish the result separately for each Ω_i considered as a new model, with traces of the filtration and probability on Ω_i and random variables restricted to this set (clearly, NA^w is inherited for such restrictions of a cone-valued process).

We apply this remark (frequent in proofs of NA criteria) to the subsets $\Omega_1 := \{\xi \notin -G_1\}$ and $\Omega_2 := \Omega^c_1$. For Ω_2, the claim is trivial. Therefore, we may assume without loss of generality that $\Omega_1 = \Omega$, i.e., $\xi\eta > 0$ for some $\eta \in L^0(G_1^*, \mathcal{F}_1)$, and normalizing, that $|\xi| = 1$ and $|\eta| = 1$.

Moreover, we need to investigate only the case where there exists a process $\tilde{Z} \in \mathcal{M}_2^{N+2}(G^* \setminus \{0\})$ such that $E\xi\tilde{Z}_2 > 0$ (otherwise, $\xi \in A_2^{N+2}$ by the induction hypothesis, and there is nothing to prove because A_2^{N+2} is a part of A_1^{N+2}).

The intersection Θ of the set $\Theta_1 := \{\xi E(\tilde{Z}_2|\mathcal{F}_1) > 0\}$ (of positive probability) with $\Theta_2 := \{E(\tilde{Z}_2|\mathcal{F}_1) \in G_1^*\}$ is a null set. Indeed, in the opposite case we would have the inequality $E\xi Z_1 > 0$ for the process $Z \in \mathcal{M}_1^{N+2}(G^*)$ with $Z_t := I_\Theta E(\tilde{Z}_{N+2}|\mathcal{F}_t)$, which is impossible in virtue of the remark at the beginning of the proof.

Thus, neglecting a null set, we may assume that $\Theta_1 \subseteq \Theta_2^c$ and, normalizing, that $|E(\tilde{Z}_2|\mathcal{F}_1)| = 1$ on the set $\Theta_2^c = \{E(\tilde{Z}_2|\mathcal{F}_1) \notin G_1^*\}$.

The following *elementary geometric fact* is obvious:

if x_1, x_2, x_3 are unit vectors in $\mathbf{R}_+^2$ with $yx_1 \geq 0$, $yx_2 \geq 0$, $yx_3 \leq 0$, and $yx_1 \geq yx_2$ for some vector y, then x_2 is a conic combination of x_1 and x_3.

Recalling that $\xi Z_1^0 \leq 0$, we obtain from this observation that on the set $\Theta_1 \cap \{\xi\eta > \xi E(\tilde{Z}_2|\mathcal{F}_1)\}$ the random vector $E(\tilde{Z}_2|\mathcal{F}_1)$ lies between η and Z_1^0, i.e., takes values in G_1^*.

Thus, by the above, $\Theta_1 \subseteq \{\xi\eta \leq \xi E(\tilde{Z}_2|\mathcal{F}_1)\}$ (a.s.). Using again the mentioned geometric fact, we get that there are nonnegative $\mathcal{F}_1$-measurable coefficients α and β such that $\eta = \alpha E(\tilde{Z}_2|\mathcal{F}_1) + \beta Z_1^0$ on Θ_1. It follows that $E\xi Z_1' > 0$ for the process $Z' \in \mathcal{M}_1^{N+2}(G^*)$ with $Z_1' = \eta I_{\Theta_1}$ and $Z_t' = (\alpha\tilde{Z}_2 + \beta Z_t^0)I_{\Theta_1}$ for $t \geq 2$. We get a contradiction that $\xi \in \Gamma_1^{N+2}$.

(b) Let $\xi \in \Gamma_1^{N+2}$ be such that $E\xi Z_1 < 0$ for every $Z \in \mathcal{M}_1^{N+2}(G^* \setminus \{0\})$. Without loss of generality we shall assume that $|\xi| = 1$. By the above we know that $\xi \in A_1^{N+2}$, i.e. $\xi = \sum_{t=1}^{N+2} \zeta_t$ with $\zeta_t \in -L^0(G_t, \mathcal{F}_t)$.

Let us consider three possible cases:

Case 1: The set $\{\xi \in -\operatorname{int} G_1\}$ is nonnull. The assertion is obvious since we can increase ξ on this set.

Case 2: The set $\Gamma := \{\xi \in \operatorname{int} G_1\}$ is nonnull. Then $\xi - \theta \in L^0(G_1, \mathcal{F}_1\}$ for some $\theta \in L^0(\mathbf{R}_+^2, \mathcal{F}_1)$ such that $\theta > 0$ on Γ and $\theta = 0$ outside. Since $\xi \in A_1^{N+2}$, we have that also $\theta \in A_1^{N+2}$, in violation of the assumed NA^w-property. So, this case is impossible.

Case 3: Complementary to the previous ones. The components of ξ have different signs, and $\xi\eta \geq 0$ for some $\eta \in L^0(G_1^*, \mathcal{F}_1)$ with $|\eta| = 1$.

We introduce the cone-valued process $\tilde{G}_1^{N+2}$ with $\tilde{G}_1 = \mathbf{R}\xi + \mathbf{R}_+^2$ and $\tilde{G}_t = G_t$, $t \geq 2$. Let us check whether NA^w holds for $\tilde{G}_1^{N+2}$. If $\tilde{G}_1^{N+2}$ satisfies NA^w, there is a bounded $\tilde{Z} \in \mathcal{M}_1^{N+2}(\tilde{G}^* \setminus \{0\})$. Let Z be an arbitrary element from $\mathcal{M}_1^{N+2}(G^* \setminus \{0\})$. Then $\xi Z_1 \leq 0$. Recalling that $\xi\eta \geq 0$ and $\xi\tilde{Z}_1 = 0$ (by the definition of $\tilde{G}_1$), we obtain using the main geometric fact that $\tilde{Z}_1$ takes values between the rays generated by Z_1 and η. It follows that $\tilde{Z}_1 \in L^\infty(G_1^*, \mathcal{F}_1)$. This means that $\tilde{Z} \in \mathcal{M}_1^{N+1}(G^* \setminus \{0\})$ but $\xi\tilde{Z}_1 = 0$, which is impossible.

Thus, NA^w does not hold for the process $\tilde{G}_1^{N+2}$, and, therefore, there exists $\epsilon \in L^0(\mathbf{R}_+^2, \mathcal{F}_{N+2})$, $\epsilon \neq 0$, such that

$$\epsilon = \sum_{t=1}^{N+2} \xi_t, \quad \xi_t \in -L^0(\tilde{G}_t, \mathcal{F}_t).$$

By definition, $\xi_1 = -\alpha\xi - \beta$, where β takes values in $\mathbf{R}_+^2$. Modifying, if necessary, ϵ and ξ_1 (by adding β), we may assume that $\xi_1 = -\alpha\xi$, where α is a scalar $\mathcal{F}_0$-measurable random variable. Moreover, we may assume without loss of generality that all ξ_i are integrable (changing, eventually, the probability measure) and α takes values -1, 0, 1.

Since G_2^{N+2} satisfies NA^w, necessarily, $\epsilon I_{\{\alpha=0\}} = 0$. Take a bounded process $Z \in \mathcal{M}_2^{N+2}(G^* \setminus \{0\})$ (existing by the induction hypothesis). Using the martingale property of Z and the duality, we have

$$E\epsilon E(Z_1|\mathcal{F}_0)I_{\{\alpha=-1\}} - E\xi E(Z_1|\mathcal{F}_0)I_{\{\alpha=-1\}} = EI_{\{\alpha=-1\}} \sum_{t=2}^{N+2} \xi_t Z_t \leq 0.$$

Since $\xi E(Z_1|\mathcal{F}_0) \leq 0$, it follows that $\epsilon E(Z_1|\mathcal{F}_0)I_{\{\alpha=-1\}} \leq 0$ (a.s.) and, therefore, $\epsilon I_{\{\alpha=-1\}} = 0$. Thus, $\epsilon I_{\{\alpha=1\}} \neq 0$. On the set $\{\alpha = 1\}$ we have the equality

$$\epsilon = -\xi + \sum_{t=2}^{n+2} \xi_t.$$

It follows that $\xi + \epsilon I_{\{\alpha=1\}} \in A_1^{N+2}$, which is a required property of ξ. □

3.2.4 Counterexamples

Example 1. *A two-asset one-period model satisfying NA^w for which A_0^1 is not closed.* Let $\Omega = \mathbf{N}$, $\mathcal{F} = 2^\Omega$, $P(k) = 2^{-k}$, $\mathcal{F}_0 = \{\emptyset, \Omega\}$, $\mathcal{F}_1 = \mathcal{F}$. Take $G_0 = \text{cone}\{2e_2 - e_1, e_1 - e_2\}$ and $G_1 = \text{cone}\{2e_1 - e_2, e_2 - e_1\}$. The vector $e_1 + e_2$ belongs to both G_0^* and G_1^*, and, hence, the constant process $Z = e_1 + e_2$ is an element of $\mathcal{M}_0^1(G^* \setminus \{0\})$. Let us check that the random variable ξ with $\xi(k) = k(e_2 - e_1)$ does not belong to the set A_0^1 but lies in the closure of the latter. Indeed, suppose that $\xi = \xi_0 + \xi_1$, where $\xi_0 \in -G_0$ and $\xi_1 \in -L^0(G_1, \mathcal{F})$. Since $Z\xi = 0$, $Z\xi_0 \leq 0$, and $Z\xi_1 = 0$, we have that $Z\xi_0 = 0$. But this means that $\xi_0 = c(e_2 - e_1)$ with some $c \geq 0$. It follows that $\xi_1(k) = (k - c)(e_2 - e_1)$, and this vector cannot belong to G_1 for $k > c$, in contradiction with the assumption. On the other hand, ξ is the limit of a sequence of random variables from A_0^1, namely, $\xi^n = \xi_0^n + \xi_1^n$, where $\xi_0^n := n(e_2 - e_1)$ and $\xi_1^n(k) := \min\{k - n, 0\}(e_2 - e_1)$.

Example 2. *A three-dimensional one-period model satisfying NA^w for which $\mathcal{M}_0^1(G^* \setminus \{0\}) = \emptyset$.* The probabilistic setting is the same as in the previous example. Take $G_0^* = \mathbf{R}_+\eta$, $G_1^* = \text{cone}\{\eta_1, \eta_2\}$, where $\eta = (3, 1, 1)$ and

$\eta_1 = (4,1,1)$ are deterministic vectors in $\mathbf{R}^3_+$, while η_2 is a random one with $\eta_2(k) = (2,1,1+1/k)$.

Clearly, $\mathcal{M}^1_0(G^* \setminus \{0\}) = \emptyset$ because one cannot find random variables $\alpha, \beta \geq 0$ to meet the conditions $E\alpha = E\beta = 1/2$ and $E\beta\gamma = 0$, where $\gamma(k) = 1/k$, needed to ensure the equality $EZ_1 = Z_0$.

Let $\xi_0 \in -G_0$ and $\xi_1 \in -L^0(G_1, \mathcal{F})$ be such that $\xi = \xi_0 + \xi_1$ takes values in $\mathbf{R}^3_+$. The latter condition implies that $\eta_1\xi \geq 0$. Since $\eta_1\xi_1 \leq 0$, we have $\eta_1\xi_0 \geq 0$. In the same way we get that $\eta_2(k)\xi_0 \geq 0$ whatever is k. But

$$\eta_1\xi_0 + \lim_k \eta_2(k)\xi_0 = 2\eta\xi_0 \leq 0,$$

and, therefore, both terms in the left-hand side are zero. So, $\eta_1\xi_0 = 0$. As a result, $\eta_1\xi = \eta_1\xi_1 \leq 0$. With ξ taking values in $\mathbf{R}^3_+$, this is possible only when $\xi = 0$ and NA^w holds.

Thus, a straightforward generalization of Theorem 3.2.15 for an arbitrary $\mathcal{C}$-valued process fails to be true already in dimension three. However, the above counterexample does not exclude that it holds in a narrower class of financial models.

Example 3. *A four-asset two-period model satisfying NA^s for which $\mathcal{M}^2_0(G^* \setminus \{0\}) = \emptyset$.* The probability space is $\Omega = \{-1,0,1,\ldots\}$, $\mathcal{F} = 2^\Omega$, $P(k) = 2^{-k-2}$, $\mathcal{F}_0 = \{\emptyset, \Omega\}$, $\mathcal{F}_1 = \mathcal{F}_2 = \mathcal{F}$. We consider a $\mathcal{C}$-valued process G corresponding to $\widehat{K}$ of the financial model parameterized by the adapted matrix-valued process $\Pi = (\Pi_t)_{t\leq 2}$ depending on the parameter $a \in (1,2)$ and given as follows:

$$\Pi_0 = \begin{pmatrix} 1 & a & a & a \\ 1 & 1 & a & a \\ a & a & 1 & a \\ a & a & 1 & 1 \end{pmatrix}; \qquad \Pi_1(-1) = \begin{pmatrix} 1 & a & a & a \\ a & 1 & a & a \\ a & a & 1 & a \\ a & a & a & 1 \end{pmatrix};$$

$$\Pi_1(0) = \begin{pmatrix} 1 & 1 & 1 & 1/2 \\ 1 & 1 & 1 & 1/2 \\ 1 & 1 & 1 & 1/2 \\ 2 & 2 & 2 & 1 \end{pmatrix}; \qquad \Pi_1(k) = \begin{pmatrix} 1 & & & k \\ & 1 & k & \\ & k^{-1} & 1 & a \\ k^{-1} & & 1 & 1 \end{pmatrix}, \quad k \geq 2;$$

the matrices $\Pi_1(1)$ and $\Pi_2(-1)$ are filled by units (this means the absence of transaction costs); at last, for $k \geq 0$, the nondiagonal elements of the matrix $\Pi_2(k)$ are $\pi_2^{ij}(k) = 2^{k+2}$; the empty spaces can be filled by arbitrary sufficiently large numbers (say, k) to avoid chains of transfers cheaper than those already specified. Recall that the cone G_t is generated by the orths e_i and the vectors $\pi_t^{ij} e_i - e_j$.

Verification of the NA^s_1-property. If $\xi = \xi_0 + \xi_1 \in L^0(G_1, \mathcal{F}_1)$, where $\xi_0 \in -G_0$ and $\xi_1 \in -L^0(G_1, \mathcal{F}_1)$, then $\xi_0 \in G_1(-1)$, which is possible only if $\xi_0 = 0$: the scalar product of any nontrivial element of $G_1(-1)$ on the vector $\mathbf{1} = (1,1,1,1)$ is strictly positive, while the scalar products of the elements of

$-G_0$ on the same vector are negative (i.e., the linear space orthogonal to $\mathbf{1}$ separates the cones $G_1(-1)$ and $-G_0$). Thus, $\xi = \xi_1 \in L^0(\mathcal{G}_1^0, \mathcal{F}_0)$.

Verification of the NA_2^s-property. Suppose that $\xi = \xi_0 + \xi_1 + \xi_2$, where $\xi_t \in -L^0(G_t, \mathcal{F}_t)$ and $\xi \in L^0(G_2, \mathcal{F}_2)$. Notice that $\mathbf{1}\xi(1) \geq 0$ and $\mathbf{1}\xi_t(1) \leq 0$ for $t = 0, 1, 2$. Therefore, $\mathbf{1}\xi_t(1) = 0$. This implies, in particular, that

$$\xi_0 = \alpha(e_1 - e_2) + \beta(e_3 - e_4), \quad \alpha, \beta \geq 0.$$

Suppose that the coefficient $\beta > 0$. For the vector $w_k = (a, 1, k, ak)$, we have

$$w_k \xi_0(k) = \alpha(a-1) - \beta k(a-1) < 0$$

when k is large. This leads to a contradiction because $w_k \xi_1(k) \leq 0$ (since $w_k \in G_1^*(k)$ as one can easily verify by multiplying w_k by the generators of $G_1(k)$) and $w_k \xi_2(k) \leq 0$, while $w_k \xi(k) \geq 0$ (since $w_k \in G_2^*(k)$). Thus, $\beta = 0$.

For any $k \geq 1$, the scalar product of the vector $f_k = (1, 1, k, k)$ with $\xi_0 = \alpha(e_1 - e_2)$ is zero; the scalar products of this vector with vectors from $G_1(k)$ are positive and strictly positive with vectors from $G_2(k) \backslash \{0\}$. It follows that $\xi_0 + \xi_1(k) + \xi_2(k) = 0$; the latter equality can be obtained also for $k = 0$ by the same arguments but with the vector $f_0 = (1, 1, 1, 1/2)$.

Finally, we have that the vector $\xi(-1) = \xi_0(-1) + \xi_1(-1) + \xi_2(-1)$ belongs to $G_2(-1)$. The vector $\xi_0 \in G_2^0(-1)$. It follows that $\xi_1(-1)$ belongs to $G_2(-1)$ and, by assumption, to $-G_1(-1)$; these two cones are separated by the subspace orthogonal to $\mathbf{1}$, that is, by $G_2^0(-1)$. So, $\xi_1(-1) \in G_2^0(-1)$. With this, we conclude that $\xi(-1) \in G_2^0(-1)$ and, hence, $\xi(-1) \in G_2^0(-1)$.

Thus, $A_0^2 \cap L^0(G_2, \mathcal{F}_2) \subseteq L^0(G_2, \mathcal{F}_2)$, i.e., the NA_2^s-property holds.

Verification that $\mathcal{M}_0^2(G^* \setminus \{0\}) = \emptyset$. Recall that $\mathcal{M}_0^2(G^* \setminus \{0\}) \neq \emptyset$ if and only if $\bar{A}_0^2 \cap L^0(\mathbf{R}_+^4, \mathcal{F}_2)$. So, it is sufficient to construct an appropriate sequence $\xi^n = \xi_0^n + \xi_1^n + \xi_2^n$ convergent to $\xi \in L^0(\mathbf{R}_+^4, \mathcal{F}_2)$, $\xi \neq 0$: this will inform us also that $A_0^2 \neq \bar{A}_0^2$.

We put

$$\xi_0^n = N(e_1 - e_2) + (e_3 - e_4),$$

$$\xi_1^n(k) = \begin{cases} 0, & k = -1, \\ -N(e_1 - e_2) - (\frac{1}{2}e_3 - e_4), & k = 0, \\ -N(e_1 - e_2) - (e_3 - e_4), & k = 1, \\ -\frac{N}{k}(ke_1 - e_4) - N(\frac{1}{k}e_3 - e_2) - (\frac{N}{k} - 1)(e_4 - e_3), & 2 \leq k \leq N, \\ -\frac{N}{k}(ke_1 - e_4) - N(\frac{1}{k}e_3 - e_2) - \frac{N}{k}(e_4 - e_3), & 2 \leq k > N, \end{cases}$$

$$\xi_2^n(k) = \begin{cases} -\xi_0^n, & k = -1, \\ 0, & k \geq 0. \end{cases}$$

It is tedious but elementary exercise to verify that $\xi_t^n \in L^0(G_t, \mathcal{F}_t)$ and

$$\xi^n = \xi_0^n + \xi_1^n + \xi_2^n = \begin{cases} 0, & k = -1, \\ \frac{1}{2}e_3, & k = 0, \\ 0, & 1 \le k \le N, \\ e_3 - e_4, & k > N. \end{cases}$$

Thus, ξ^n is a required sequence.

3.2.5 A Complement: The Rásonyi Theorem

Unlike other proofs, the arguments used to establish the Grigoriev theorem do not rely upon the closedness of the set A_0^T in L^0 and a separation theorem in an infinite-dimensional space. It was shown by Rásonyi that they can be extended to get the following interesting result which can be considered as a complement to Theorem 3.2.2.

Let $G = (G_t)$ be an adapted cone-valued process, i.e., $G_t(\omega)$ are closed cones in $\mathbf{R}^d$, and the sets $\{(\omega, x) : \ x \in G_t(\omega)\}$ are in $\mathcal{F}_t \otimes \mathcal{B}(\mathbf{R}^d)$. Let

$$A_0^T := -\sum_{t=0}^{T} L^0(G_t, \mathcal{F}_t).$$

Suppose that $G_t^0 := G_t \cap (-G_t) = \{0\}$ or, equivalently, $\operatorname{int} G_t^* \neq \emptyset$ for every t (efficient friction condition in the context of financial modeling). In accordance with our previous terminology, we say that the NA^s-property holds if $A_0^t \cap L^0(G_t, \mathcal{F}_t) = \{0\}$ for $t = 0, 1, \dots, T$.

Theorem 3.2.17 *Suppose that $G_t + G_{t+1}$ are closed cones for $t = 0, \dots, T-1$. Then $NA^s \Leftrightarrow \mathcal{M}_0^T(\operatorname{int} G^*) \neq \emptyset$.*

Note that the condition of the theorem is fulfilled when the cones G_t are polyhedral.

The proof of the result needs some prerequisites from finite-dimensional convex analysis and measurable selections. Further information, in particular on expectations of set-valued mappings, can be found in the Appendix 5.4. To work comfortably, we assume that all σ-algebras here are complete.

1. If B is a closed convex set with $\operatorname{int} B \neq \emptyset$, then B is a closure of $\operatorname{int} B$ and, for any closed convex set A,

$$\operatorname{int} A \cap B \neq \emptyset \quad \Longleftrightarrow \quad \operatorname{int} (A \cap B) \neq \emptyset.$$

If also $\operatorname{int} A \neq \emptyset$, then

$$\operatorname{int} A \cap \operatorname{int} B = \operatorname{int} (A \cap B).$$

2. Let $\omega \mapsto K(\omega)$ be a measurable mapping the values of which are nonempty convex closed subsets of $\mathbf{R}^d$. Then the mapping $\omega \mapsto \operatorname{int} K(\omega)$ is also measurable.

Suppose that the values of K are closed convex cones. Let $\omega \mapsto K_1(\omega)$ be another measurable mapping the values of which are nonempty open subsets of $\mathbf{R}^d$ and such that $K \cap K_1 = \emptyset$ on a set Γ. Then K and K_1 can be separated on Γ in a measurable way, i.e., there is $\xi \in L^0(\mathbf{R}^d, \mathcal{F})$ with $\xi = \xi I_\Gamma$ such that $\sup \eta\xi \leq 0$ for every $\eta \in L^0(K, \mathcal{F})$ and $\zeta\xi > 0$ on Γ for every ζ in $L^0(K_1, \mathcal{F})$.

3. Let $\omega \mapsto C(\omega)$ be a measurable mapping the values of which are closed convex subsets of the unit ball $\bar{\mathcal{O}}_1(0)$ in $\mathbf{R}^d$. Let $\mathcal{G}$ be a sub-σ-algebra of $\mathcal{F}$. There exists a $\mathcal{G}$-measurable mapping, denoted $E(C|\mathcal{G})$, the values of which are closed convex subsets of the unit ball and such that

$$L^0(E(C|\mathcal{G}), \mathcal{G}) = \{E(\eta|\mathcal{G}) : \ \eta \in L^0(C, \mathcal{F})\}.$$

Lemma 3.2.18 *Suppose that* $0 \in C$ *and* $\operatorname{int} C \neq \emptyset$ *(a.s.). Then*

$$\begin{aligned}\{E(\eta|\mathcal{G}) : \ \eta \in L^0(\operatorname{int} C, \mathcal{F})\} &\subseteq \{L^0(\operatorname{int} E(C|\mathcal{G}), \mathcal{G})\} \\ &\subseteq \{E(\eta|\mathcal{G}) : \ \eta \in L^0(2 \operatorname{int} C, \mathcal{F})\}.\end{aligned}$$

The first inclusion implies that $\operatorname{int} E(C|\mathcal{G}) \neq \emptyset$ *(a.s.).*

Proof. Let $\eta \in L^0(\operatorname{int} C, \mathcal{F})$. Then $\rho := 2\text{dist}\,(\eta, \partial C)$ is a strictly positive random variable, and $\eta + \rho\bar{\mathcal{O}}_1(0) \subseteq C$. Hence, $E(\eta|\mathcal{G}) + E(\rho|\mathcal{G})\bar{\mathcal{O}}_1(0) \subseteq E(C|\mathcal{G})$ and $E(\eta|\mathcal{G}) \in L^0(\operatorname{int} E(C|\mathcal{G}), \mathcal{G})\}$, i.e., the first inclusion holds.

To check the second inclusion, we fix an arbitrary $\vartheta_0 \in L^0(\operatorname{int} C, \mathcal{F})$ and put $\tilde{\vartheta}_0 := E(\vartheta_0|\mathcal{G})$. Let $\eta \in L^0(\operatorname{int} E(C|\mathcal{G}), \mathcal{G})$. Let ϵ denote the distance of η from the boundary of $E(C|\mathcal{G})$. It is a $\mathcal{G}$-measurable random variable taking values in $]0, 1]$. We have $\eta - \epsilon\tilde{\vartheta}_0 \in E(C|\mathcal{G})$ and, by the definition of the set-valued conditional expectation, $\eta - \epsilon\tilde{\vartheta}_0 = E(\vartheta_1|\mathcal{G})$ for some $\vartheta_1 \in L^0(C, \mathcal{F})$. Putting $\vartheta := \vartheta_1 + \epsilon\vartheta_0$, we get that $\eta = E(\vartheta|\mathcal{G})$. Since

$$\vartheta = (1+\epsilon)\left[\frac{1}{(1+\epsilon)}\vartheta_1 + \frac{\epsilon}{(1+\epsilon)}\vartheta_0\right]$$

and the expression in the square bracket defines an element from $L^0(\operatorname{int} C)$, the random variable ϑ is a selector of the set $(1+\epsilon)\operatorname{int} C$. The latter is contained in the set $2 \operatorname{int} C$ due to the assumption $0 \in C$, which we use only here. □

Proof of Theorem 3.2.17. We need to establish only the "difficult" implication $(\Rightarrow)$. To this aim define by backward induction the adapted set-valued process $C = (C_t)$ with $C_T := G_T^* \cap \bar{\mathcal{O}}_1(0)$ and $C_t := E(C_{t+1}|\mathcal{F}_t) \cap G_t^*$ for $t \leq T-1$.

In the case where all ($\mathcal{F}_t$-measurable) random sets $\operatorname{int} C_t \neq \emptyset$, the needed martingale is $Z = Z^T$, where Z^T is obtained by the following procedure. Take an arbitrary $Z_0^0 \in L^0(\operatorname{int} C_0, \mathcal{F}_0)$. Suppose that we already constructed $Z^t \in \mathcal{M}_0^t(\operatorname{int} C)$. In virtue of Lemma 3.2.18, for the element

$Z_t^t \in L^0(\operatorname{int} C_t, \mathcal{F}_t)$, we can find $\eta \in L^0(2\operatorname{int} C_{t+1}, \mathcal{F}_{t+1})$ such that $Z_t^t = E(\eta|\mathcal{F}_t)$. The process $Z^{t+1} := \frac{1}{2}(Z_0^t, \dots, Z_t^t, \eta)$ is an element of $\mathcal{M}_0^{t+1}(\operatorname{int} C)$.

So, the theorem will be proven if we show that NA^s implies that $\operatorname{int} C_t \neq \emptyset$ for all t. Suppose that this is not the case and there is s such that $\operatorname{int} C_t \neq \emptyset$ for every $t \geq s+1$, but $\operatorname{int} C_s = \emptyset$ on a nonnull set $\Gamma_s \in \mathcal{F}_s$; note that $s \leq T-1$ as $\operatorname{int} C_T \neq \emptyset$. Since $\operatorname{int} C_s = \operatorname{int}\big(E(C_{s+1}|\mathcal{F}_s) \cap G_s^*\big) = \emptyset$ on Γ_s and, by assumption, $\operatorname{int} G_s^* \neq \emptyset$, one can separate $\operatorname{int} E(C_{s+1}|\mathcal{F}_0)$ and G_s^* on Γ_s in an $\mathcal{F}_s$-measurable way, that is, to find $\xi \in L^0(\mathbf{R}^d, \mathcal{F}_s)$ vanishing outside Γ_s and such that $\eta\xi \leq 0$ for every $\eta \in L^0(G_s^*, \mathcal{F}_s)$ (therefore, $\xi \in -L^0(G_s, \mathcal{F}_s)$) and $\zeta\xi > 0$ on Γ_s for any selector ζ of $\operatorname{int} E(C_{s+1}|\mathcal{F}_1)$. According to the lemma below, this contradicts to NA^s. □

Lemma 3.2.19 *Assume that* $\operatorname{int} C_t \neq \emptyset$ *for* $t \geq s+1$. *Suppose that there exists* $\xi \in A_s^T$ *such that* $\zeta\xi \geq 0$ *whatever is* $\zeta \in L^0(\operatorname{int} E(C_{s+1}|\mathcal{F}_s), \mathcal{F}_s)$ *and this inequality is strict on a nonnull set (hence,* $\xi \neq 0$*). Then* NA^s *fails.*

Proof. If $s = T-1$, then ξ is a weak arbitrage opportunity already for A_{T-1}^T. Indeed, let $B := \{\xi \notin G_T\}$ be a nonnull set. Then one can find $\nu \in L^0(G_T^*, \mathcal{F}_T)$ with $|\nu| \leq 1$ such that $\nu\xi < 0$ on B. Thus,

$$E(\nu I_B|\mathcal{F}_{T-1})\xi = E(\nu\xi I_B|\mathcal{F}_{T-1}) \leq 0,$$

and the inequality is strict on a nonnull set. This contradicts to the assumption of the lemma that the inequality $\zeta\xi \geq 0$ holds for all random vectors $\zeta \in L^0(E(C_T|\mathcal{F}_{T-1}), \mathcal{F}_{T-1})$. Hence, ξ are in G_T, and NA^s fails to be true.

Going backward, let us establish the assertion for an arbitrary s assuming that it holds for $s+1, \dots, T-1$. Since the behavior before s does not matter, we may assume for the notational convenience that $s = 0$ and also that the cone $G_0 = -\mathbf{R}\xi$, i.e., $G_0^* = \{y:\ y\xi \leq 0\}$. Consider the partition of Ω into the following three $\mathcal{F}_1$-measurable subsets:

$$\Omega_1 := \big\{\operatorname{int} E(C_2|\mathcal{F}_1) \cap \operatorname{int} G_1^* \cap G_0^* \neq \emptyset\big\},$$

$$\Omega_2 := \big\{\operatorname{int} E(C_2|\mathcal{F}_1) \cap G_1^* \cap G_0^* \neq \emptyset\big\} \cap \big\{\operatorname{int} E(C_2|\mathcal{F}_1) \cap G_1^* \cap G_0^* \subseteq \partial G_1^*\big\},$$

$$\Omega_3 := \big\{\operatorname{int} E(C_2|\mathcal{F}_1) \cap G_1^* \cap G_0^* = \emptyset\big\}.$$

Suppose that $P(\Omega_1) > 0$. On Ω_1, the intersection of $\operatorname{int}(E(C_2|\mathcal{F}_1) \cap G_1^*)$ and $\operatorname{int} G_0^*$ is nonempty, and hence $C_1 \cap \{y:\ y\xi < 0\} \neq \emptyset$. Let ν be a selector of the latter set extended by zero outside Ω_1. Then $\eta := E(\nu|\mathcal{F}_0)$ belongs to $E(C_1|\mathcal{F}_0)$, and $\eta\xi \leq 0$ with the strict inequality on a nonnull set. This contradiction with the hypothesis of the lemma means that $P(\Omega_1) = 0$.

Suppose now that $P(\Omega_2) > 0$. Put $H := \partial G_0^* = \{y:\ y\xi = 0\}$. Note that the assumption of the lemma implies that (a.s.)

$$\operatorname{int} E(C_2|\mathcal{F}_1) \cap G_1^* \cap G_0^* = \operatorname{int} E(C_2|\mathcal{F}_1) \cap G_1^* \cap H.$$

Indeed, on the subset of Ω where the left-hand side is empty, this identity is obvious. On the complementary subset,

$$\operatorname{int} E(C_2|\mathcal{F}_1) \cap G_1^* \cap \operatorname{int} G_0^* \neq \emptyset,$$

and, hence, $\operatorname{int}(E(C_2|\mathcal{F}_1) \cap G_1^*) \cap \operatorname{int} G_0^* \neq \emptyset$, implying that $C_1 \cap \operatorname{int} G_0^* \neq \emptyset$. Using this, we arrive at a contradiction by the same separation argument as was done for the set Ω_1.

Arguing with convex sets in the subspace H, we easily get that

$$\begin{aligned}
&\{\operatorname{int} E(C_2|\mathcal{F}_1) \cap G_1^* \cap G_0^* \neq \emptyset\} \cap \{\operatorname{int} G_1^* \cap H \neq \emptyset\} \\
&\quad = \{(\operatorname{int} E(C_2|\mathcal{F}_1) \cap H) \cap (G_1^* \cap H) \neq \emptyset,\ \operatorname{int} G_1^* \cap H \neq \emptyset\} \\
&\quad \subseteq \{\operatorname{int} E(C_2|\mathcal{F}_1) \cap \operatorname{int} G_1^* \cap H \neq \emptyset\} \subseteq \Omega_1.
\end{aligned}$$

It follows that $\Omega_2 \cap \{\operatorname{int} G_1^* \cap H \neq \emptyset\} = \emptyset$. So, Ω_2 is the union of two sets $\Omega_2 \cap \{G_1^* \subseteq G_0^*\}$ and $\Omega_2 \cap \{G_1^* \subseteq -G_0^*\}$. The first has zero probability because, in virtue of the assumption of the lemma, the inclusion $C_1 \subseteq \{y :\ y\xi \le 0\}$ may hold only on a null-set. Thus, $P(\Omega_2) = P(\Omega_2 \cap \{G_1^* \subseteq -G_0^*\})$. This means that $\xi \in G_1$ on Ω_2. If $P(\Omega_2) = 1$, then the NA^s-property fails. It remains to consider the case $P(\Omega_2) < 1$.

So, suppose that $P(\Omega_3) > 0$. In this case we can separate $\operatorname{int} E(C_2|\mathcal{F}_1)$ and $G_1^* \cap G_0^*$ on Ω_3, i.e., find an $\mathcal{F}_1$-measurable random vector $\nu = I_{\Omega_3}\nu$ equal to zero outside Ω_3 and such that $\nu\zeta > 0$ for any $\zeta \in L^0(\operatorname{int} E(C_2|\mathcal{F}_1), \mathcal{F}_1)$ on Ω_3 and $\nu\zeta \le 0$ for any $\zeta \in L^0(G_1^* \cap G_0^*, \mathcal{F}_1)$. Since $(G_1^* \cap G_0^*)^* = G_0 + G_1$ in virtue of the assumption on closedness of $G_0 + G_1$, the second property means that $\nu = \alpha\xi - \xi_1$, where $\alpha \in L^0(\mathbf{R}_+, \mathcal{F}_1)$ and $\xi_1 \in L^0(G_1, \mathcal{F}_1)$. If $\alpha > 0$ on Ω_3, we can divide the identity by α and claim the existence of ν with the above properties having the form $\nu = \xi - \xi_1$. In this case, $\nu \in A_0^1$, and by the induction hypothesis NA^s fails. If $\tilde{\Omega}_3 := \{\alpha = 0\} \cap \Omega_3$ is a nonnull set, we arrive at the same conclusion by applying the induction hypothesis with $\nu = \xi_1 I_{\tilde{\Omega}_3}$. □

3.2.6 Arbitrage Opportunities of the Second Kind

On some markets satisfying the NA^w-property, it may happen that an investor with an initial endowment outside the solvency cone may run a portfolio to get rid of debts for sure. As the following example shows, sometimes it is sufficient just to wait.

Example. Let us consider the two-asset model with $S_0^1 = S_0^2 = 1$, where the first asset is the numéraire, i.e., $S_1^1 = 1$, while S_1^2 takes values $1+\varepsilon$ and $1-\varepsilon > 0$ with probabilities $1/2$. The filtration is generated by S. Suppose that K_0^* is the cone generated by the vectors $(1,2)$ and $(1,1/2)$ and that $K_1^* = \mathbf{R}_+\mathbf{1}$, i.e., there are no transaction costs at the date $T=1$. Then $\widehat{K}_1^*$ is the ray generated by the vector S_1. The process Z with $Z_0 = (1,1)$ and $Z_1 = S_1$ is a strictly consistent price system, so the NA^w-property holds. Let v be a point from the dual C of the cone generated by the vectors $(1, 1+\varepsilon)$ and $(1, 1-\varepsilon)$. It lies in the solvency cone $\widehat{K}_1$ a.s. But for $\varepsilon \in]0, 1/2[$, this dual C is strictly larger than the solvency cone $\widehat{K}_0 = K_0$. The investor having $v \in C \setminus K_0$ as the initial endowment became solvent at time $T = 1$, though he was not solvent at the date zero. Clearly, we can modify the model by introducing small transaction costs at time $T = 1$ to get the same conclusion for a model with efficient friction.

We say that the model G admits *arbitrage opportunities of the second kind* if there exist $s \le T-1$ and an $\mathcal{F}_s$-measurable d-dimensional random variable for which $\Gamma := \{\xi \notin G_s\}$ is not a null-set and such that

$$(\xi + A_s^T) \cap L^0(G_T, \mathcal{F}_T) \neq \emptyset,$$

i.e., $\xi = \xi_s + \cdots + \xi_T$ for some $\xi_t \in L^0(G_t, \mathcal{F}_t)$, $s \le t \le T$. If such ξ does exist, then, in the financial context where $G = \widehat{K}$, an investor having $I_\Gamma \xi$ as the initial endowments at time s may use the strategy $(I_\Gamma \xi_t)_{t \ge s}$ and get rid of all debts at time T.

So, the model has *no arbitrage opportunities of the second kind* (abbreviation: has the *NA*2-property) if, for every date s and for $\xi \in L^0(\mathbf{R}^d, \mathcal{F}_s)$, the intersection $(\xi + A_s^T) \cap L^0(G_T, \mathcal{F}_T)$ is nonempty only if $\xi \in L^0(G_s, \mathcal{F}_s)$. Alternatively, the *NA*2-property can be expressed in the following way:

$$L^0(\mathbf{R}^d, \mathcal{F}_s) \cap (-A_s^T) = L^0(G_s, \mathcal{F}_s) \quad \forall s \le T.$$

Note that in the original paper by Rásonyi the *NA*2-property was called the *no sure gain in liquidation value* property (NGV) or, in earlier version, *no sure profits* property with the abbreviation *NSP*. We use the terminology consistent with that developed in the theory of large financial markets, see [121].

Theorem 3.2.20 *Suppose that the efficient friction condition is fulfilled and $\mathbf{R}_+^d \subseteq G_t$ for all t. Then the following conditions are equivalent:*

(a) *NA*2;
(b) $L^0(\mathbf{R}^d, \mathcal{F}_s) \cap L^0(G_{s+1}, \mathcal{F}_{s+1}) \subseteq L^0(G_s, \mathcal{F}_s)$ *for all* $s = 0, \dots, T-1$;
(c) $\operatorname{coneint} E(G_{s+1}^* \cap \bar{\mathcal{O}}_1(0) | \mathcal{F}_s) \supseteq \operatorname{int} G_s^*$ *(a.s.) for all* $s = 0, \dots, T-1$;
(d) *for all* $s \le T-1$ *and* $\eta \in L^1(\operatorname{int} G_s^*, \mathcal{F}_s)$, *there is* $Z \in \mathcal{M}_s^T(\operatorname{int} G^*)$ *such that* $Z_s = \eta$.

Proof. (a) $\Rightarrow$ (b). It follows from the inclusions

$$L^0(G_{s+1}, \mathcal{F}_s) \subseteq L^0(G_{s+1}, \mathcal{F}_{s+1}) \subseteq -A_s^T.$$

(b) $\Rightarrow$ (c). Put $H_s := \operatorname{coneint} E(G_{s+1}^* \cap \bar{\mathcal{O}}_1(0) | \mathcal{F}_s)$. Suppose that for some $t \le T-1$, the set of $\Gamma := \{\operatorname{int} G_t^* \setminus H_t \neq \emptyset\}$ has strictly positive probability. The convex sets $H_t(\omega) \setminus \{0\}$ are open. Hence, $\Gamma = \{\operatorname{int} G_t^* \setminus \bar{H}_t \neq \emptyset\}$. Using measurable selection, we can find an $\mathcal{F}_t$-measurable d-dimensional random variable $\zeta = \zeta I_\Gamma$ such that $\zeta \in \operatorname{int} G_t^* \setminus \bar{H}_t$ a.s. on the set Γ. By the measurable version of the separation theorem, we find a d-dimensional random variable $\tilde{\xi} = \tilde{\xi} I_\Gamma$ satisfying the following two properties:

(i) $\tilde{\xi}\eta \geq 0$ for all $\eta \in L^0(\bar{H}_t, \mathcal{F}_t)$;
(ii) $\tilde{\xi}\zeta < 0$ a.s. on Γ.

Normalizing, we may assume that $|\tilde{\xi}| = I_\Gamma$.

Property (i) implies that $\tilde{\xi} \in L^0(G_{t+1}, \mathcal{F}_t)$ (a.s.). Indeed, a random variable $\xi \in L^1(\mathbf{R}^d, \mathcal{F}_t)$ takes values in G_{t+1} if and only if $E\xi E(\eta|\mathcal{F}_t) \geq 0$ for every $\eta \in L^0(G^*_{t+1}, \mathcal{F}_{t+1})$ with $|\eta| \leq 1$. But the set of the conditional expectations $E(\eta|\mathcal{F}_t)$ for such η coincides with the set $L^0(E(G^*_{t+1} \cap \bar{\mathcal{O}}_1(0)|\mathcal{F}_t), \mathcal{F}_t)$.

Property (ii) means that $\tilde{\xi}$ does not take values in G_t on the set Γ of strictly positive probability. The existence of such $\tilde{\xi}$ contradicts (b).

(c) $\Rightarrow$ (d). Let $Z_t \in L^1(G^*_t, \mathcal{F}_t)$. Since

$$\big\{(a, y) \in]0, \infty[\times \operatorname{int} E\big(G^*_{t+1} \cap \bar{\mathcal{O}}_1(0)|\mathcal{F}_t\big)(\omega) : \ ay = Z_t(\omega)\big\} \neq \emptyset \quad \text{a.s.},$$

we get, using measurable selection, that $Z_t = \alpha_t Y_t$, where $\alpha_t \in L^0(]0, \infty[, \mathcal{F}_t)$ and $Y_t \in L^0(\operatorname{int} E(G^*_{t+1} \cap \bar{\mathcal{O}}_1(0)|\mathcal{F}_t), \mathcal{F}_t)$. By Lemma 3.2.18, $Y_t \in E(\tilde{Z}_{t+1}|\mathcal{F}_t)$ for some $\tilde{Z}_{t+1} \in L^0(2 \operatorname{int}(G^*_{t+1} \cap \bar{\mathcal{O}}_1(0)), |\mathcal{F}_{t+1})$. Put $Z_{t+1} := \alpha_t \tilde{Z}_{t+1}$. Then $E(|\tilde{Z}_{t+1}||\mathcal{F}_t)) \leq 2\alpha < \infty$ and $E(\tilde{Z}_{t+1}|\mathcal{F}_t) = Z_t$. Since the process G^* evolves in $\mathbf{R}^d_+$, we have that $E(\alpha_t Z^i_{t+1}) = EZ^i_t < \infty$, i.e., Z_{t+1} is integrable. Repeating successively these arguments starting from $Z_s = \eta$, we obtain a martingale with the required property. Note that, without the assumption that $\mathbf{R}^d_+ \subseteq G_t$, by this construction we could obtain only a generalized martingale.

(d) $\Rightarrow$ (a). Let us suppose that $\xi \in L^0(\mathbf{R}^d, \mathcal{F}_s)$ admits the representation $\xi = \xi_s + \dots + \xi_T$ with $\xi_t \in L^0(G_t, \mathcal{F}_t)$, $s \leq t \leq T$, but the set $\Gamma := \{\xi \notin G_s\}$ is of strictly positive probability. Without loss of generality we may assume that ξ is bounded. For each $\omega \in \Gamma$, one can find $\eta(\omega) \in \operatorname{int} G^*_s(\omega)$ such that $\eta(\omega)\xi(\omega) < 0$. Using measurable selection, we can find $\eta \in L^0(\operatorname{int} G^*_s, \mathcal{F}_s)$ such that the latter inequality holds a.s. Condition (*d*) ensures the existence of $Z \in \mathcal{M}^T_s(\operatorname{int} G^*)$ such that $Z_s = \eta$. Then $EZ_T\xi I_\Gamma = EZ_s\xi I_\Gamma < 0$, and, in virtue of Lemma 3.3.2 (next section),

$$EZ_T \sum_{t=s}^{T} \xi_t I_\Gamma \geq 0,$$

a contradiction. □

Remark. As we could see, the criteria investigated in this chapter relate certain properties of financial models with the existence of a consistent price system, i.e., a martingale which is a selector of a set-valued process (for the considered models, in the dual to the process of solvency cones in physical units). It was observed by Rokhlin that the latter problem can be placed in a more general framework of the *martingale selection problem*, which is not only of mathematical but also of financial interest because it is useful also for models with liquidity constraints. The martingale selection problem suggests that there is a set-valued adapted process $G = (G_t)$, and the question is whether there exist a probability $Q \sim P$ and a Q-martingale $Z \in \mathcal{M}(G)$. For the development in this direction, we refer to the papers [198, 199, 196, 197].

3.3 Hedging of European Options

3.3.1 Hedging Theorem: Finite Ω

Let C be an $\mathbf{R}^d$-valued $\mathcal{F}_T$-measurable random variable, interpreted as a *contingent claim* of values of corresponding assets.

Our aim now is to describe the set of all initial endowments starting from which one can "super-replicate," in the sense of the partial ordering, the contingent claim C by the terminal value of a self-financing portfolio.

The formal description of the convex set of *hedging endowments* (in values or in physical units since we use a convention that all $S_0^i = 1$) is as follows:

$$\Gamma := \{v \in \mathbf{R}^d : \ \exists B \in \mathcal{B} \text{ such that } \ v + V_T^B \succeq_T C\}.$$

It is easy to see that

$$\Gamma = \{v \in \mathbf{R}^d : \ \widehat{C} \in v + \widehat{A}_0^T\}.$$

We also introduce the closed convex set

$$D := \Big\{v \in \mathbf{R}^d : \ \sup_Z E(Z_T\widehat{C} - Z_0 v) \le 0\Big\} = \bigcap_Z \{v \in \mathbf{R}^d : \ Z_0 v \ge EZ_T\widehat{C}\},$$

where Z runs over the set $\mathcal{M}_0^T(\widehat{K}^* \setminus \{0\})$ assumed to be nonempty. Having in mind the approximation of Z_T by $Z_T^\varepsilon := (1-\varepsilon)Z_T + \varepsilon\tilde{Z}_T$, where $\tilde{Z}$ belongs to $\mathcal{M}_0^T(\widehat{K}^* \setminus \{0\})$, we can, of course, take above the supremum over $Z \in \mathcal{M}_0^T(\widehat{K}^*)$.

Theorem 3.3.1 *Let Ω be finite, and let $\mathcal{M}_0^T(\widehat{K}^* \setminus \{0\}) \neq \emptyset$. Then $\Gamma = D$.*

Proof. Take $\xi = \sum_{t=0}^T \xi_t$ with $\xi_t \in -L^0(\widehat{K}_t, \mathcal{F}_t)$. For any $Z \in \mathcal{M}_0^T(\widehat{K}^* \setminus \{0\})$, we have

$$EZ_T\widehat{C} \le EZ_T\left(v + \sum_{t=0}^T \xi_t\right) = Z_0 v + \sum_{t=0}^T EZ_t\xi_t \le Z_0 v,$$

and the "easy" inclusion $\Gamma \subseteq D$ holds.

Take now $v \notin \Gamma$. To show that $v \notin D$, it is sufficient to find $Z \in \mathcal{M}_0^T(\widehat{K}^*)$ with $Z_0 v < EZ_T\widehat{C}$. Since $\widehat{C} \notin v + \widehat{A}_0^T$ and the latter set, being a shift of a polyhedral cone, is closed, the separation theorem for a finite-dimensional space implies that

$$\sup_{\xi \in v + \widehat{A}_0^T} E\eta\xi < E\eta\widehat{C} \tag{3.3.1}$$

for some d-dimensional random variable η. Define the martingale $Z_t := E(\eta|\mathcal{F}_t)$. It follows that $EZ_t\xi_t \ge 0$ for all $\xi_t \in L^0(\widehat{K}_t, \mathcal{F}_t)$, implying that $Z \in \mathcal{M}_0^T(\widehat{K}^*)$. Taking in (3.3.1) $\xi = v$ and using the martingale property, we get the desired inequality $EZ_0 v < E\eta\widehat{C}$. $\square$

Financial interpretation. We want to attract the reader's attention to the financial interpretation of the obtained result and the role of consistent price systems. The theorem asserts that a contingent claim $\widehat{C}$ (in physical units) can be super-replicated starting from an initial endowment v by a self-financing portfolio if and only if the "value" $Z_0 v$ of this initial endowment is not less than the expected "value" of the contingent claim $EZ_T\widehat{C}$, whatever is the consistent price system Z (we write "value" in quotation marks to emphasize its particular meaning in the present context). In other words, consistent price systems allow the option seller to relate benefits from possessing v at time $t=0$ and the liabilities $\widehat{C}$ at time $t=T$ and provide information whether there is a portfolio ending up on the safe side.

3.3.2 Hedging Theorem: Discrete Time, Arbitrary Ω

Now we present a hedging result extending the Theorem 3.3.1 to the case of arbitrary Ω. There is no need to change the definition of the set of Γ of initial endowments from which one can start a portfolio process the terminal values of which dominate the contingent claim. However, its dual description, that is, the definition of the set D requires some precautions needed to ensure the existence of expectations involved. Moreover, the techniques used requires the closedness of the set of replicable claims. That is why we shall assume that the set $\mathcal{M}_0^T(\operatorname{ri} G^*)$ is nonempty.

We present the result in the abstract setting of the $\mathcal{C}$-valued process G dominating the constant process $\mathbf{R}^d_+$; as we can indicated earlier, such a setting is not only more mathematically transparent but covers various financial models with proportional transaction costs besides our basic one.

So, we fix a d-dimensional random variable ζ (which corresponds in financial context to $\widehat{C}$, the contingent claim expressed in physical units). Define the set

$$\Gamma = \{v \in \mathbf{R}^d : \ \zeta \in v + A_0^T\}.$$

Let $\mathcal{Z}$ be the set of martingales from $\mathcal{M}_0^T(\operatorname{ri} G^*)$ such that $E(Z_T\zeta)^- < \infty$. Put

$$D := \Big\{v \in \mathbf{R}^d : \ \sup_{Z\in\mathcal{Z}} E(Z_T\zeta - Z_0 v) \le 0\Big\}.$$

The following simple assertion is a key to understanding the role of the integrability assumption involved in the definition of $\mathcal{Z}$.

Lemma 3.3.2 *Let Z be an $\mathbf{R}^d$-valued martingale, and let $\Sigma_T := Z_T\sum_{s=0}^T \xi_s$, where $\xi_s \in L^0(\mathbf{R}^d,\mathcal{F}_s)$ are such that $Z_s\xi_s \le 0$. If $E\Sigma_T^- < \infty$, then all products $Z_s\xi_s$ are integrable, Σ_T is integrable, and $E\Sigma_T \le 0$.*

Proof. For $T=0$, there is nothing to prove. Assume that the claim is true for $T-1$. Clearly,

$$Z_T\sum_{s=0}^{T-1}\xi_s \ge -\Sigma_T^- - Z_T\xi_T \ge -\Sigma_T^-.$$

By conditioning the resulting inequality we get that

$$Z_{T-1}\sum_{s=0}^{T-1}\xi_s \geq -E(\Sigma_T^-|\mathcal{F}_{T-1}).$$

Since the left-hand side is Σ_{T-1}, we have the bound $\Sigma_{T-1}^- \leq E(\Sigma_T^-|\mathcal{F}_{T-1})$, implying that $E\Sigma_{T-1}^- \leq E\Sigma_T^- < \infty$. By the induction hypothesis, Σ_{T-1} is integrable, and $E\Sigma_{T-1} \leq 0$. We have the same properties for Σ_T because $Z_T\xi_T \leq 0$. □

Theorem 3.3.3 *Suppose that* $\mathcal{M}_0^T(\operatorname{ri} G^*) \neq \emptyset$. *Then* $\Gamma = D$.

Proof. The arguments follow the same line as in the proof of Theorem 3.3.1. The inclusion $\Gamma \subseteq D$ is clear: if $\zeta = v + \sum_{s=0}^T \xi_s$ with $\xi_s \in -L^0(G_s, \mathcal{F}_s)$, then $EZ_T\zeta \leq Z_0 v$ for any martingale from $\mathcal{Z}$ in virtue of the above lemma.

To check the opposite inclusion, we take a point $v \notin \Gamma$ and show that $v \notin D$. It is sufficient to find $Z \in \mathcal{Z}$ such that $Z_0 v < EZ_T\zeta$. Consider a measure $\tilde{P} \sim P$ with bounded density ρ such that $\zeta \in L^1(\tilde{P})$. The convex set $\tilde{A}^1 := A_0^T \cap L^1(\tilde{P})$ is closed and does not contain the point $\zeta - v$ and, hence, can be separated from the latter by a functional η from L^∞. This means that

$$\sup_{\xi\in\tilde{A}^1} E\rho\eta\xi < E\eta\rho(\zeta - v).$$

The bounded martingale $E(\rho\eta|\mathcal{F}_t)$ satisfies the needed inequality and belongs to $\mathcal{M}_0^T(G^*)$. Adding to it the martingale $\epsilon\tilde{Z}$ where $\tilde{Z} \in \mathcal{M}_0^T(\operatorname{ri} G^*)$ and taking $\varepsilon > 0$ small enough, we get Z with all needed properties. □

3.4 Hedging of American Options

3.4.1 American Options: Finite Ω

We consider again the abstract setting where the model is given $\mathcal{C}$-valued process $G = (G_t)$, $t = 0, 1, \ldots, T$, dominating the constant process $\mathbf{R}_+^d$. Recall that a particular case of this setting serves as a model of financial market with transactions costs specified in the "hat" terms, i.e., the assets are counted in physical units. The pay-off process $Y = (Y_t)$ is now $\mathbf{R}^d$-valued. Our aim is to describe the set $\Gamma = \Gamma(Y)$ of all $v \in \mathbf{R}^d$ such that there is a portfolio process $X = (X_t)$ starting from zero for which $v + X_t \geq_{G_t} Y_t$, i.e., the process $v + X$ dominates Y in the sense of partial orderings generated by G. More formally, we denote by $\mathcal{X}^0$ the set of $X = (X_t)$ with $X_{-1} = 0$ and $\Delta X_t \in -L^0(G_t, \mathcal{F}_t)$ for $t = 0, 1, \ldots, T$ and put

$$\Gamma := \{v \in \mathbf{R}^d : \exists X \in \mathcal{X}^0 \text{ such that } v + X_t - Y_t \in G_t,\ t = 0, 1, \ldots, T\}.$$

We also introduce the set $A_0^T(.)$ of hedgeable American claims consisting of all processes Y which can be dominated, in the above sense, by a portfolio process with zero initial capital. Clearly, if $Y \in A_0^T(.)$, then $Y_t \in A_0^t$ for all $t \geq 0$. Note that the adapted processes can be viewed as measurable functions on the measure space $(\tilde{\Omega}, \tilde{\mathcal{F}}, P^N)$, where $\tilde{\Omega} = \Omega \times \{0, 1, \dots, T\}$, the σ-algebra $\tilde{\mathcal{F}}$ is generated by the adapted processes themselves,[1] and $dP^N = dPdN$ with N the counting measure on integers (i.e., $N(\{t\}) = 1$). The averaging with respect to P^N will be denoted by E^N.

By analogy with the result available for frictionless market (Proposition 2.1.14) and having in mind just established hedging theorems for European-type options under transaction costs, one may guess that, at least for the case of finite Ω,

$$\Gamma = \{v \in \mathbf{R}^d : \ Z_0 v \geq E Z_\tau Y_\tau \ \forall Z \in \mathcal{M}(G^*), \ \tau \in \mathcal{T}\}. \tag{3.4.1}$$

Surprisingly, in general, this equality, as we show later, fails to be true. To formulate the correct result, we introduce the notation

$$\bar{Z}_t := \sum_{r=t}^{T} E(Z_r | \mathcal{F}_t)$$

and define the set of adapted bounded processes

$$\mathcal{Z}(G^*, P) := \{Z : \ Z_t, \bar{Z}_t \in L^\infty(G_t^*, \mathcal{F}_t), \ t = 0, 1, \dots, T\}.$$

Clearly, all bounded martingales from $\mathcal{M}(G^*, P)$ belong to $\mathcal{Z}(G^*, P)$.

Theorem 3.4.1 *Suppose that Ω is finite. Then*

$$\Gamma = \{v \in \mathbf{R}^d : \ \bar{Z}_0 v \geq E^N Z Y \ \forall Z \in \mathcal{Z}(G^*, P)\}. \tag{3.4.2}$$

Proof. The inclusion $\subseteq$ is easy. Indeed, let $v \in \Gamma$, and let X be a value process which dominates $Y - v$. Then, for $Z \in \mathcal{Z}(G^*, P)$, we have

$$E^N Z Y = \sum_{t=0}^{T} E Z_t Y_t \leq \sum_{t=0}^{T} E Z_t (v + X_t) = \bar{Z}_0 v + \sum_{t=0}^{T} E Z_t X_t \leq \bar{Z}_0 v$$

since

$$\sum_{t=0}^{T} E Z_t X_t = \sum_{t=0}^{T} \sum_{r=0}^{t} E Z_t \Delta X_r = \sum_{t=0}^{T} E \left(\sum_{r=t}^{T} Z_r \right) \Delta X_t = \sum_{t=0}^{T} E \bar{Z}_t \Delta X_t \leq 0$$

because ΔX_t and $\bar{Z}_t$ take values in $-G_t$ and G_t^*, respectively.

To prove the reverse inclusion, we follow the usual pattern. Suppose that $v \notin \Gamma$, i.e., $Y - v$ does not belong to the closed convex cone $A_0^T(.)$ in the finite-dimensional Euclidean space $L^2(\mathbf{R}^d; \tilde{\Omega}, \tilde{F}, P^N)$. The separation theorem provides us an element Z from this space (which is simply an adapted process)

[1] That is, $\tilde{\mathcal{F}}$ is the discrete-time analogue of the optional σ-algebra of the general theory of processes.

such that

$$\sup_{X \in A_0^T(.)} E^N ZX < E^N Z(Y - v). \tag{3.4.3}$$

Since $A_0^T(.)$ is a cone, the supremum is zero. Thus, for every $X \in A_0^T(.)$,

$$\sum_{t=0}^{T} EZ_tX_t \le 0, \qquad \sum_{t=0}^{T} E\bar{Z}_t\Delta X_t \le 0 \tag{3.4.4}$$

(the left-hand sides of these equalities are the same according to the above calculation). It follows that $EZ_t\xi \le 0$ and $E\bar{Z}_t\xi \le 0$ whatever is $\xi \in -L^0(G_t, \mathcal{F}_t)$. Therefore, $Z \in \mathcal{Z}(G^*, P)$. As $E^N Z(Y - v) > 0$, the point v does not belong to the set in the right-hand side of (3.4.2), and we conclude. □

Remark. It is easy to see that the hedging theorem for American options remains true if we replace the counting measure by an arbitrary probability measure ν on the set $\{0, 1, \dots, T\}$ with $\nu(\{t\}) > 0$ for all $t = 0, 1, \dots, T$. Of course, $\bar{Z}_t$ should be replaced by

$$\bar{Z}_t^\nu = \sum_{r=t}^{T} E(Z_r|\mathcal{F}_t)\nu(\{r\}) = \int_{[t,T]} E(Z_r|\mathcal{F}_t)\nu(dr).$$

This gives a hint to the corresponding result in the continuous-time setting.

Financial interpretation. As we shall see below, even in a very simple discrete-time model, consistent price systems form a class which is too narrow to evaluate American claims correctly. The phenomenon appears because one cannot prohibit the option buyer to toss a coin and take a decision to exercise the option at time t or not, in dependence of the outcome. It happens that the expected "value" of an American claim is the mathematical expectation of the weighted average of "values" of assets obtained by the option holder for a variety of exercise dates. This expected "value" should be compared with the "value" of the initial endowment. The main question is what is the class of price systems which should be involved to calculate "values" to be compared. The above theorem shows that the comparison can be done with the systems for which the expected weighted average of future prices knowing the past is again a price system. The structure of such price systems is coherent with the option buyer actions. We shall call them *coherent price systems* and use the abbreviation CoPS.

3.4.2 American Options: Arbitrary Ω

Assuming a condition which guarantees the closedness of $A_0^T(.)$ (see Proposition 3.4.3 below), we can easily establish the following result, which holds for an arbitrary probability space.

Theorem 3.4.2 *Suppose that NA^r-property holds. Then*

$$\Gamma = \{v \in \mathbf{R}^d : \bar{Z}_0 v \geq E^N ZY \ \forall Z \in \mathcal{Z}(G^*, P) \ \textit{with} \ E^N|ZY| < \infty\}. \quad (3.4.5)$$

Proof. The arguments to establish the inclusion $\subseteq$ remain the same as in the previous theorem. To check the opposite inclusion, take an arbitrary $v \notin \Gamma$. So, $Y - v \notin A_0^T(.)$. Choose an auxiliary probability measure $\tilde{P}$ such that the density process $\rho = (\rho_t)_{t \leq T}$ is bounded and $\tilde{E}^N|Y| < \infty$. Now $Y - v \in L^1(\tilde{P}^N)$ and does not belong to the closed convex cone $A_0^T(.) \cap L^1(\tilde{P}^N)$. Let $\tilde{Z}$ denote the separating functional. The bounded process $Z := \rho\tilde{Z}$ satisfies (3.4.3) and (3.4.4), and we conclude exactly in the same way as above. □

Proposition 3.4.3 *If NA^r-property holds, then the set $A_0^T(.)$ is closed in $L^0(P^N)$.*

Proof. We know already that the claim holds for the one-step model. Arguing by induction, we suppose that it is true for $T-1$. Let us consider a sequence of processes $Y^n \in A_0^T(.)$ converging to some Y. By definition, there is a sequence of portfolio processes X^n (i.e., with $\Delta X_t^n \in -G_t$) such that $X_t^n - Y_t^n \in G_t$. By the standard reduction it is sufficient to consider the following two cases.

The first, easy, case: X_0^n converges to a limit $X_0 \in L^0(-G_0, \mathcal{F}_0)$. It follows that $X_0 - Y_0 \in L^0(G_0, \mathcal{F}_0)$. The process $\tilde{Y}_t^n := Y_t^n - \Delta X_0^n$, $t \geq 1$, is an element of $A_1^T(.)$. By the induction hypothesis, the limit of the sequence of processes $\tilde{Y}^n$ belongs to $A_1^T(.)$. This means that $\tilde{X}_t - \tilde{Y}_t \in L^0(G_t, \mathcal{F}_t)$ for some portfolio process defined for $t \geq 1$. Put $X_t := \tilde{X}_t + X_0$ for $t \geq 1$. Then $X = (X_t)_{t \geq 0}$ is a portfolio process dominating Y on the whole time range, and, hence, $Y \in A_0^T(.)$.

The second case: $|X_0^n|$ diverges to infinity. Using the lemma on subsequences, we can assume that $X_0^n/|X_0^n|$ converges to an element $\tilde{X}_0$ from $L^0(G_0^0, \mathcal{F}_0)$, where $G_0 := G_0 \cap (-G_0)$. By the induction hypothesis, the set $A_1^T(.)$ is closed. We get from here that the constant process $-\tilde{X}_0$ is an element of $A_1^T(.)$, and, hence, the zero process is an element of $A_0^T(.)$ dominated by some process $\tilde{X}$ with $|\tilde{X}_0| = 1$. The domination property means that $\tilde{X}_t = \sum_{r=0}^t \Delta\tilde{X}_r$ takes values in G_t for each $t \geq 0$. Recalling that the equality

$$\sum_{r=0}^{t-1} \Delta\tilde{X}_r + (\Delta\tilde{X}_t - \tilde{X}_t) = 0$$

is fulfilled only if the summands are elements of the corresponding linear spaces $L^0(G_r^0, \mathcal{F}_r)$ (see Lemma 3.2.12), we obtain that $\Delta\tilde{X}_t$ and $\tilde{X}_t$ belong to $L^0(G_t^0, \mathcal{F}_t)$ for every $t \geq 0$. The existence of such a process allows us to make a step of the Gauss-type elimination algorithm to diminish the number of nonzero components of X_0^n, and we accomplish the proof in the same way as in Lemma 3.2.8. □

3.4.3 Complementary Results and Comments

Now we examine the question under which condition on the market the identity $\Gamma(Y) = D(Y)$ holds for every pay-off process Y; here $D(Y)$ is the set in the right-hand side of (3.4.1).

Put

$$c_t(x) := \inf\{\lambda \in \mathbf{R} : \lambda e_1 - x \in G_t\}, \quad x \in \mathbf{R}^d.$$

In financial context, $c_t(x)$ is the number of units of the first asset needed to acquire, at date t, the portfolio x; if the first asset is the numéraire, $c_t(x)$ is a *constitutional value* of x.

Proposition 3.4.4 *Let $T \geq 1$. Suppose that there is $x \in \mathbf{R}^d$ such that the following two conditions are fulfilled:*

(i) *if $y - c_0(x)e_1 \in G_0^0$, then either $y - x \in G_0^0$ or $P(y - x \in G_1) < 1$;*
(ii) $x - c_0(x)e_1 \notin G_0$.

Then there exists $Y = (Y_t)$ such that $\Gamma(Y) \neq D(Y)$.

Proof. The process $Y_t = c_0(x)e_1 I_{\{t=0\}} + xI_{\{t>0\}}$ has the needed property. For any $Z \in \mathcal{M}(G^*)$ and an arbitrary stopping time $\tau \leq T$, we have

$$\begin{aligned} E\big(Z_\tau Y_\tau - Z_0 c_0(x)e_1\big) &= E\big(Z_\tau x - Z_0 c_0(x)e_1\big)I_{\{\tau>0\}} \\ &= Z_0\big(x - c_0(x)e_1\big)P(\tau > 0). \end{aligned}$$

The right-hand side being negative, $c_0(x)e_1 \in D(Y)$. If $c_0(x)e_1 \in \Gamma(Y)$, there is a portfolio process X such that $c_0(x)e_1 + X$ dominates Y. In particular, $X_0 \in G_0$. But $X_0 = \Delta X_0 \in -G_0$. According to condition (i), we have two possibilities. The first one: $c_0(x)e_1 + X_0 - x \in G_0^0$; then $c_0(x)e_1 - x \in G_0^0$, and this is impossible due to (ii). The second possibility is also impossible because the domination property $c_0(x)e_1 + X_0 + \Delta X_1 - x \in G_1$ implies that $c_0(x)e_1 + X_0 - x \in G_1$ (a.s.). Therefore, $v \notin \Gamma(Y)$. □

Remark. Note that always $x - c_0(x)e_1 \in -G_0$. Thus, in the case $G_0^0 = \{0\}$, condition (ii) holds for all x except $x = c_0(x)e_1$.

Example. Let us consider the two-dimensional model with $T = 1$, trivial filtration,

$$G_t := \{x : p_t x \geq 0, hx \geq 0\}, \quad t = 0, 1,$$

$p_t = (1, 1 + \lambda_t)$, $\lambda_1 > \lambda_0 \geq 0$, and $h^1 > h^2 > 0$. That is, G_t are sharpened sectors containing the first quadrant; the upper boundary is taken to be common. For $x = e_2$, condition (ii) holds obviously. Let $y = c_0(e_2)e_1$, i.e., y is the projection on the x-axis of the point $y - x = c_0(e_2)e_1 - e_2$ lying on the intersection point of the lower boundary of G_0 with the line parallel to the x-axis and containing $-e_2$; note that $c_1(e_2) > 1$. The lower boundary of G_1 lies above that of G_0. Thus, $y - x \notin G_1$, condition (i) holds, and $\Gamma(Y) \neq D(Y)$.

3.5 Ramifications

3.5.1 Models with Incomplete Information

Models with transaction costs where the investor has an uncomplete (for example, delayed) information not only necessitate important changes in the description of value processes but also appropriate modifications of the main concepts. In particular, one cannot work on the level of portfolio positions represented by a point in $\mathbf{R}^d$ but has to remain on the primary level of the investor's decisions (orders), i.e., in the space of much higher dimension.

Example 1. Let us consider the barter market which is described by an $\mathbf{F}$-measurable conversion ("bid-ask") process $\Pi = (\pi_t^{ij})$ taking values in the set of strictly positive $d \times d$ matrices such that $\pi_t^{ij}\pi_t^{ji} \geq 1$. The entry π_t^{ij} stands for the number of units of the ith asset needed to exchange, at time t, for one unit of the jth asset. The above inequality means that exchanging one unit of the ith asset for $1/\pi_t^{ij}$ units of the jth asset with simultaneous exchange back of the latter quantity results in decrease of the ith position.

In the case of fully informed investor, the portfolio process is generated by an $\mathbf{F}$-adapted process (η_t^{ij}) with values in the set $\mathbf{M}_+^d$ of positive $d\times d$ matrices; the entry $\eta_t^{ij} \geq 0$ is the investor's order to increase the position j by η_t^{ij} units by converting a certain number of units of the ith asset. The investor has a precise idea about this "certain number": it is $\pi_t^{ij}\eta_t^{ij}$. The situation is radically different when the information available is given by a smaller filtration $\mathbf{G}$, i.e., η_t^{ij} is only $\mathcal{G}_t$-measurable. The decrease of the ith asset implied by such an order, being $\mathcal{F}_t$-measurable, is unknown to the investor. However, one can easily imagine a situation where the latter is willing to control the lower level of investments in some assets in his portfolio. This can be done by using the $\mathbf{G}$-adapted order process $(\tilde{\eta}_t^{ij})$ with the element $\tilde{\eta}_t^{ij}$ representing the number of units of the ith asset to be exchanged for the jth asset—the result of this transaction yields an increase of the jth position in $\tilde{\eta}_t^{ij}/\pi_t^{ij}$ units, and, in general, now this quantity is unknown to the investor at time t. Of course, orders of both types, "to get" and "to send", can be used simultaneously. In other words, the investor's orders form a $\mathbf{G}$-adapted process $[(\eta_t^{ij}), (\tilde{\eta}_t^{ij})]$ taking values in the set of positive rectangular matrices $\mathbf{M}_+^{d\times 2d} = \mathbf{M}_+^d \times \mathbf{M}_+^d$. The dynamics of the portfolio processes is given by the formula

$$\Delta\widehat{V}_t = \widehat{\Delta B}_t^1 + \widehat{\Delta B}_t^2, \tag{3.5.1}$$

where the coordinates of $\widehat{\Delta B}_t^1$ and $\widehat{\Delta B}_t^2$ are

$$\widehat{\Delta B}_t^{1,i} := \sum_{j=1}^d \big[\eta_t^{ji} - \pi_t^{ij}\eta_t^{ij}\big],$$

$$\widehat{\Delta B}_t^{2,i} := \sum_{j=1}^d \big[\tilde{\eta}_t^{ji}/\pi_t^{ji} - \tilde{\eta}_t^{ij}\big].$$

Let $(e^{ij}) \in \mathbf{M}^d_+$ be the matrix with all zero entries except the entry (i,j) which is equal to unity. The union of the elementary orders $[(e^{ij}),0]$ and $[0,(e^{ji})]$ forms a basis in $\mathbf{M}^{d\times 2d}$. The execution of the order $[(e^{ij}),(e^{ji})]$ (buying a unit of the jth asset in exchange for the ith asset and then exchanging it back) leads to a certain loss in the ith position, while others remain unchanged, i.e., $\Delta\widehat{V}^i_t \le 0$, $\Delta\widehat{V}^j_t = 0$, $j \ne i$. This observation will be used further, in the analysis of the NA^r-property.

Example 2. Let us turn back to our basic model which is defined by a price process $S=(S_t)$ (describing the evolution of prices of units of assets in terms of some numéraire, e.g., the euro) and an $\mathbf{M}^d_+$-valued process $\Lambda = (\lambda^{ij}_t)$ of transaction cost coefficients. This model admits a formulation in terms of portfolio positions in physical units: one can introduce the matrix Π by setting

$$\pi^{ij}_t = \left(1+\lambda^{ij}_t\right)S^j_t/S^i_t, \quad 1\le i,j\le d.$$

In the full information case the difference between two models is only in parameterizations: one can introduce in the barter market "money" by taking as the price process S an arbitrary one evolving in the duals to the solvency cones and nonvanishing and defining λ^{ij}_t from the above relations. On the other hand, from the perspective of partial information, the setting based on price quotes is more flexible and provides a wider range of possible generalizations.

Again, assume that the investor's information is described by a smaller filtration $\mathbf{G}$, while S and Λ are $\mathbf{F}$-adapted (note that these processes may be adapted with respect to different filtrations).

In contrast to the barter market, the investor now may communicate orders of four types: in addition to the orders (η^{ij}_t) and $(\tilde{\eta}^{ij}_t)$, one can imagine also similar orders, "to get" and "to send", but expressed in units of the numéraire and given by $\mathbf{G}$-adapted matrix-valued processes (α^{ij}_t) and $(\tilde{\alpha}^{ij}_t)$ with positive components. The entry α^{ij}_t is the increment of value in the position j due to diminishing the position i, while the entry $\tilde{\alpha}^{ij}_t$ is a value of the ith asset ordered to be exchanged for the jth asset.

The dynamics of value processes in such a model, in physical units, is given by the formula

$$\Delta\widehat{V}_t = \widehat{\Delta B}^1_t + \widehat{\Delta B}^2_t + \widehat{\Delta B}^3_t + \widehat{\Delta B}^4_t, \tag{3.5.2}$$

where $\widehat{\Delta B}^{3,i}_t := \Delta B^{3,i}_t/S^i_t$, $\widehat{\Delta B}^{4,i}_t := \Delta B^{4,i}_t/S^i_t$ with

$$\Delta B^{3,i}_t := \sum_{j=1}^d \alpha^{ji}_t - \sum_{j=1}^d (1+\lambda^{ij}_t)\alpha^{ij}_t,$$

$$\Delta B^{4,i}_t := \sum_{j=1}^d \frac{\tilde{\alpha}^{ji}_t}{1+\lambda^{ji}_t} - \sum_{j=1}^d \tilde{\alpha}^{ij}_t.$$

Of course, in this case the dynamics can be expressed also in values, that is, in units of the numéraire (using the relation $X^i = \widehat{X}^i S^i$).

Thus, in both cases the set of "results" (for portfolios with zero initial endowments) consists of the d-dimensional random variables

$$\xi = \sum_{t=0}^{T} \mathcal{L}_t \zeta_t, \quad \zeta_t \in O_t := L^0(\mathbf{M}_+^{d\times m}, \mathcal{G}_t), \tag{3.5.3}$$

where m is either $2d$ or $4d$, and $\mathcal{L}_{\omega,t} : \mathbf{M}^{d\times m} \to \mathbf{R}^d$ are linear operators such that the mappings $\omega \mapsto \mathcal{L}_{\omega,t}$ are measurable with respect to the σ-algebra $\mathcal{F}_t$. We shall denote this set by $\widehat{R}_T$ or, when needed, by $\widehat{R}_T(\mathcal{L})$ to show the dependence on the defining operator-valued random process. As usual, we define the set of hedgeable claims $\widehat{A}_T(\mathcal{L}) := \widehat{R}_T(\mathcal{L}) - L^0(\mathbf{R}^d_+)$.

Let us associate with the random linear operator $\mathcal{L}_t$ (acting on elements of $\mathbf{M}^{d\times m}$) the linear operator $\mathbf{L}_t$ acting on $\mathbf{M}^{d\times m}$-valued random variables, $\mathbf{L}_t : L^0(\mathbf{M}^{d\times m}, \mathcal{G}_t) \to L^0(\mathbf{R}^d, \mathcal{F}_t)$, by setting $(\mathbf{L}_t\zeta)(\omega) = \mathcal{L}_{\omega,t}\zeta(\omega)$. With this notation,

$$\widehat{R}_T = \sum_{t=0}^{T} \mathbf{L}_t(O_t).$$

Sometimes, it is convenient to view $\mathbf{M}^{d\times m}$ as the set of linear operators defined by the corresponding matrices.

Unlike the case of frictionless market, the set R_T, in general, is not closed even for models with complete information: see Example 1 in Sect. 3.2.4, where the set $\widehat{R}_1 = \widehat{A}_1$ is not closed though the NA^w-condition is satisfied. However, similarly to the models with complete information, we have the following result.

Proposition 3.5.1 *The sets $\mathbf{L}_t(O_t)$ are closed in probability.*

Proof. The arguments being standard, we only sketch them. In a slightly more general setting, consider the sequence of random vectors $\zeta^n = \sum_{i=1}^N c_i^n g_i$ in a finite-dimensional Euclidean space, where g_i are $\mathcal{G}$-measurable random vectors, and $c_i^n \in L^0_+(\mathcal{G})$. Let $\mathcal{L}$ be an $\mathcal{F}$-measurable random linear operator. Knowing that the sequence $\xi^n = \mathcal{L}\zeta^n$ converges to ξ, we want to show that $\xi = \mathcal{L}\zeta$ for some $\zeta = \sum_{i=1}^N c_i g_i$. Supposing that the result holds for $N-1$ (for $N=1$, it is obvious), we extend it to N. Indeed, it is easy to see, recalling, as usual, the lemma on random subsequences, that we may assume without loss of generality that all sequences c_i^n converge to infinity and, moreover, the normalized sequences $\tilde{c}_i^n := c_i^n/|c^n|$, where $|c^n|$ is the sum of c_i^n, converge to some $\mathcal{G}$-measurable random variables $\tilde{c}_i$. For the random vector $\tilde{\zeta} := \sum_{i=1}^N \tilde{c}_i g_i$, we have that $\mathcal{L}\tilde{\zeta} = 0$. Put $\alpha^n := \min_i\{c_i^n/\tilde{c}_i : \tilde{c}_i > 0\}$. Note that the random variable $\bar{c}_i^n := c_i^n - \alpha^n\tilde{c}_i \geq 0$ and, for each ω, at least one of $\bar{c}_i^n(\omega)$ vanishes. For $\bar{\zeta}^n = \sum_{i=1}^N \bar{c}_i^n g_i$, we have that $\mathcal{L}\bar{\zeta}^n$ also tends to ξ. Considering the partition of Ω by N disjoint $\mathcal{G}$-measurable subsets Γ_i constructed from the covering of Ω by sets $\{\liminf_n \bar{c}_i^n = 0\}$ and replacing on Γ_i the coefficients $\bar{c}_i^n$ by zero (without affecting the limit ξ), we obtain a reduction to the case with $N-1$ generators. □

3.5.2 No Arbitrage Criteria: Finite Ω

The definition of the NA^w-property remains the same as in the model with full information: $\widehat{R}_T \cap L^0(\mathbf{R}^d_+, \mathcal{F}_T) = \{0\}$ or $\widehat{A}_T \cap L^0(\mathbf{R}^d_+, \mathcal{F}_T) = \{0\}$.

As always, criteria in the case of finite Ω are easy to establish using the finite-dimensional separation theorem.

Proposition 3.5.2 *Let Ω be finite. The following conditions are equivalent:*

(a) *NA^w;*
(b) *there exists $Z \in \mathcal{M}(\operatorname{int} \mathbf{R}^d_+, \mathbf{F})$ such that $E(Z_t \mathcal{L}_t \zeta | \mathcal{G}_t) \le 0$ for any $\zeta \in O_t$.*

Proof. (a) $\Rightarrow$ (b). Note that $\widehat{A}_T$ is a finite-dimensional polyhedral (thus, closed) cone containing $-L^0(\mathbf{R}^d_+)$. The NA^w-property implies that nonzero elements of $L^0(\mathbf{R}^d_+)$ can be separated from $\widehat{A}_T$ in a strict sense. Using a classical argument, we construct an $\mathbf{F}$-martingale $Z = (Z_t)$ with strictly positive components such that $EZ_T\xi \le 0$ for every $\xi \in \widehat{A}_T$. Namely, we can take Z_T equal to the sum of functionals negative on $\widehat{A}_T$ and strictly positive on $e_i I_\Gamma$ with the summation index Γ running through the family of atoms of $\mathcal{F}_T$ and $i = 1, 2, \dots, d$. It follows that $E(Z_t \mathcal{L}_t \zeta_t) \le 0$ for any $\zeta_t \in O_t$, implying the assertion.

(b) $\Rightarrow$ (a). This implication is obvious because for ζ admitting the representation (3.5.3), we have that

$$EZ_T\xi = \sum_{t=0}^{T} E\big[E(Z_t \mathcal{L}_t \zeta_t | \mathcal{G}_t)\big] \le 0,$$

and, therefore, ξ cannot be an element of $L^0(\mathbf{R}^d_+, \mathcal{F}_T)$ other than zero. □

As we know, even in the case of full information, a straightforward generalization of the above criterion to an arbitrary Ω fails to be true. To get "satisfactory" theorems, one needs either to impose extra assumptions or to modify the concept of absence of arbitrage. We investigate here an analog of the NA^r-condition starting from the simple case where Ω is finite.

First, we establish a simple lemma which holds in a "very abstract" setting where the word "premodel" instead of "model" means that we do not suggest any particular properties of $(\mathcal{L}_t)$.

Fix a subset $\mathcal{I}_t$ of O_t. The elements of $\mathcal{I}_t$ will be interpreted later, in a more specific "financial" framework, as the reversible orders.

We say that the premodel has the *NA^r-property* if the NA^w-property holds for the premodel based on an $\mathbf{F}$-adapted process $\mathcal{L}' = (\mathcal{L}'_t)$ such that

(i) $\mathcal{L}'_t\zeta \ge \mathcal{L}_t\zeta$ componentwise for every $\zeta \in O_t$;
(ii) $\mathbf{1}\mathcal{L}'_t\zeta \ne \mathbf{1}\mathcal{L}_t\zeta$ if $\zeta \in O_t \setminus \mathcal{I}_t$ (i.e., the above inequality is not identity).

Lemma 3.5.3 *Let Ω be finite. If a premodel has the NA^r-property, then there is a process $Z \in \mathcal{M}(\operatorname{int} \mathbf{R}^d_+, \mathbf{F})$ such that $E(Z_t\mathcal{L}_t\zeta|\mathcal{G}_t) \le 0$ for every $\zeta \in O_t$ and, if $\zeta \in O_t \setminus \mathcal{I}_t$,*

$$\zeta I_{\{E(Z_t\mathcal{L}_t\zeta|\mathcal{G}_t)=0\}} \in \mathcal{I}_t. \tag{3.5.4}$$

Proof. According to Proposition 3.5.2 applied to the premodel based on the process $\mathcal{L}'$ from the definition of NA^r, there exists $Z \in \mathcal{M}(\operatorname{int} \mathbf{R}^d_+, \mathbf{F})$ such that $E(Z_t\mathcal{L}'_t\zeta|\mathcal{G}_t) \le 0$ for any $\zeta \in O_t$. Hence, $E(Z_t\mathcal{L}_t\zeta|\mathcal{G}_t) \le 0$ by virtue of (i). Again by (i) we have, for $\zeta \in O_t \setminus \mathcal{I}_t$, that

$$Z_t\mathcal{L}'_t\zeta I_{\{E(Z_t\mathcal{L}_t\zeta|\mathcal{G}_t)=0\}} \ge Z_t\mathcal{L}_t\zeta I_{\{E(Z_t\mathcal{L}_t\zeta|\mathcal{G}_t)=0\}}.$$

If the order $\zeta I_{\{E(Z_t\mathcal{L}_t\zeta|\mathcal{G}_t)=0\}}$ is not in $\mathcal{I}_t$, this inequality is strict on a nonnull set. Thus, taking the expectation, we obtain

$$EZ_t\mathcal{L}'_t\zeta I_{\{E(Z_t\mathcal{L}_t\zeta|\mathcal{G}_t)=0\}} > 0,$$

which is contradiction. □

Now we give a precise meaning to the word "model" by imposing an *assumption* on the generating process (fulfilled in both our examples) and specifying the sets $\mathcal{I}_t$.

Namely, we suppose that in $\mathbf{M}^{d\times m}$ there is a basis formed by the union of two families of vectors $\{f_i\}$ and $\{\tilde{f}_i\}$, $1 \le i \le md/2$, belonging to $\mathbf{M}^{d\times m}_+$ and such that, componentwise,

$$\mathcal{L}_t f_i + \mathcal{L}_t\tilde{f}_i \le 0, \tag{3.5.5}$$

while $\mathcal{I}_t$ is the cone of (matrix-valued) random variables having the form $\sum_i(\eta_i f_i + \tilde{\eta}_i\tilde{f}_i)$ with $\eta_i, \tilde{\eta}_i \in L^0_+(\mathcal{G}_t)$ and such that $\mathcal{L}_t\sum_i(\eta_i + \tilde{\eta}_i)(f_i + \tilde{f}_i) = 0$.

Note that the latter equality implies that $\mathbf{L}_t(\mathcal{I}_t) \subseteq \mathbf{L}_t(O_t)\cap(-\mathbf{L}_t(O_t))$. It is clear that the set $\mathcal{I}_t$ is stable under multiplication by elements of $L^0(\mathbf{R}_+, \mathcal{G}_t)$. This implies that (3.5.4) for $\zeta \in \mathcal{I}_t$ always holds (cf. the formulations of Lemma 3.5.3 and the theorems below).

Inequality (3.5.5) means that the elementary transfers in opposite directions cannot lead to gains. The orders from $\mathcal{I}_t$, even symmetrized, do not incur losses.

For the models, in the definition of the NA^r the words "premodel" are replaced by "models," i.e., we require that property (3.5.5) should also hold for the dominating process $\mathcal{L}'$.

Theorem 3.5.4 *Let Ω be finite. Then the following properties of the model are equivalent:*

(a) *NA^r;*
(b) *there is $Z \in \mathcal{M}(\operatorname{int} \mathbf{R}^d_+, \mathbf{F})$ such that $E(Z_t\mathcal{L}_t\zeta|\mathcal{G}_t) \le 0$ for every $\zeta \in O_t$ and, if $\zeta \in O_t$,*

$$\zeta I_{\{E(Z_t\mathcal{L}_t\zeta|\mathcal{G}_t)=0\}} \in \mathcal{I}_t.$$

Proof. To check the remaining implication (b) ⇒ (a), we put $\mathcal{L}'_t\zeta := \mathcal{L}_t\zeta - \bar{\mathcal{L}}_t\zeta$ defining the action of $\bar{\mathcal{L}}_t$ on the element $\zeta = \sum_i(\eta_i f_i + \tilde{\eta}_i\tilde{f}_i)$ by the formula $\bar{\mathcal{L}}_t\zeta := \sum_i(\eta_i + \tilde{\eta}_i)\theta_i$, where $\theta_i = \theta_i(t)$ has the components

$$\theta_i^k := \max\left\{\frac{1}{2}\left[\mathcal{L}_t(f_i + \tilde{f}_i)\right]^k, \frac{1}{d}\frac{E(Z_t\mathcal{L}_t f_i|\mathcal{G}_t)}{E(Z_t^k|\mathcal{G}_t)}, \frac{1}{d}\frac{E(Z_t\mathcal{L}_t\tilde{f}_i|\mathcal{G}_t)}{E(Z_t^k|\mathcal{G}_t)}\right\}.$$

The values $\theta_i^k(t)$ being negative, condition (i) holds. Inequality (3.5.5) for $\mathcal{L}'_t$ is obviously fulfilled due to the first term in the definition of $\theta_i^k(t)$. Now let ζ be an element of $O_t \setminus \mathcal{I}_t$. This means that, for some k and i, the set

$$\Gamma := \left\{(\eta_i + \tilde{\eta}_i)\left[\mathcal{L}_t(f_i + \tilde{f}_i)\right]^k < 0\right\} = \left\{(\eta_i + \tilde{\eta}_i)Z_t^k\left[\mathcal{L}_t(f_i + \tilde{f}_i)\right]^k < 0\right\}$$

is nonnull. From elementary properties of conditional expectations it follows that $(\eta_i + \tilde{\eta}_i)E(Z_t^k[\mathcal{L}_t(f_i + \tilde{f}_i)]^k|\mathcal{G}_t) < 0$ on Γ. Property (ii) holds because on Γ both $E(Z_t\mathcal{L}_t f_i|\mathcal{G}_t)$ and $E(Z_t\mathcal{L}_t\tilde{f}_i|\mathcal{G}_t)$ are strictly negative, as follows from the coincidence of sets

$$\left\{E(Z_t\mathcal{L}_t f_i|\mathcal{G}_t) < 0\right\} = \left\{E(Z_t\mathcal{L}_t\tilde{f}_i|\mathcal{G}_t) < 0\right\} = \left\{E\big(Z_t\mathcal{L}_t(f_i + \tilde{f}_i)|\mathcal{G}_t\big) < 0\right\},$$

which can be easily established. Indeed, $f_i I_{\{E(Z_t\mathcal{L}_t f_i|\mathcal{G}_t)=0\}} \in \mathcal{I}_t$ and, by the definition of $\mathcal{I}_t$,

$$I_{\{E(Z_t\mathcal{L}_t f_i|\mathcal{G}_t)=0\}}\mathcal{L}_t\tilde{f}_i = -I_{\{E(Z_t\mathcal{L}_t f_i|\mathcal{G}_t)=0\}}\mathcal{L}_t f_i.$$

Multiplying this identity by Z_t and taking the conditional expectation with respect to $\mathcal{G}_t$, we get that

$$I_{\{E(Z_t\mathcal{L}_t f_i|\mathcal{G}_t)=0\}}E(Z_t\mathcal{L}_t\tilde{f}_i|\mathcal{G}_t) = 0.$$

Similarly,

$$I_{\{E(Z_t\mathcal{L}_t\tilde{f}_i|\mathcal{G}_t)=0\}}E(Z_t\mathcal{L}_t f_i|\mathcal{G}_t) = 0.$$

These two equalities imply the coincidence of sets where the conditional expectations (always negative) are zero, i.e., the required assertion.

Finally, we check the NA^w-property of $(\mathcal{L}'_t)$ using Proposition 3.5.2. For any $\zeta = \sum_i(\eta_i f_i + \tilde{\eta}_i\tilde{f}_i)$ from O_t, we have

$$\begin{aligned} E(Z_t\mathcal{L}'_t\zeta|\mathcal{G}_t) &= E(Z_t\mathcal{L}_t\zeta|\mathcal{G}_t) - E\left(\sum_i(\eta_i + \tilde{\eta}_i)\sum_{k=1}^d Z_t^k\theta_i^k\middle|\mathcal{G}_t\right) \\ &\le E(Z_t\mathcal{L}_t\zeta|\mathcal{G}_t) - \sum_i \eta_i E(Z_t\mathcal{L}_t f_i|\mathcal{G}_t) - \sum_i \tilde{\eta}_i E(Z_t\mathcal{L}_t\tilde{f}_i|\mathcal{G}_t) = 0. \end{aligned}$$

It follows that $EZ_T\xi \le 0$ for every $\xi \in \widehat{R}_T(\mathcal{L}') \cap L^0(\mathbf{R}_+^d)$, excluding arbitrage opportunities for the model based on $\mathcal{L}'$.

The theorem is proven. □

Remark 1. One might find it convenient to view $\mathbf{M}^{d\times m}$ as the set of linear operators defined by corresponding matrices and consider the adjoint operators $\mathcal{L}^*_{\omega,t} : \mathbf{R}^d \to (\mathbf{M}^{d\times m})^*$. This gives a certain flexibility of notation, e.g., the property "$E(Z_t\mathcal{L}_t\zeta|\mathcal{G}_t) \le 0$ for every $\zeta \in O_t$" can be formulated as "the operator $E(\mathcal{L}^*_t Z_t|\mathcal{G}_t)$ is negative" (in the sense of partial ordering induced by $\mathbf{M}^{d\times m}_+$), the inclusion $f_i \in \operatorname{Ker} E(\mathcal{L}^*_t Z_t|\mathcal{G}_t)$ can be written instead of the equality $E(Z_t\mathcal{L}_t f_i|\mathcal{G}_t) = 0$, and so on. However, the current notation has the advantage of being easier adjustable for the more general situation where $\mathcal{L}_t$ is a concave positive homogeneous mapping from $\mathbf{M}^{d\times m}_+$ into $L^0(\mathbf{R}^d, \mathcal{F}_t)$.

Remark 2. The hypothesis on the structure of invertible claims may not be fulfilled for Examples 1 and 2. For the investor having access to full information, the set of all assets can be split into classes of equivalence within which one can do frictionless transfers though not necessary in one step. Our assumption means that all transfers within each class are frictionless, a hypothesis which does not lead to a loss of generality as a fully informed "intelligent" investor will not lose money making charged transfers within an equivalence class. However, in the context of restricted information, it seems that such an assumption means that the information on equivalence classes is available to the investor.

3.5.3 No Arbitrage Criteria: Arbitrary Ω

In the general case the assertion of Proposition 3.5.2 fails to be true, though with a suitable modification, its condition (b) remains sufficient for the NA^w-property. Namely, we have:

Proposition 3.5.5 *The NA^w-property holds if there exists $Z \in \mathcal{M}(\operatorname{int}\mathbf{R}^d_+, \mathbf{F})$ such that all conditional expectations $E(|Z_t||\mathcal{L}_t f_i||\mathcal{G}_t)$ and $E(|Z_t||\mathcal{L}_t \tilde{f}_i||\mathcal{G}_t)$ are finite and $E(Z_t\mathcal{L}_t\zeta|\mathcal{G}_t) \le 0$ for any $\zeta \in O_t$.*

This result is an obvious corollary of the following technical lemma dealing with integration issues.

Lemma 3.5.6 *Suppose that $\Sigma_T = Z_T\sum_{t=0}^T \xi_t$, where $Z \in \mathcal{M}(\mathbf{R}^d_+, \mathbf{F})$ and $\xi_t \in L^0(\mathbf{R}^d, \mathcal{F}_t)$ are such that $E(|Z_t||\xi_t||\mathcal{G}_t) < \infty$ and $E(Z_t\xi_t|\mathcal{G}_t) \le 0$. Put $\bar{\Sigma}_T := E(\Sigma_T|\mathcal{G}_T)$. If $\bar{\Sigma}^-_T \in L^1$, then $\bar{\Sigma}_T \in L^1$ and $E\bar{\Sigma}_T \le 0$.*

Proof. We proceed by induction. The claim is obvious for $T = 0$. Suppose that it holds for $T-1$. Clearly,

$$Z_T\sum_{t=0}^{T-1} \xi_t = \Sigma_T - Z_T\xi_T.$$

By the martingale property $E(Z^i_T|\xi_t||\mathcal{G}_t) = E(Z^i_t|\xi_t||\mathcal{G}_t) < \infty$, implying that $E(|Z_T||\xi_t||\mathcal{G}_t) < \infty$ for any $t \le T$. Thus, $\bar{\Sigma}_T$ is well defined and finite. Taking the conditional expectation with respect to $\mathcal{G}_T$ in the above identity, we get, using the martingale property, that

$$E(\Sigma_{T-1}|\mathcal{G}_T) = E\left(Z_T \sum_{t=0}^{T-1} \xi_t \Big| \mathcal{G}_T\right) = \bar{\Sigma}_T - E(Z_T\xi_T|\mathcal{G}_T) \geq \bar{\Sigma}_T.$$

Therefore, the negative part of $E(\Sigma_{T-1}|\mathcal{G}_T)$ is dominated by the negative part of $\bar{\Sigma}_T$, which is integrable. Using Jensen's inequality, we have

$$\begin{aligned}\bar{\Sigma}_{T-1}^- &= \left[E\big(E(\Sigma_{T-1}|\mathcal{G}_T)|\mathcal{G}_{T-1}\big)\right]^- \\ &\leq E\big(\left[E(\Sigma_{T-1}|\mathcal{G}_T)\right]^-|\mathcal{G}_{T-1}\big) \leq E\big(\bar{\Sigma}_T^-|\mathcal{G}_{T-1}\big).\end{aligned}$$

Thus, $\bar{\Sigma}_{T-1}^- \in L^1$ and, by virtue of the induction hypothesis, $\bar{\Sigma}_{T-1} \in L^1$ and $E\bar{\Sigma}_{T-1} \leq 0$. In the representation $\bar{\Sigma}_T = E(\bar{\Sigma}_{T-1}|\mathcal{G}_T) + E(\bar{\Sigma}_T^-|\mathcal{G}_{T-1})$, the first term is integrable and has negative expectation, while the second is negative. Thus, $E\bar{\Sigma}_T \leq 0$ and, automatically, $E\bar{\Sigma}_T^+ < \infty$. □

The NA^r-criterion, suitably modified, remains true without any restriction on the probability space. Of course, in its formulation one needs to take care about the existence of the involved conditional expectations. This can be done as in the next result.

Theorem 3.5.7 *The following conditions are equivalent:*

(a) *NA^r;*
(b) *there is $Z \in \mathcal{M}(\operatorname{int}\mathbf{R}^d_+, \mathbf{F})$ such that all random variables $E(Z_t\mathcal{L}_t f_i|\mathcal{G}_t)$ and $E(Z_t\mathcal{L}_t\tilde{f}_i|\mathcal{G}_t)$ are finite, $E(Z_t\mathcal{L}_t\zeta|\mathcal{G}_t) \leq 0$ for every $\zeta \in O_t$, and, if $\zeta \in O_t$,*

$$\zeta I_{\{E(Z_t\mathcal{L}_t\zeta|\mathcal{G}_t)=0\}} \in \mathcal{I}_t. \tag{3.5.6}$$

We have no trouble with the implication (b) ⇒ (a): an inspection of the arguments given in the case of finite Ω shows that they work well until the concluding step, which now can be done just by reference to Lemma 3.5.6.

The proof of the "difficult" implication (a) ⇒ (b) follows the same line of ideas as in the case of full information.

Lemma 3.5.8 *Suppose that the equality*

$$\sum_{t=0}^{T} \mathcal{L}_t\tilde{\zeta}_t - \tilde{r} = 0 \tag{3.5.7}$$

with $\tilde{\zeta}_t \in O_t$ and $\tilde{r}_t \in L^0(\mathbf{R}^d_+)$ holds only if $\tilde{\zeta}_t \in \mathcal{I}_t$ and $\tilde{r} = 0$. Then $\widehat{A}_T$ is closed in probability.

Proof. For $T = 0$, the arguments are exactly the same as those used for Proposition 3.5.1 with obvious changes caused by the extra term describing the funds withdrawals. Namely, the difference is that for the limiting normalized order $\tilde{\zeta} := \sum_{i=1}^{N} \tilde{c}_i g_i$, we get the equality $\mathcal{L}\tilde{\zeta} - \tilde{r} = 0$ where $\tilde{r} \in L^0(\mathbf{R}^d_+, \mathcal{F}_T)$ is the limit of normalized funds withdrawals. By hypothesis, $\tilde{r} = 0$, and we can complete the proof using the same Gauss-type reduction procedure.

Arguing by induction, we suppose that $\widehat{A}_{T-1}$ is closed and consider the sequence of order processes $(\zeta_t^n)_{t \le T}$ such that $\sum_{t=0}^{T} \mathcal{L}_t \zeta_t^n - r^n \to \eta$. There is an obvious reduction to the case where at least one of "elementary" orders at time zero tends to infinity. Normalizing and using the induction hypothesis, we obtain that there exists an order process $(\tilde{\zeta}_t)_{t \le T}$ with nontrivial $\tilde{\zeta}_0$ such that $\sum_{t=0}^{T} \mathcal{L}_t \tilde{\zeta}_t - \tilde{r} = 0$, and we can use the assumption of the lemma. It ensures that $\tilde{r} = 0$ and there are $\zeta'_t \in O_t$ such that $\mathcal{L}_t \zeta'_t = -\mathcal{L}_t \tilde{\zeta}_t$. This allows us to reduce a number of nonzero coefficients (i.e., "elementary" orders) at the initial order by putting $\bar{\zeta}_0^n = \zeta_0^n - \alpha^n \tilde{\zeta}_0$, as in the proof of Proposition 3.5.1, and $\bar{\zeta}_t^n = \zeta_t^n + \alpha^n \zeta'_t$ for $t \ge 1$. □

Lemma 3.5.9 *The NA^r-condition implies the hypothesis of the above lemma.*

Proof. Of course, $\tilde{r} = 0$ (otherwise, $(\tilde{\zeta}_t)$ is an arbitrage opportunity, i.e., even NA^w is violated). For the process $(\mathcal{L}'_t)$, from the definition of NA^r we have that componentwise

$$\sum_{t=0}^{T} \mathcal{L}'_t \tilde{\zeta}_t \ge \sum_{t=0}^{T} \mathcal{L}_t \tilde{\zeta}_t = 0$$

and $\mathbf{1} \sum_{t=0}^{T} \mathcal{L}'_t \tilde{\zeta}_t > 0$ with strictly positive probability if at least one of $\tilde{\zeta}_t$ does not belong to $\mathcal{I}_t$. This means that $(\tilde{\zeta}_t)$ is an arbitrage opportunity for the model based on $(\mathcal{L}'_t)$. □

Lemma 3.5.10 *Assume that the hypothesis of Lemma 3.5.8 holds. Then for any "elementary" order f and every $t \le T$, one can find a bounded process $Z = Z^{(t,f)} \in \mathcal{M}(\operatorname{int} \mathbf{R}^d_+, \mathbf{F})$ such that:*

(1) *$E(|Z_s||\mathcal{L}_s g|) < \infty$ and $E(Z_s \mathcal{L}_s g | \mathcal{G}_s) \le 0$ for all $s \le T$ and all "elementary" orders g,*
(2) *$f I_{\{E(Z_t \mathcal{L}_t f | \mathcal{G}_t) = 0\}} \in \mathcal{I}_t$.*

Proof. We may assume without loss of generality that all portfolio increments $\mathcal{L}_s g$ corresponding to the elementary orders g are integrable (otherwise we can pass to an equivalent measure P' with bounded density ρ, find a process Z' with the needed properties under P', and take $Z = \rho \tilde{Z}'$).

Let $\mathcal{Z}$ be the set of all bounded processes $Z \in \mathcal{M}(\mathbf{R}^d_+, \mathbf{F})$ such that $EZ_T \xi \le 0$ for all $\xi \in \widehat{A}^1_T := \widehat{A}_T \cap L^1$. Let

$$c_t := \sup_{Z \in \mathcal{Z}} P\big(E(Z_t \mathcal{L}_t f | \mathcal{G}_t) < 0\big). \tag{3.5.8}$$

Let Z be an element for which the supremum is attained (one can take as Z a countable convex combination of any uniformly bounded sequence along which the supremum is attained).

If (2) fails, then the random vector $\mathcal{L}_t(f+\tilde{f})I_{\{E(Z_t\mathcal{L}_tf|\mathcal{G}_t)=0\}}$ (all components of which are negative) is not zero. This implies that the element $-\mathcal{L}_t\tilde{f}I_{\{E(Z_t\mathcal{L}_tf|\mathcal{G}_t)=0\}}$ does not belong to $\widehat{A}^1_T$. Indeed, in the opposite case we would have the identity

$$\sum_{s=0}^{T}\mathcal{L}_s\zeta_s=-\mathcal{L}_t\tilde{f}I_{\{E(Z_t\mathcal{L}_tf|\mathcal{G}_t)=0\}}.$$

The assumption of Lemma 3.5.8 ensures that the order $\tilde{f}I_{\{E(Z_t\mathcal{L}_tf|\mathcal{G}_t)=0\}}+\zeta_t$ is in $\mathcal{I}_t$. Thus, for the symmetrized order, we have that

$$\mathcal{L}_t(f+\tilde{f})I_{\{E(Z_t\mathcal{L}_tf|\mathcal{G}_t)=0\}}+\mathcal{L}_t(\zeta+\tilde{\zeta})=0.$$

Since the second term is also componentwise negative, both should be equal to zero, and we get a contradiction.

By the Hahn–Banach theorem one can separate $\varphi:=-\mathcal{L}_t\tilde{f}I_{\{E(Z_t\mathcal{L}_tf|\mathcal{G}_t)=0\}}$ and $\widehat{A}^1_T$: that is, we may find $\eta\in L^\infty(\mathbf{R}^d)$ such that

$$\sup_{\xi\in\widehat{A}^1_T}E\eta\xi<E\eta\varphi.$$

Since $\widehat{A}^1_T$ is a cone containing $-L^1(\mathbf{R}^d_+)$, the supremum above is equal to zero, $\eta\in L^1(\mathbf{R}^d_+)$, and $E\eta\varphi>0$. The latter inequality implies that, for $Z^\eta_t=E(\eta|\mathcal{G}_t)$, we have $EE(Z^\eta_t\mathcal{L}_tf|\mathcal{G}_t)I_{\{E(Z_t\mathcal{L}_tf|\mathcal{G}_t)=0\}}<0$. Therefore, for the martingale $Z':=Z+Z^\eta$, we have that

$$P\big(E(Z'_t\mathcal{L}_tf|\mathcal{G}_t)<0\big)>P\big(E(Z_t\mathcal{L}_tf|\mathcal{G}_t)<0\big)=c_t.$$

This contradiction shows that (2) holds.

The process Z constructed in this way may be not in $\mathcal{M}(\operatorname{int}\mathbf{R}^d_+,\mathbf{F})$. However, it can be easily "improved" to meet the latter property. To this end, fix $i\le d$ and consider, in the subset of $\mathcal{Z}$ on which the supremum c_t in (3.5.8) is attained, a process Z with maximal probability $P(Z^i_T>0)$ (such a process does exist). Then $P(\bar{Z}^i_T>0)=1$. Indeed, in the opposite case, the element $e_iI_{\{Z^i_T=0\}}\in L^1(\mathbf{R}^d_+)$ is not zero and, therefore, does not belong to $\widehat{A}^1_T$. So it can be separated from the latter set. The separating functional generates a martingale $Z'\in\mathcal{Z}$. Since $P(\bar{Z}_T+Z'_T>0)>P(\bar{Z}_T>0)$, we arrive at a contradiction with the definition of $\bar{Z}$. The set of $Z\in\mathcal{Z}$ satisfying (1) and (2) is convex, and, hence, a convex combination of d processes obtained in this way for each coordinate has the required properties. □

The implication (a) ⇒ (b) of the theorem follows from the lemmas above. Indeed, by virtue of Lemmas 3.5.9–3.5.10, NA^r ensures the existence of processes $Z^{(t,f)}$ satisfying (1) and (2) of Lemma 3.5.10. One can take as a required martingale Z the process $Z := \sum_{t,f} Z^{(t,f)}$, where $t = 0, 1, \dots, T$, and f runs through the set of "elementary" orders. An arbitrary order $\zeta \in O_t$ is a linear combination of elementary orders with positive $\mathcal{G}_t$-measurable coefficients. The condition $E(Z_t \mathcal{L}_t \zeta | \mathcal{G}_t) \le 0$ follows from property (1) of Lemma 3.5.10. To prove the inclusion (3.5.6), we note that $I_{\{\Sigma \xi_i = 0\}} = \prod I_{\{\xi_i = 0\}}$ when $\xi_i \le 0$. With this observation, the required inclusion is an easy corollary of property (2) of Lemma 3.5.10 and the stability of $\mathcal{I}_t$ under multiplication by positive $\mathcal{G}_t$-measurable random variables.

Remark. In the above proof we get from NA^r a condition which looks stronger than (b), with bounded Z and integrable random variables $|Z_t||\mathcal{L}_t f|$, but, in fact, it is equivalent to (b).

3.5.4 Hedging Theorem

Thanks to the previous development, hedging theorems in the model with partial information do not require new ideas. For the case of finite Ω the result can be formulated in our "very abstract" setting without additional assumptions on the structure of the sets $\mathcal{I}_t$.

We fix a d-dimensional random variable $\widehat{C}$, a contingent claim expressed in physical units. Define the set

$$\Gamma = \{v \in \mathbf{R}^d : \widehat{C} \in v + \widehat{A}_T\}.$$

Let $\mathcal{Z}$ be the set of martingales $Z \in \mathcal{M}_T(\mathbf{R}^d_+, \mathbf{F})$ such that $E(Z_t \mathcal{L}_t \zeta_t | \mathcal{G}_t) \le 0$ for every $\zeta_t \in O_t$. Put

$$D := \Big\{ v \in \mathbf{R}^d : \sup_{Z \in \mathcal{Z}} E(Z_T \widehat{C} - Z_0 v) \le 0 \Big\}.$$

Proposition 3.5.11 *Let Ω be finite, and let $\mathcal{Z} \ne \emptyset$. Then $\Gamma = D$.*

In this theorem the inclusion $\Gamma \subseteq D$ is obvious, while the reverse inclusion is an easy exercise on the finite-dimensional separation theorem. We leave it to the reader.

In the case of general Ω we should take care about the integrability and closedness of the set $\widehat{A}_T$. To this end we shall work with the model in the "narrow" sense of the preceding sections assuming the NA^r-property. Now $\mathcal{Z}$ is the set of bounded martingales $Z \in \mathcal{M}_T(\mathbf{R}^d_+, \mathbf{F})$ such that $E(Z_t \mathcal{L}_t f_i | \mathcal{G}_t)$ and $E(Z_t \mathcal{L}_t \tilde{f}_i | \mathcal{G}_t)$ are finite, $E(Z_t \mathcal{L}_t \zeta | \mathcal{G}_t) \le 0$, and $E(Z_T \widehat{C})^- < \infty$. The definitions of the sets Γ and D remain the same.

Theorem 3.5.12 *Suppose that NA^r holds. Then $\Gamma = D$.*

Proof. The inclusion $\Gamma \subseteq D$ follows from the inequality

$$Z_T(\widehat{C} - v) \le Z_T \sum_{t=0}^{T} \mathcal{L}_t \zeta_t, \quad \zeta_t \in O_t,$$

and Lemma 3.5.6.

To check the inclusion $D \subseteq \Gamma$, we take a point $v \notin \Gamma$ and show that $v \notin D$. It is sufficient to find $Z \in \mathcal{Z}$ such that $Z_0 v < EZ_T\widehat{C}$. Consider a measure $\tilde{P} \sim P$ with bounded density ρ such that $\widehat{C}$ and all $|\mathcal{L}_t||f_i|$ and $|\mathcal{L}_t||f_i|$ belong to $L^1(\tilde{P})$. Under NA^r, the convex set $\tilde{A}^1 := A_0^T \cap L^1(\tilde{P})$ is closed and does not contain the point $\widehat{C} - v$. Thus, we can separate the latter by a functional η from L^∞. This means that

$$\sup_{\xi \in \tilde{A}^1} E\rho\eta\xi < E\eta\rho(\widehat{C} - v).$$

It is clear that the bounded martingale $Z_t := E(\rho\eta|\mathcal{F}_t)$ satisfies the required properties. □

3.6 Hedging Theorems: Continuous Time

3.6.1 Introductory Comments

In this section we develop two continuous-time models, both with efficient market friction, and suggest versions of hedging theorems for European and American options.

The first, "X-model" is based on the right-continuous adapted value processes of bounded variation. Our hedging theorem for European options requires the continuity of the price processes and transaction cost coefficients and involves also some technical hypotheses. Fortunately, all of them are fulfilled in the reference case of Brownian filtration and constant transaction coefficients.

It would be natural to proceed after exposing a corresponding theory for frictionless market. However, the latter, being based on the extensive use of advanced stochastic calculus, is much more demanding mathematically. Rather surprisingly, in the considered setting we need only very elementary facts from the general theory of stochastic processes. The explanation of this difference is simple. In contrast with the classical theory, in our transaction cost model one can consider only the strategies which are processes of bounded variation, while already in the Black–Scholes model one needs to consider a more general class, and the replicating strategy is not of bounded variation. Having in mind this specific feature of models with transaction costs, we believe that a preliminary knowledge of continuous-time results, though desirable, is not necessary.

Using many ideas which appeared already in the study of discrete-time models, we shall try to circumvent technicalities related with the structure of continuous-time set-valued processes. That is why we postulate simply in our abstract formulation that the dual $\mathcal{C}$-valued process is generated by a countable family of continuous vector-valued processes. The main difficulty of the theory originates from the necessity to exclude doubling strategies.

We consider, as admissible, strategies generating the value processes which, being expressed in physical units of assets, are bounded from below in the sense of partial ordering. Our assumptions allow us to verify the closedness property of the set of hedgeable claims, which leads, as one might expect, to a super-replication result.

The continuity assumption on price processes plays an essential role in our proof. It happens that it is a principal condition which cannot be relaxed in the considered framework: the Rásonyi counterexample shows that a straightforward generalization of the hedging theorem fails to be true if the price process has jumps.

The situation is better with the "Y-model" suggested by Campi and Schachermayer, where the value processes are predictable, of bounded variations, and have right and left limits. Of course, the trajectories of such processes are not distribution functions of (vector-valued) measures, and working with them requires more efforts. Fortunately, this class is more flexible, consistent with a financial intuition, and the dual description of the set of hedging endowments for a European option is of the same type as in the case of the X-model.

We complete the section by a hedging theorem for American options in the framework of the Y-model and deduce from it a corresponding result for the X-model assuming, of course, the continuity hypothesis for the latter.

3.6.2 Model Specification

All processes are given on a fixed stochastic basis $(\Omega, \mathcal{F}, (\mathcal{F}_t)_{t\le T}, P)$ (i.e., a filtered probability space satisfying the "usual conditions" of stochastic calculus); a finite time horizon T is fixed; the initial σ-algebra is *trivial* (in the sense: generated by the null sets), and $\mathcal{F} = \mathcal{F}_T$.

A continuous-time version of our basic model can be described as follows. The price process $S = (S^1, \ldots, S^d)$ is a *continuous* semimartingale with strictly positive components and $S_0 = \mathbf{1}$. The adapted matrix-valued process $\Lambda = (\lambda^{ij})$ representing the transaction cost coefficients also is assumed to be continuous, with $\lambda^{ij} \ge 0$ and $\lambda^{ii} = 0$. The definitions of the solvency cone K and the cone M remain the same as in the discrete-time model. We shall assume that at each instant at least one $\lambda^{ij} \ne 0$ and, therefore, $K_t = M_t$. Again, the agent's positions at time t can be described either by the vector of their values V_t or by the vector of "physical" quantities $\widehat{V}_t$, where $\widehat{V}_t^i := V_t^i / S_t^i$.

To describe the portfolio dynamics we define the class of strategies $\mathcal{B}$ as the set of all *right-continuous* d-dimensional adapted processes B of bounded variation such that, for any stopping time $\tau \le T$,

$$\dot{B}_\tau \in -K_\tau. \tag{3.6.1}$$

The symbol $\dot{B}$ denotes the Radon–Nikodym derivative of B with respect to the (scalar) total variation $\|B\|$ (we prefer this notation to more common $\operatorname{Var} B$).

More precisely, $\dot{B} = (\dot{B}_t)$ is the optional process the trajectories $\dot{B}_t(\omega)$ of which, for almost all ω, are the Radon–Nikodym derivatives of $B_.(\omega)$ with respect to $\|B\|_.(\omega)$. The reader can find a proof of the existence of such a process in any treatise on stochastic calculus, see, e.g., Proposition I.3.13 in [102]. By convention, $B_{0-} := 0$, but B_0 is not necessarily equal to zero (this allows us to interpret the value ΔB_0 as the initial transfer). It should be noted that the definitions of $\|B\|$ may differ in dependence of the chosen norm in $\mathbf{R}^d$; for the absolute norm, the quantity $\|B\|_0$ is just the sum of $|\Delta B_0^i|$. Since all norms in a finite-dimensional space are equivalent, property (3.6.1) is invariant with respect to the choice of a particular one.

For $v \in \mathbf{R}^d$ and $B \in \mathcal{B}$, we define the process $V = V^{v,B}$, the value of a self-financing portfolio with the strategy B and initial endowment v, by

$$V^i = v^i + \widehat{V}^i_- \cdot S^i + B^i, \tag{3.6.2}$$

where we use the abbreviated notation for stochastic integrals, i.e.,

$$\widehat{V}^i_- \cdot S^i_t := \int_{[0,t]} \widehat{V}^i_{u-}\, dS^i_u.$$

By financial consideration it is obvious that the dynamics in the "physical domain" is much simpler:

$$\widehat{V}^i_t = v^i + \frac{1}{S} \cdot B^i = v^i + \int_{[0,t]} \frac{dB^i_u}{S^i_u} \tag{3.6.3}$$

(the changes in number of units of assets does not depend on the price evolution). This relation implies the representation

$$V^i_t = S^i_t v^i + S^i_t \int_{[0,t]} \frac{dB^i_u}{S^i_u},$$

which is nothing but the (stochastic) Cauchy formula for the solution of linear equation. Indeed, putting $\tilde{S}^i := (1/S^i_-) \cdot S^i$, we may rewrite the V-dynamics as the linear equation

$$V^i = v^i + V^i_- \cdot \tilde{S}^i + B^i, \tag{3.6.4}$$

solve it, and obtain that

$$V^i_t = \mathcal{E}_t(\tilde{S}^i) v^i + \mathcal{E}_t(\tilde{S}^i) \int_{[0,t]} \frac{dB^i_s}{\mathcal{E}_s(\tilde{S}^i)}, \tag{3.6.5}$$

where the process $\mathcal{E}(\tilde{S}^i)$ is the Doléans exponential coinciding with S^i.

Remark. This model can be described using the family of increasing adapted processes L^{ij}, $1 \le i, j \le d$, $i \ne j$, representing the accumulated net values transferred from the position j to the position i. Of course, an L-process generates the B process, and the converse to this assertion is an exercise on measurable selection arguments.

As in the discrete-time case, it is easier to analyze the hedging problem in the "physical units domain." The definition of the cones $\widehat{K}_t$ remains the same. The portfolio processes $\widehat{V}$ are of bounded variation, right-continuous, and such that $d\widehat{V}/d\|\widehat{V}\| \in -\widehat{K}$ (up to a null set).

The reader acquainted with continuous-time market models without friction knows well that one should make precautions to avoid doubling strategies. In the classical theory, a standard constraint imposed on the strategies is the so-called *admissibility*, requiring the boundedness of portfolio processes from below. That is why it seems natural to use the same idea and consider as admissible the strategies with V-processes bounded from below, in the sense of partial orderings induced by solvency cones. Unfortunately, this does not work. To get reasonable theorems we need to permit strategies with $\widehat{V}$-processes bounded from below. The latter class includes strategies allowing to keep fixed negative quantities of certain assets (i.e., "buy-and-hold" with eventual short positions). We give the precise definition in the following abstract framework.

3.6.3 Hedging Theorem in Abstract Setting

We are given a *$\mathcal{C}$-valued process* $G = (G_t)_{t\in[0,T]}$ dominating $\mathbf{R}^d_+$ and defined by a countable sequence of adapted d-dimensional processes $\xi^k = (\xi^k_t)$ such that, for every t and every ω, only a finite but nonzero number of $\xi^k_t(\omega)$ are different from zero and $G_t(\omega) := \operatorname{cone}\{\xi^k_t(\omega),\ k \in \mathbf{N}\}$, i.e., $G_t(\omega)$ is a polyhedral cone generated by the finite set $\{\xi^k_t(\omega),\ k \in \mathbf{N}\}$.

The class of $\mathcal{C}$-valued processes is stable under intersections, sums, and duality (the operations are understood in the sense of set-valued mappings). It is also stable under linear transforms defined by adapted matrix-valued processes.

Efficient friction hypothesis. Throughout this section we shall assume that all cones G_t are proper, i.e., $G_t \cap (-G_t) = \{0\}$ or, equivalently, $\operatorname{int} G^*_t \neq \emptyset$.

We suppose that the generators of G are continuous processes and add to this the following assumption on the continuity of the generators of G^*:

G. There is a countable family of continuous adapted processes (ζ^k) such that, for each ω, only a finite number of vectors ζ^k are different from zero and $G^*_t = \operatorname{cone}\{\zeta^k_t :\ k \in \mathbf{N}\}$ for every t.

Clearly, this hypothesis is fulfilled for financial models with constant coefficients (i.e., with constant K and K^*) and continuous price process S.

Remark. It is important to note that the continuity of generators does not imply the continuity of the cone-valued processes. The following simple example gives the idea: $G_t = \operatorname{cone}\{\xi^1_t, \xi^2_t\}$, where $\xi^1_t = e_1$, $\xi^1_t = (t-1)^+ e_2$.

We denote by $\mathcal{D} = \mathcal{D}(G)$ the subset of $\mathcal{M}_0^T(\operatorname{int} G^*)$ formed by martingales Z such that not only $Z_t \in L^0(\operatorname{int} G_t^*, \mathcal{F}_t)$ but also $Z_{t-} \in L^0(\operatorname{int} G_t^*, \mathcal{F}_t)$ for all $t \in [0,T]$.

Let $\mathcal{X}^0$ be the set of all càdlàg processes X of bounded variation with $X_0 = 0$ and such that $\dot{X}_\tau \in L^0(-G_\tau, \mathcal{F}_\tau)$ for all $\tau \le T$, and let $\mathcal{X}^x = x + \mathcal{X}^0$, $x \in \mathbf{R}^d$. We denote by $\mathcal{X}_b^x$ the subset of $\mathcal{X}^x$ formed by the processes X such that $X_t + \kappa_X \mathbf{1} \in L^0(G_t, \mathcal{F}_t)$, where $\kappa_X \in \mathbf{R}$. One can say that such a process X is bounded from below by a constant vector-valued process $-\kappa_X \mathbf{1}$ in the sense of the partial orderings $\ge_G$ defined in the obvious way. At last, we put $\mathcal{X}_b^x(T) := \{X_T : X \in \mathcal{X}_b^x\}$.

The following lemma gives a useful alternative description of the class $\mathcal{X}^0$.

Lemma 3.6.1 *An adapted right-continuous process X of bounded variation belongs to $\mathcal{X}^0$ if and only if the processes $\zeta^k \cdot X$, $k \in \mathbf{N}$, are decreasing.*

Proof. If $\zeta^k \cdot X$ are decreasing, then, outside a P-null set, $\zeta_t^k(\omega)\dot{X}_t(\omega) \le 0$ for all k and $t \in [0,T]$ except a set of dates having $d\|X\|(\omega)$-measure zero. Replacing $\dot{X}_t$ by zero on the set where at least one of the inequalities fails to be true, we obtain an optional version of the Radon–Nikodym derivative which evolves in G. The "only if part" is obvious. □

Lemma 3.6.2 *If $Z \in \mathcal{M}_0^T(G^*)$ and $X \in \mathcal{X}_b^x$, then ZX is a supermartingale, and*

$$E(-Z\dot{X}) \cdot \|X\|_T \le Z_0 x - EZ_T X_T. \tag{3.6.6}$$

Proof. By the product formula we have that

$$ZX = Z_0 X_0 + X_- \cdot Z + Z\dot{X} \cdot \|X\|. \tag{3.6.7}$$

The second term in the right-hand side is a local martingale, the third is a negative decreasing process because $Z\dot{X} \le 0$. Thus,

$$Z_0 x + X_- \cdot Z \ge Z_0 X_0 + X_- \cdot Z + Z \ge ZX \ge -\kappa_X Z\mathbf{1}$$

due to the definition of the set $\mathcal{X}_b^x$. The (scalar) local martingale $X_- \cdot Z$, bounded from below by a martingale, is a supermartingale. The terminal value of the decreasing process in (3.6.7), namely, $Z\dot{X} \cdot \|X\|_T \le 0$, being bounded from below by an integrable random variable, is also integrable. Hence, ZX is a supermartingale. The bound (3.6.6) follows from (3.6.7) and the supermartingale property of $X_- \cdot Z$. □

Let $\xi \in L^0(\mathbf{R}^d, \mathcal{F}_T)$ be such that $\xi + \kappa\mathbf{1} \in L^0(G_T, \mathcal{F}_T)$, where κ is a constant.

Define the convex set $\Gamma := \Gamma_\xi := \{x \in \mathbf{R}^d : \ \xi \in \mathcal{X}_b^x(T)\}$ and the closed convex set

$$D := D_\xi := \{x \in \mathbf{R}^d : \ Z_0 x \ge EZ_T\xi \ \ \forall Z \in \mathcal{M}_0^T(G^*)\}.$$

Obviously, $\Gamma \subseteq D$: if $x \in \Gamma$, then there is $\xi = x + X_T$ where $X \in \mathcal{X}_b^0$. For any $Z \in \mathcal{M}_0^T(G^*)$, we have $Z_0 x \geq E Z_T \xi$ in virtue of the supermartingale property (Lemma 3.6.2).

The next hypothesis is a requirement that the set $\mathcal{M}_0^T(G^*)$ is rich enough. It will serve us to propagate the property of boundedness from below from the terminal date to the whole time interval.

B. Let $\xi \in L^0(\mathbf{R}^d, \mathcal{F}_t)$. If the scalar product $Z_t \xi \geq 0$ for all $Z \in \mathcal{M}_0^T(G^*)$, then $\xi \in L^0(G_t, \mathcal{F}_t)$.

Condition **B** is fulfilled for the model with constant transaction costs admitting an equivalent martingale measure, i.e., when K^* is constant and there exists a strictly positive martingale ρ such that all ρS^i are martingales. Indeed, for any $w \in K^*$, the process Z^w with the coordinates $\rho w^i S^i$ is an element of $\mathcal{M}_0^T(\widehat{K}^*)$. Recall that, in our notation, G corresponds to $\widehat{K}$. The condition $Z_t^w \xi \geq 0$ means that $w \eta_t \geq 0$, where $\eta_t(\omega) = \phi_t^{-1}(\omega)\xi(\omega)$, and the diagonal operator ϕ_t is defined in (3.1.1). We have the latter inequality for all $w \in K^*$, and, hence, $\phi_t^{-1}\xi \in K$ and $\xi \in \widehat{K}_t$.

Our main result for the considered model is the following:

Theorem 3.6.3 *Assume that $\mathcal{D}(G) \neq \emptyset$ and that* **G** *and* **B** *hold. Then $\Gamma = D$.*

Its proof is given in the next subsection.

3.6.4 Hedging Theorem: Proof

We need some auxiliary results. The following one is very important. It expresses the fact that, under the efficient friction condition, the value processes cannot oscillate too much if their terminal values are bounded from below by a constant vector.

Lemma 3.6.4 *Assume that $\mathcal{D}(G) \neq \emptyset$ and that* **G** *holds. Let A be a subset of $\mathcal{X}_b^0$. Suppose that there is a constant κ such that $X_T + \kappa \mathbf{1} \in L^0(G_T, \mathcal{F}_T)$ for all $X \in A$. Then there exists a probability measure $Q \sim P$ with bounded density such that $\sup_{X \in A} E_Q \|X\|_T < \infty$.*

Proof. Let $\tilde{Z} \in \mathcal{D}(G)$. Since $(-\tilde{Z}_T X_T) \leq \kappa \tilde{Z}_T \mathbf{1}$, it follows from (3.6.6) that

$$\sup_{X \in A} E(-\tilde{Z}\dot{X}) \cdot \|X\|_T \leq \sup_{X \in A} E(-\tilde{Z}_T X_T) \leq \kappa E \tilde{Z}_T \mathbf{1} < \infty.$$

Let us consider the random variable

$$\alpha := \inf_{t \leq T} \inf_{x \in G_t,\ |x|=1} \tilde{Z}_t x = \inf_{t \leq T} \tilde{Z}_t x_t,$$

where x_t is the point on the unit sphere at which the interior infimum is attained. If $t_n \to t_0$ and the sequence x_{t_n} tends to some x_0, then $x_0 \in G_{t_0}$, because the continuity of the generators ζ^k imposed in **G** ensures that the inequalities $x_{t_n}\zeta^k_{t_n} \geq 0$ imply that $x_{t_0}\zeta^k_{t_0} \geq 0$. The infimum in t can be obtained either on a decreasing sequence of t_n (in this case, $\alpha = \tilde{Z}_{t_0}x_{t_0}$) or on an increasing one (in this case, $\alpha = \tilde{Z}_{t_0-}x_{t_0}$). The assumption on $\tilde{Z}$ guaranties that α is strictly positive.

It is easily seen that

$$E(-\tilde{Z}\dot{X}) \cdot \|X\|_T \geq E\alpha|\dot{X}| \cdot \|X\|_T = E\alpha\|X\|_T \geq E\alpha e^{-\alpha}\|X\|_T.$$

With these observations, the assertion becomes obvious: one can take Q with the density $dQ/dP = c\alpha e^{-\alpha}$. □

Recall that a sequence a^n is *Césaro convergent* if $\bar{a}^n := n^{-1}\sum_{k=1}^n a^k$ converges; in general, a type of convergence must be specified. The Komlós theorem asserts that if ξ^n are random variables with $\sup_n E|\xi^n| < \infty$, then there exist $\xi \in L^1$ and a subsequence $\xi^{n'}$ such that all its subsequences are Césaro convergent to ξ a.s.

Let $\mathcal{V}_T$ be the space of positive finite measures on $[0,T]$ equipped by the topology of weak convergence (in probabilistic sense). An optional random measure is simply a $\mathcal{V}_T$-valued random variable μ such that the process $\mu_t(\omega) := \mu(\omega, [0,t])$ is adapted.

The following result can be easily obtained from the Komlós theorem by the diagonal procedure.

Lemma 3.6.5 *Let μ^n be optional random measures with $\sup_n E\mu^n_T < \infty$. Then there exist an optional random measure μ with $\mu_T \in L^1$ and a subsequence $\mu^{n'}$ such that all its further subsequences are Césaro convergent in $\mathcal{V}_T$ to μ a.s.*

Proof. Let Q be a countable set dense in $[0,T]$ with $T \in$ Q. Using the Komlós theorem and the diagonal procedure, we find a subsequence such that $\mu^{n'}_r$ is Césaro convergent a.s. to $\mu^o_r \in L^1$ (with all further subsequences) for all $r \in$ Q . We can always choose a common null set N and assume that $\mu^o_r(\omega)$ is increasing in r for each $\omega \notin N$. The process $\mu_t := \limsup_{r\downarrow t} \mu^o_r$ (where $r \in$ Q and $r > t$) is such that $\bar{\mu}^{n'}_t(\omega) \to \mu_t(\omega)$ for all t, where $\mu_.(\omega)$ is continuous for $\omega \notin N$; it defines the random measure μ we need. □

We say that a sequence of $\mathbf{R}^d$-valued random variables U^n is *Fatou-convergent* to U if $U^n \to U$ a.s. and $U^n + \kappa\mathbf{1} \in L^0(G_T, \mathcal{F}_T)$ for some κ (in other words, $U^n \succeq_{G_T} -\kappa\mathbf{1}$). A set A is *Fatou-closed* if it contains all limits of its Fatou convergent sequences. A subset A_0 is Fatou-dense in A if any element of A is a limit of a Fatou-convergent sequence of elements from A_0.

Lemma 3.6.6 *Assume that $\mathcal{D}(G) \neq \emptyset$. Let* **G** *and* **B** *hold. Then $\mathcal{X}^x_b(T)$ is Fatou-closed.*

Proof. Let $X^n_T \in \mathcal{X}^x_b(T)$ be a sequence with $X^n_T + \kappa\mathbf{1} \in L^0(G_T, \mathcal{F}_T)$ converging to U a.s. Applying Lemmas 3.6.4 and 3.6.5, we may assume that each component of X^n converges a.s. in $\mathcal{V}_T$ to a certain process X of bounded variation. All the processes $\zeta^k \cdot X^n$ are decreasing. It follows from the continuity of the generators ζ^k that the processes $\zeta^k \cdot X$ are also decreasing. According to Lemma 3.6.1, $X \in \mathcal{X}^x$. It remains to check that $X_t + \kappa\mathbf{1} \in L^0(G_t, \mathcal{F}_t)$ for all $t \le T$. Let $Z \in \mathcal{M}(G)$. In virtue of Lemma 3.6.2, the prelimit processes $Z(X^n + \kappa\mathbf{1})$ are supermartingales, positive at the terminal date. It follows that $EI_\Gamma Z_t(X_t^n + \kappa\mathbf{1}) \ge 0$ whatever is $\Gamma \in \mathcal{F}_t$. Therefore, $Z_t(X_t^n + \kappa\mathbf{1}) \ge 0$. Condition **B** implies that $X_t^n(\omega) + \kappa\mathbf{1} \in G_t(\omega)$ for all t a.s. (that is, outside a P-null set). Thus, (again, outside a P-null set) $X_t(\omega) + \kappa\mathbf{1} \in G_t(\omega)$ for the points of continuity of the trajectory $X_.(\omega)$ and, due to the regularity of trajectories, for all points $t \in [0, T]$. □

Lemma 3.6.7 *The set $\mathcal{X}_b^x(T) \cap L^\infty$ is Fatou-dense in $\mathcal{X}_b^x(T)$.*

Proof. Let us consider a random variable $U = X_T$ where $X \in \mathcal{X}_b^x$. Then there is a constant $\kappa \ge 0$ such that $X_T + \kappa\mathbf{1}$ belongs to $L^0(G_T, \mathcal{F}_T)$. The bounded random variables $U^n := UI_{\{|U|\le n\}} - \kappa\mathbf{1}I_{\{|U|>n\}}$ belong to $\mathcal{X}_b^x(T)$ (since $U - U^n = (U + \kappa\mathbf{1})I_{\{|U|>n\}}$ is in $L^0(G_T, \mathcal{F}_T)$) and form a sequence Fatou-convergent to U. □

Lemma 3.6.8 *Assume that $\mathcal{D}(G) \ne \emptyset$ and that* **G** *and* **B** *hold. Then*

$$\mathcal{X}_b^x(T) = \{\xi \in L_b^0(\mathbf{R}^d) : \ E\eta\xi \le f(\eta) \ \forall \eta \in L^1(G_T^*, \mathcal{F}_T)\}, \tag{3.6.8}$$

where $f(\eta) = \sup_{\zeta \in \mathcal{X}_b^x(T)} E\eta\zeta$.

Proof. This is just a corollary of the bipolar theorem given in the Appendix 5.5: its hypotheses hold in virtue of the preceding lemmas. □

Proof of Theorem 3.6.3. To verify the remaining inclusion $D \subseteq \Gamma$, we take arbitrary $x \notin \Gamma$. This means that $\xi \notin \mathcal{X}_b^x(T)$. In virtue of (3.6.8), there exists $\eta \in L^1(G_T^*, \mathcal{F}_T)$ such that $E\eta\xi > f(\eta)$. Automatically, $f(\eta) < \infty$, and, therefore, $E\eta\zeta \le 0$ for every ζ from the cone $\mathcal{X}_b^0(T)$ containing $-L^\infty(G_t, \mathcal{F}_t)$, $t \le T$. Let us consider the martingale $Z_t = E(\eta|\mathcal{F}_t)$. It belongs to $\mathcal{M}(G^*)$, and $EZ_T\xi > EZ_0 x$. Thus, $x \notin D$, and we get the result. □

3.6.5 Rásonyi Counterexample

Let us examine now the question whether one can extend the hedging theorem replacing the continuity assumption on generators in hypothesis **G** by their right-continuity. The reader acquainted with the theory of frictionless market may expect that it is the case.

Surprisingly, the answer is negative. We provide an example due to Rásonyi showing that, already in a two-asset financial model with constant transaction cost coefficients but a discontinuous price process, this theorem fails to be true.

We start with a result which holds, as it is easily seen from its proof, in much more general situations including one where the generators of the cone-valued process G are only càdlàg processes. Let us introduce the following hypothesis.

H. For every d-dimensional random variable ξ such that $\xi+\kappa_\xi \mathbf{1} \in L^0(G_T)$, the equality $\Gamma_\xi = D_\xi$ holds.

Lemma 3.6.9 *If* **H** *holds, then the set* $\mathcal{X}_b^0(T)$ *is Fatou-closed.*

Proof. Let $U^n \in \mathcal{X}_b^0(T)$ be such that $U^n + \kappa \mathbf{1} \in G_T$ and $U^n \to U$ a.s. For any $Z \in \mathcal{M}_0^T(G^*)$, the sequence $Z_T U^n$ is bounded from below by an integrable random variable (namely, by a multiple of $|Z_T|$). For each n, the point zero belongs to Γ_{U^n}, the set coinciding with D_{U^n} by virtue of hypothesis **H**. Thus, $EZ_TU^n \le 0$ and, by the Fatou lemma, $EZ_TU \le 0$, that is, $0 \in D_U$. Applying again hypothesis **H**, we infer that $0 \in \Gamma_U$, which means that $U \in \mathcal{X}_b^0(T)$. □

As we already observed, **H** holds for models with efficient friction when the transaction costs are constant (i.e., K is constant) and the price process S a continuous martingale.

In the following example with discontinuous martingale S, the set $\mathcal{X}_b^0(T)$ is not Fatou-closed, and, therefore, the property **H**, i.e., the assertion of Theorem 3.6.3 fails to be true.

Let $(\Omega, \mathcal{F}, P)$ be a probability space, and let η, η_i, $i \ge 1$, be independent random variables on it such that

$$P(\eta = -1) = P(\eta = 1) = 1/2,$$
$$P(\eta_i = -1) = 1 - p_i, \qquad P(\eta_i = a_i) = p_i,$$

where $p_i = \exp\{-2^{-i}\}$ and $a_i = 1/p_i - 1$; clearly, $a_i \in]0,1[$ and $E\eta_i = 0$.

We consider, on the time interval $[0,1]$, the two-asset model with the price process $S = (S^1, S^2)$ where $S^1 \equiv 1$ and

$$S^2 = 1 + \frac{1}{2}\sum_{i=1}^{\infty} \frac{1}{2^i}\eta_i I_{[t_i,1]} + \frac{1}{4}\eta I_{\{1\}},$$

$t_i = 1 - 1/(i+1)$. The piecewise constant process S^2 with independent increments is a martingale with respect to its natural filtration. Note that $1/4 \le S^2 \le 7/4$.

Let $\lambda^{12} = \lambda^{21} = 1/2$. The vectors $3e_2 - 2e_1$ and $3e_1 - 2e_2$ are generators of the constant solvency cone K.

The random variable

$$\zeta := I_{\{\eta_i = a_i,\ \forall i \ge 1\}}\left(2e_1 - \frac{3}{S_{1-}^2}e_2\right)$$

is in the Fatou-closure of $\mathcal{X}_b^0(T)$ and is the limit of the bounded sequence

$$\widehat{V}_1^n = I_{\{\eta_i=a_i,\ \forall i\le n\}}\left(2e_1 - \frac{3}{S_{t_n}^2}e_2\right),$$

where $\widehat{V}_1^n$ are the terminal values of portfolios corresponding to a single transfer of funds at the date t_n, namely, with $\Delta B_{t_n} = I_{\{\eta_i=a_i,\ \forall i\le n\}}(2e_1 - 3e_2)$.

To check that $\zeta \notin \mathcal{X}_b^0(T)$, we observe that the set

$$A := \{\eta_i = a_i,\ \forall i \ge 1,\ \eta = 1\}$$

is of positive probability. On A the trajectories S^2 increase on $[0,1[$ from unit to the value $S_{1-}^2 = 3/2$ but do not attend it jumping downwards to $5/4$ at the terminal point. Inspecting the evolution of the boundary rays of the cones $\widehat{K}_t$ for $\omega \in A$, we note that, for all $t \in [0,T]$,

$$\widehat{K}_t(\omega) \subseteq J := \operatorname{cone}\{e_2 - e_1, 3e_1 - 2e_2\} \setminus R,$$

where the ray $R := \mathbf{R}_+(e_2 - e_1) = \{v : \ v = \lambda(e_2 - e_1),\ \lambda > 0\}$.

It is easily seen that the integral $\int_0^T dB_t$ for $\dot{B}$ evolving in the cone $-J$ (though not closed) is also an element of $-J$. It follows that on the set A we have $\widehat{V}_1 \in -J$ whatsoever is a portfolio process $\widehat{V}$, i.e., $\widehat{V}_1$ cannot take values on the ray $-R$, while $\zeta = 2(e_1 - e_2)$ on A.

3.6.6 Campi–Schachermayer Model

We discuss, following the lines suggested by Campi and Schachermayer, a modification of the abstract setting which permits to extend the hedging theorem to the case including models with discontinuous evolution of prices and transaction costs. The contents of this and the following subsections requires a bit more knowledge from the general theory of stochastic processes, and we send the reader to the major textbooks on this subject, [59, 161], or others.

Efficient friction. All cones G_t and G_{t-} are proper and contain $\mathbf{R}_+^d$.

As we already mentioned, the regularity of trajectories of the generators, and even their continuity, does not imply the regularity of the cone-valued processes (so the definition G_{t-} should not be interpreted as the left limit; moreover, it depends on the choice of generating process). As we shall see further, some arguments require certain regularity properties of the latter. We formulated them as an assumption on G introducing first some notation.

Let $G_{s,t}(\omega)$ denote the closure of cone$\{G_r(\omega) : \ s \le r < t\}$, and let

$$G_{s,t+}(\omega) := \bigcap_{\varepsilon>0} G_{s,t+\varepsilon}(\omega), \qquad G_{s-,t}(\omega) := \bigcap_{\varepsilon>0} G_{s-\varepsilon,t}(\omega),$$

with an obvious change for $s = 0$.

Regularity hypothesis. We assume that $G_{t,t+} = G_t$, $G_{t-,t} = G_{t-}$, and $G_{t-,t+} = \operatorname{cone}\{G_{t-}, G_t\}$ for all t.

It is easy to see that these regularity conditions are fulfilled for the case where the cones G_t and G_{t-} are proper and generated by a finite number of generators of unit length. Indeed, let $G_t = \operatorname{cone}\{\xi_t^k : k \le n\}$ with $|\xi_t^k| = 1$ for all t. Since the dependence on ω here is not important, we may argue for the deterministic case. Let $x \notin G_t$. The proper closed convex cones $\mathbf{R}_+ x$ and G_t intersect each other only at the origin, so the intersections of the interiors of $(-\mathbf{R}_+ x)^*$ and G_t^* are nonempty (see Lemma 5.1.2 in the Appendix). It follows that there is $y \in \mathbf{R}^d$ such that x belongs to the open half-space $\{z : yz < 0\}$, while the balls $\{z : |z - \xi_t^k| < \delta\}$ for sufficiently small $\delta > 0$ lie in the complementary half-space. Since ξ^k are right-continuous, the cones $G_{s,t+\varepsilon}$ for all sufficiently small $\varepsilon > 0$ also lay in the latter. Thus, $x \notin G_{t,t+}$ and $G_{t,t+} \subseteq G_t$. The opposite inclusion is obvious. In the same way we get other two identities.

In the case of financial models,

$$G_t = \widehat{K}_t = \operatorname{cone}\{\pi_t^{ij} e_i - e_j,\ e_i,\ 1 \le i, j \le d\},$$

where $\pi_t^{ij} = (1 + \lambda_t^{ij}) S_t^j / S_t^i$. The above remark shows that the standing hypothesis is fulfilled in the case of efficient friction with càdlàg processes π^{ij} and the additional assumption that the cones $\widehat{K}_{t-}$ are proper. Of course, the latter property may fail even if S is continuous. For instance, this will be in the case where all $\lambda_{t-}^{ij} = 0$ at some date t. Note that, under the efficient friction hypothesis, the generators e_i may be skipped from the set of generators of $\widehat{K}_t$ without changing the latter, but the cones $\widehat{K}_{t-}$ will be different.

The notion of consistent price system also should be adapted to the considered setting. We denote now by $\mathcal{D} = \mathcal{D}(G)$ the subset of $\mathcal{M}_0^T(\operatorname{int} G^*)$ formed by martingales Z such that not only $Z_t \in L^0(\operatorname{int} G_t^*, \mathcal{F}_t)$ but also $Z_{t-} \in L^0(\operatorname{int}(G_{t-})^*, \mathcal{F}_t)$ for all $t \in [0, T]$. The elements of $\mathcal{D}$ are called *consistent price systems.*

The most radical modification is in the definition of the portfolio processes. Recall that previously the processes X (corresponding to $\widehat{V}$ in the financial context) were always assumed to be adapted càdlàg processes of bounded variation. This was not justified by modeling reasons but by a will to work comfortably in the standard framework of stochastic calculus.

Let Y be a d-dimensional predictable process of bounded variation starting from zero and having trajectories with left and right limits (French abbreviation: *làdlàg*). Put $\Delta Y := Y - Y_-$, as usual, and $\Delta^+ Y := Y_+ - Y$, where $Y_+ = (Y_{t+})$. Define the right-continuous processes

$$Y_t^d = \sum_{s \le t} \Delta Y_s, \qquad Y_t^{d,+} = \sum_{s \le t} \Delta^+ Y_s$$

(the first is predictable, while the second is only adapted) and, at last, the continuous one:

$$Y^c := Y - Y^d - Y_-^{d,+}.$$

Recall that $\dot{Y}^c$ denotes the optional version of the Radon–Nikodym derivative $dY^c/d\|Y^c\|$.

Let $\mathcal{Y}$ be the set of processes Y satisfying the following conditions:

(1) $\dot{Y}^c \in -G\,dP\,d\|Y^c\|$-a.e.;
(2) $\Delta^+ Y_\tau \in -G_\tau$ a.s. for all stopping times $\tau \le T$;
(3) $\Delta Y_\sigma \in -G_{\sigma-}$ a.s. for all predictable[2] stopping times $\sigma \le T$.

Let $\mathcal{Y}^x := x + \mathcal{Y}$, $x \in \mathbf{R}^d$. We denote by $\mathcal{Y}^x_b$ the subset of $\mathcal{Y}^x$ formed by the processes Y bounded from below in the sense of partial ordering, i.e., such that $Y_t + \kappa_Y \mathbf{1} \in L^0(G_t, \mathcal{F}_t)$, $t \le T$, for some $\kappa_Y \in \mathbf{R}$. In the financial context (where $G = \widehat{K}$) the elements of $\mathcal{Y}^x_b$ are *admissible* portfolio processes.

To use classical stochastic calculus, we shall operate with the right-continuous adapted process of bounded variation

$$Y_+ := Y^c + Y^d + Y^{d,+}$$

and use the relation $Y_+ = Y + \Delta^+ Y$. Since the generators are right-continuous, the process Y_+ inherits the boundedness from below of Y (by the same constant process $\kappa_Y \mathbf{1}$). Note that $\|Y_+\|_t = \|Y\|_{t-} + |\Delta Y_t + \Delta^+ Y_t|$.

In the sequel we shall use a larger set of portfolio processes depending on $Z \in \mathcal{M}^T_0(G^*)$, namely,

$$\mathcal{Y}^x_b(Z) := \big\{ Y \in \mathcal{Y}^x : \text{ there is a scalar martingale } M \text{ such that } ZY \ge M \big\}.$$

The inclusion $\mathcal{Y}^x_b \subseteq \mathcal{Y}^x_b(Z)$ is obvious: if $Y_t + \kappa \mathbf{1} \in L^0(G_t, \mathcal{F}_t)$ for all $t \le T$, then $ZY \ge M$ with $M = -\kappa Z \mathbf{1}$.

Lemma 3.6.10 *If $Z \in \mathcal{M}^T_0(G^*)$ and $Y \in \mathcal{Y}^x_b(Z)$, then both processes ZY_+ and ZY are supermartingales, and*

$$E\Big(- Z\dot{Y}^c \cdot \|Y^c\|_T - \sum_{s \le T} Z_{s-} \Delta Y_s - \sum_{s<T} Z_s \Delta^+ Y_s \Big) \le Z_0 x - E Z_T Y_T. \quad (3.6.9)$$

Proof. With the right-continuous process Y_+ (having the same left limits as Y), the standard product formula is readily applied:

$$Z_t Y_{t+} = Z_0 x + Y_- \cdot Z_t + Z\dot{Y}^c \cdot \|Y^c\|_t + \sum_{s \le t} Z_s \Delta Y_s + \sum_{s \le t} Z_s \Delta^+ Y_s.$$

Taking into account that $Y = Y_- + \Delta Y$, we rewrite this identity as

$$Z_t Y_{t+} = Z_0 x + Y \cdot Z_t + Z\dot{Y}^c \cdot \|Y^c\|_t + \sum_{s \le t} Z_{s-} \Delta Y_s + \sum_{s \le t} Z_s \Delta^+ Y_s.$$

[2] Since Y is a predictable process, the set $\{\Delta Y \ne 0\}$ can be represented as a disjoint union of graphs of predictable stopping times. Hence, (3) implies that $\Delta Y_\tau \in -G_{\tau-}$ a.s. for all stopping times $\tau \le T$.

Since $Y_+ = Y + \Delta^+ Y$, we obtain from here the product formula for ZY (which is "nonstandard" since Y may not be càdlàg):

$$Z_t Y_t = Z_0 x + Y \cdot Z_t + Z\dot{Y}^c \cdot \|Y^c\|_t + \sum_{s \le t} Z_{s-} \Delta Y_s + \sum_{s<t} Z_s \Delta^+ Y_s.$$

By virtue of requirements on Y, the stochastic integral $Y \cdot Z$ is a local martingale, while the last three terms define decreasing processes (by our standing assumption $Z_{s-} \in (G_{s-})^*$). Recalling that the process ZY is bounded from below by a martingale, we deduce from here that the local martingale $Y \cdot Z$ is bounded from below by a martingale and, hence, is a supermartingale and thus integrable. It follows that the terminal values of the mentioned decreasing processes are integrable. Therefore, ZY is a supermartingale. By the Fatou lemma, its right-continuous limit, i.e., the process ZY_+ is a supermartingale. Finally, taking the expectation of the last identity above and using the inequality $Y \cdot Z_T \le 0$, we get the required bound (3.6.9). □

Lemma 3.6.11 *Suppose that $Y^n \in \mathcal{Y}$ and, for all ω (except of a null set), $\lim_n Y_t^n(\omega) = Y_t(\omega)$ for all $t \in [0,T]$, where Y is a process of bounded variation. Then the process Y belongs to $\mathcal{Y}$.*

This assertion follows immediately from the alternative description of $\mathcal{Y}$ given in the lemma below.

Lemma 3.6.12 *Let Y be a predictable process of bounded variation. Then*

$$Y \in \mathcal{Y} \iff Y_\sigma - Y_\tau \in L^0(G_{\sigma,\tau}) \quad \text{for all stopping times } \sigma, \tau,\ \sigma \le \tau \le T.$$

Proof. ($\Rightarrow$) Follows obviously from the representation

$$Y_\tau - Y_\sigma = \int_\sigma^\tau \dot{Y}_r^c \, d\|Y^c\|_r + \sum_{\sigma < r \le \tau} \Delta Y_r + \sum_{\sigma \le r < \tau} \Delta^+ Y_r.$$

($\Leftarrow$) First, we provide an "explicit" formula for $\dot{Y}^c$ using the classical approach due to Doob, see [59], V.5.58. For the reader's convenience, we recall the idea. Put $t_k = t_k^n = k2^{-n}T$, fix ω (omitted in the notation), and consider the sequence of functions

$$X_n(t) = \sum_k \frac{Y_{t_{k+1}^n +} - Y_{t_k^n +}}{\|Y_+\|_{t_{k+1}^n} - \|Y_+\|_{t_k^n}} I_{]t_k^n, t_{k+1}^n]}(t), \quad [0,T].$$

This sequence is a bounded martingale with respect to the dyadic filtration on $[0,T]$ and the finite measure $d\|Y_+\|$. So, it converges (almost everywhere with respect to this measure) to a limit X_∞, which is the Radon–Nikodym derivative $dY_+/d\|Y_+\|$ and which may serve also as the Radon–Nikodym derivative $dY^c/d\|Y^c\|$.

Thus,

$$\dot{Y}^c = \limsup_n \sum_k \frac{Y_{t^n_{k+1}+} - Y_{t^n_k+}}{\|Y_+\|_{t^n_{k+1}} - \|Y_+\|_{t^n_k}} I_{]t^n_k, t^n_{k+1}]} \quad dP\, d\|Y^c\|\text{-a.e.}$$

It follows that $-\dot{Y}^c_t \in G_{t-,t+}$ (a.e.). By assumption, $G_{t-,t+} = \operatorname{cone}\{G_{t-}, G_t\}$. Recall that the generators ξ^k are càdlàg processes. Thus, for each ω, the set $\{t:\ G_t(\omega) \neq G_{t-}(\omega)\}$ is at most countable, and property 1) in the definition of $\mathcal{Y}$ is fulfilled.

For a stopping time τ, we put $\tau^n := \tau + 1/n$. Then $\tau_n \downarrow \tau$ and

$$\Delta^+ Y_\tau = \lim_n (Y_{\tau^n} - Y_\tau) \in -G_{\tau,\tau+} = -G_\tau.$$

For a predictable stopping time σ, one can find an announcing sequence of stopping times $\sigma^n \uparrow \sigma$ with $\sigma^n < \sigma$ on the set $\{\sigma > 0\}$. Thus, on this set,

$$\Delta Y_\sigma = \lim_n (Y_\sigma - Y_{\sigma^n}) \in -G_{\sigma-,\sigma} = -G_{\sigma-}.$$

The lemma is proven. □

The following assertion is a variant of Lemma 3.6.4. In its proof we use the regularity property of G.

Lemma 3.6.13 *Let $Z \in \mathcal{D}$. Let A be a subset of $\mathcal{Y}^0_b(Z)$ for which there is a constant κ such that $Y_T + \kappa \mathbf{1} \in L^0(G_T, \mathcal{F}_T)$ for all $Y \in A$. Then there exists a probability measure $Q \sim P$ such that* $\sup_{Y \in A} E_Q \|Y\|_T < \infty$.

Proof. Fix $Z \in \mathcal{D}$ and consider the random variable

$$\alpha := \inf_{t \le T} \inf_{x \in G_t,\ |x|=1} Z_t x = \inf_{t \le T} Z_t x_t,$$

where $x_t = x_t(\omega)$ is the point on the unit sphere at which the interior infimum is attained. If $t_n \downarrow t_0$ and the sequence x_{t_n} tends to some x_0, then the point $x_0 \in \bigcap_{\varepsilon > 0} G_{t_0, t_0+\varepsilon} = G_{t_0,t_0+}$. By our assumption, $G_{t_0,t_0+} = G_{t_0}$. If $t_n \uparrow t_0$ and the sequence x_{t_n} tends to some x_0, then $x_0 \in G_{t_0-}$ by virtue of a similar argument. On various ω the infimum in t can be obtained either on a decreasing sequence of t_n (in this case, $\alpha = Z_{t_0} x_{t_0}$) or on a increasing one (in this case, $\alpha = Z_{t_0-} x_{t_0}$). The assumption on Z guaranties that in both cases the values of α are strictly positive.

It is easily seen that the left-hand side of (3.6.9) dominates

$$E\alpha\Big(|\dot{Y}^c| \cdot \|Y^c\|_T + \sum_{s \le T} |\Delta Y_s| + \sum_{s < T} |\Delta^+ Y_s|\Big) = E\alpha \|Y\|_T,$$

and, therefore,

$$E\alpha e^{-\alpha} \|Y\|_T \le E\alpha \|Y\|_T \le Z_0 x - E Z_T Y_T \le Z_0 x + \kappa E Z_T \mathbf{1}.$$

It follows that the measure Q with the density $dQ/dP = \alpha e^{-\alpha}/(E\alpha e^{-\alpha})$ is the required one. □

The Komlós-type result given by Lemma 3.6.5 also need a modification.

Lemma 3.6.14 *Let A^n be a sequence of predictable increasing processes starting from zero and with $\sup_n EA^n_T < \infty$. Then there is an increasing process A with $A_T \in L^1$ and a subsequence $A^{n'}$ which is Césaro convergent to A pointwise at every point of $[0,T]$ for all ω except a P-null set.*

Proof. Let $\mathbf{T} := \{k2^{-n}T : \ k = 0, \dots, 2^n, \ n \in \mathbf{N}\}$. Using the Komlós theorem and the diagonal procedure, we find a subsequence such that $A^{n'}_r$ is Césaro convergent a.s. to $A^o_r \in L^1$ (with all further subsequences) for all $r \in \mathbf{T}$. We can always choose a common null set Ω_0 and assume that $A^o_r(\omega)$ is increasing in r for each $\omega \notin \Omega_0$. Let us consider its left-continuous envelope defined on the whole interval, i.e., the process $A_t := \liminf_{r \uparrow t} A^o_t$ ($r \in \mathbf{T}$ and $r < t$). By the same argument as in the theory of weak convergence of probability distribution functions, we conclude that if the sequence of functions $A^{n'}(\omega)$ converges at all points of $\mathbf{T}$ in Césaro sense to $A^o(\omega)$, then it converges, in the same sense, to the function $A(\omega)$ at all points of continuity of the latter. The crucial observation is that one can sacrifice the left-continuity of A but "improve" the convergence property. To this aim let us consider a sequence of stopping times τ_k exhausting the jumps of the process A (i.e., such that $\{\Delta^+ A > 0\} \subseteq \bigcup_k [\tau_k]$, where $[\tau_k]$ is the graph of τ_k). Refining the subsequence of $A^{n'}$, we may assume that each sequence of random variables $A^{n'}_{\tau_k}$ also converges in Césaro sense. Replacing A_{τ_k} by these limiting values, we obtain the required process, which is a pointwise Césaro limit of a certain subsequence of A^n (thus, predictable). □

Remark. The above lemma is one of the key elements of the proof. It is worthy to make a look at its deterministic counterpart, which is just a version of the Helly theorem. The latter is usually formulated for left-continuous (or, more frequently, for right-continuous) monotone functions. The proof is easy: combining the Bolzano–Weierstrass theorem and the diagonal procedure, one defines a monotone function A^o on $\mathbf{T}$ and a subsequence $A^{n'}$ convergent to A^o on $\mathbf{T}$. Let A be the left envelope of A^o. Due to monotonicity, the same subsequence will converge to A at all points of $[0,T]$, where A is continuous, and this gives the standard version of the Helly theorem. Of course, the convergence may fail at the denumerable set where A is discontinuous. Repeating the arguments, one can find a further subsequence having limits also at each point of discontinuity of A. Replacing the values of A by these limits, we get an increasing function approximated by the refined subsequence at all points of the interval. The proofs in the stochastic setting (respectively, of Lemmas 3.6.5 and 3.6.14) follows the same lines, but the classical compactness argument is replaced by a reference to the Komlós theorem.

Fix $\xi \in L^0(\mathbf{R}^d, \mathcal{F}_T)$ with $\xi + \kappa \mathbf{1} \in L^0(G_T, \mathcal{F}_T)$, where κ is constant, and define the convex set $\Gamma := \{x \in \mathbf{R}^d : \ \xi \in \mathcal{Y}^x_b(T)\}$ and the closed convex set

$$D := \{x \in \mathbf{R}^d : \ Z_0 x \geq EZ_T \xi \ \forall Z \in \mathcal{M}^T_0(G^*)\}.$$

Again, it is clear that $\Gamma \subseteq D$. Indeed, if $x \in \Gamma$, then $\xi = x + Y_T$, where $Y \in \mathcal{Y}_b^0$. For any $Z \in \mathcal{M}_0^T(G^*)$, we have $Z_0 x \geq EZ_T\xi$ in virtue of the supermartingale property of ZY.

Theorem 3.6.15 *Assume that $\mathcal{D}(G) \neq \emptyset$ and* **B** *holds. Then $\Gamma = D$.*

Lemma 3.6.16 *Assume that $\mathcal{D}(G) \neq \emptyset$ and* **B** *holds. Then the set $\mathcal{Y}_b^x(T)$ is Fatou-closed.*

Proof. Let $Y_T^n \in \mathcal{Y}_b^x(T)$ be a sequence with $Y_T^n + \kappa\mathbf{1} \in L^0(G_T, \mathcal{F}_T)$ converging to U a.s. Applying Lemmas 3.6.13 and 3.6.14, we may assume that each component of Y^n converges point-wise to a certain process Y of bounded variation. In virtue of Lemma 3.6.11, $Y \in \mathcal{Y}^x$. It remains to check that $Y_t^n + \kappa\mathbf{1} \in L^0(G_t, \mathcal{F}_t)$ for all $t \leq T$. Let $Z \in \mathcal{M}(G^*)$. In virtue of Lemma 3.6.10, the processes $Z(Y^n + \kappa\mathbf{1})$ are supermartingales, positive at the terminal date. It follows that $EI_\Gamma Z_t(Y_t^n + \kappa\mathbf{1}) \geq 0$ whatever is $\Gamma \in \mathcal{F}_t$. Therefore, $Z_t(Y_t^n + \kappa\mathbf{1}) \geq 0$. Condition **B** implies that $Y_t^n + \kappa\mathbf{1} \in G_t$, and, therefore, $Y_t + \kappa\mathbf{1} \in G_t$. □

With this property, we obtain as in Sect. 3.6.4, without any changes, the following assertions needed to complete the proof of the hedging theorem.

Lemma 3.6.17 *The set $\mathcal{Y}_b^x(T) \cap L^\infty$ is Fatou-dense in $\mathcal{Y}_b^x(T)$.*

Lemma 3.6.18 *Assume that $\mathcal{D}(G) \neq \emptyset$. Then*

$$\mathcal{Y}_b^x(T) = \{\xi \in L_b^0(\mathbf{R}^d) : \ E\eta\xi \leq f(\eta) \ \forall \eta \in L^1(G_T^*, \mathcal{F}_T)\}, \tag{3.6.10}$$

where $f(\eta) = \sup_{\xi \in \mathcal{Y}_b^x(T)} E\eta\xi$.

3.6.7 Hedging Theorem for American Options

First, we introduce some notation and concepts which are natural generalizations of those used for the dual description of the set of hedging endowments in the discrete-time model.

Coherent price systems. Let ν be a finite measure on the interval $[0,T]$, and let $\mathcal{N}$ denote the set of all such measures. For an $\mathbf{R}_+^d$-valued process Z, we denote by $\bar{Z}^\nu$ the optional projection of the process $\int_{[t,T]} Z_s\nu(ds)$, i.e., an optional process such that, for every stopping time $\tau \leq T$, we have

$$\bar{Z}_\tau^\nu = E\Big(\int_{[\tau,T]} Z_s\nu(ds)\big|\mathcal{F}_\tau\Big).$$

The process $\bar{Z}^\nu$ can be represented as the difference of a martingale and a left-continuous process with increasing components:

$$\bar{Z}_t^\nu = E\Big(\int_{[0,T]} Z_s\nu(ds)\big|\mathcal{F}_t\Big) - \int_{[0,t[} Z_s\nu(ds).$$

We associate with ν the product-measure $P^\nu(d\omega, dt) = P(d\omega)\nu(dt)$ on the space $(\Omega \times [0,T], \mathcal{F} \times \mathcal{B}_{[0,T]})$; the average with respect to this measure is denoted by E^ν.

Let $\mathcal{Z}(G^*, P, \nu)$ denote the set of adapted càdlàg processes $Z \in L^1(P^\nu)$ such that $Z_t, \bar{Z}^\nu_t \in L^0(G^*_t, \mathcal{F}_t)$ for all $t \le T$. We call the elements of this set *coherent price systems.* In the case where Z is a martingale, $\bar{Z}^\nu_\tau = \nu([\tau, T])Z_\tau$, and, hence, $\mathcal{M}_0^T(G^*) \subseteq \mathcal{Z}(G^*, P, \nu)$.

Define the convex set

$$\Gamma := \{x \in \mathbf{R}^d : \exists Y \in \mathcal{Y}_b^x \text{ such that } Y \ge_G U\} = \{x \in \mathbf{R}^d : U - x \in A_0^T(.)\}$$

and the closed convex set

$$D := D(P) := \{x \in \mathbf{R}^d : \bar{Z}^\nu_0 x \ge E^\nu ZU \quad \forall Z \in \mathcal{Z}(G^*, P, \nu),\ \forall \nu \in \mathcal{N}\}.$$

It is easy to check that $D(P) = D(\tilde{P})$ if $\tilde{P} \sim P$. Indeed, let $x \in D(P)$, $\nu \in \mathcal{N}$, and $\tilde{Z} \in \mathcal{Z}(G^*, \tilde{P}, \nu)$. Define $\rho_t = E(d\tilde{P}/dP|\mathcal{F}_t)$ and consider the process $Z_t = \rho_t \tilde{Z}_t$. It is in $\mathcal{Z}(G^*, P, \nu)$, and

$$E\rho_T \int_{[0,T]} \tilde{Z}_t U_t \nu(dt) = E\int_{[0,T]} \rho_t \tilde{Z}_t U_t \nu(dt) = E\int_{[0,T]} Z_t U_t \nu(dt) \le x E^\nu Z.$$

Since $E^\nu Z = \tilde{E}^\nu \tilde{Z}$, it follows that $x \in D(\tilde{P})$.

Proposition 3.6.19 $\Gamma \subseteq D$.

Proof. Let $x \in \Gamma$. Then there exists Y in $\mathcal{Y}_b^0$ such that the process $x + Y$ dominates U, i.e., $x + Y_t - U_t \in L^0(G_t, \mathcal{F}_t)$ for all $t \in [0,T]$. It follows that $x + Y_{t+} - U_t \in L^0(G_t, \mathcal{F}_t)$. By duality, for any $Z \in \mathcal{Z}(G^*, P, \nu)$ and $\nu \in \mathcal{N}$, we have that

$$E\int_{[0,T]} Z_t x \nu(dt) + E\int_{[0,T]} Z_t Y_{t+} \nu(dt) \ge E\int_{[0,T]} Z_t U_t \nu(dt).$$

It remains to verify that $E^\nu Z Y_+ \le 0$. Using the Fubini theorem and the property of the optional projection given by Theorem VI.2.57 in [59], we have

$$\begin{aligned} E\int_{[0,T]} Z_t Y_{t+} \nu(dt) &= E\int_{[0,T]} Z_t \Big(\int_{[0,t]} \dot{Y}_{+s} d\|Y_+\|_s\Big) \nu(dt) \\ &= E\int_{[0,T]} \dot{Y}_{+s} \Big(\int_{[s,T]} Z_t \nu(dt)\Big) d\|Y_+\|_s \\ &= E\int_{[0,T]} \dot{Y}_{+s} \bar{Z}^\nu_s \, d\|Y_+\|_s. \end{aligned}$$

It is easy to see that

$$\int_{[0,T]} \dot{Y}_{+s} \bar{Z}^\nu_s \, d\|Y_+\|_s = \int_{[0,T]} \dot{Y}^c_s \bar{Z}^\nu_s \, d\|Y^c\|_s + \sum_{s \le T} \bar{Z}^\nu_s \Delta Y_s + \sum_{s \le T} \bar{Z}^\nu_s \Delta^+ Y_s.$$

Since $\dot{Y}^c_s$ and $\Delta^+ Y_s$ take values in the cone $-G_s$ and $\bar{Z}^\nu_s$ takes values in G^*_s, the first and third terms of the above identity are negative. The increment ΔY_s takes values in $-G_{s-}$. If $G_{s-} = G_s$ for all s, the second term is also negative, and we conclude. As we do not assume the continuity of the process G, the proof requires a bit more work.

Let us suppose for a moment that the random variable $||Y^d||_T$ is bounded. To get the needed inequality $E^\nu Z Y_+ \leq 0$, it is sufficient to check that the expectation of the second term is negative. We proceed as follows. Recall that $\bar{Z}^\nu = M^\nu - R$, where M^ν is a martingale, and the process

$$R_t = \int_{[0,t[} Z_u \nu(du)$$

is left-continuous. The last property implies that $\Delta \bar{Z}^\nu = \Delta M^\nu$. It follows that

$$\sum_{s \leq T} \bar{Z}^\nu_s \Delta Y_s = \sum_{s \leq T} \bar{Z}^\nu_{s-} \Delta Y_s + \sum_{s \leq T} \Delta M^\nu_s \Delta Y_s.$$

The first sum in the right-hand side is obviously negative, while the expectation of the second one is zero. This follows from the classical property (see, e.g., [102], Lemma I.3.12): if M is a positive martingale and B is a predictable increasing process starting from zero, then

$$E M_T B_T = E \int_{[0,T]} M_s \, dB_s = E \int_{[0,T]} M_{s-} \, dB_s.$$

We can easily remove the condition of the boundedness of $||Y^d||$. Indeed, a finite predictable increasing process is locally bounded, see [59], Chap. VIII.11. Hence, there is a sequence of stopping times τ^n increasing stationary to T (i.e., with $P(\tau_n = T) \to 1$) such that $||Y^d||_{\tau^n} \leq C_n$. Let U^n be the process coinciding with U on $[0, \tau^n[$ and taking the value $x + Y_{\tau^n}$ on $[\tau^n, T]$. It follows from the above arguments that $\bar{Z}^\nu x \geq E^\nu Z U^n$, and the result follows from the Fatou lemma. □

Theorem 3.6.20 *Suppose that $\mathcal{D} \neq \emptyset$. Then $\Gamma = D$.*

Proof. We fix $\tilde{Z} \in \mathcal{D}$ and define the set of hedging endowments corresponding to portfolios with the "relaxed" admissibility property, namely, we put

$$\Gamma(\tilde{Z}) := \{x \in \mathbf{R}^d : \ \exists Y \in \mathcal{Y}^x_b(\tilde{Z}) \text{ such that } Y \succeq_G U\}.$$

Since $\mathcal{Y}^x_b(\tilde{Z}) \supseteq \mathcal{Y}^x_b$, this set is larger than Γ. On the other hand, if a portfolio Y dominates U, it is bounded from below. Hence, $\Gamma(\tilde{Z}) = \Gamma$.

Let $\mathbf{T}^m := \{t_k = t^m_k : \ t_k = k2^{-m}T, \ k = 0, \dots, 2^m\}$; then $\mathbf{T} = \bigcup_{m \geq 1} \mathbf{T}^m$. Define the convex set $A_{\mathbf{T}^m}(.)$ of American options W which can be hedged at the dates from $\mathbf{T}^m$ by a portfolio belonging to the class $\mathcal{Y}^0_b(\tilde{Z})$, i.e., such that $Y_t - W_t \in G_t$, $t \in \mathbf{T}^m$, for some $Y \in \mathcal{Y}^0_b(\tilde{Z})$. Let us consider $A_{\mathbf{T}^m}(.)$ as a subset of the space $L^0(P \otimes \nu^m) := L^0(\Omega \times [0,T], \mathcal{F} \times \mathcal{B}_{[0,T]}, P \otimes \nu^m)$, where the probability measure ν^m is the uniform distribution on $\mathbf{T}^m$, i.e., it charges only the points of $\mathbf{T}^m$ with weights $1/(2^m+1)$. From the point of view of this space, W is just the random vector $(W_0, W_{1/2^m}, \dots, W_T)$ (the components of the latter are d-dimensional). For such random vectors (with fixed $m \geq 1$), we extend the concept of the Fatou-convergence in the same spirit as it was developed in the problem of hedging of European options. Note that $A_{\mathbf{T}^m}(.)$, in general, depends on $\tilde{Z}$.

We say that a sequence W^n is *Fatou-convergent in* $L^0(P \otimes \nu^m)$ to W if there is a constant κ such that $W^n_r + \kappa\mathbf{1} \in L^0(G_r, \mathcal{F}_r)$ (i.e., $W^n_r \succeq_{G_r} -\kappa\mathbf{1}$) for all $r \in \mathbf{T}^m$, $n \geq 1$, and $W^n_r \to W_r$ a.s., $n \to \infty$, for all $r \in \mathbf{T}^m$. The subsequent definitions of *Fatou-closed* and *Fatou-dense* are obvious.

Lemma 3.6.21 *The set $A_{\mathbf{T}^m}(.)$ is Fatou-closed in $L^0(P \otimes \nu^m)$.*

Proof. Let $W^n \in A_{\mathbf{T}^m}(.)$ be a sequence Fatou-converging to W, and let Y^n be a corresponding sequence of dominating elements from $\mathcal{Y}^0_b(\tilde{Z})$. Our aim is to show that W also can be dominated by some element $\mathcal{Y}^0_b(\tilde{Z})$ at the points of $\mathbf{T}^m$. Using the preceding results (see Lemmas 3.6.13 and 3.6.14), we can replace W^n and Y^n by appropriate sequences of arithmetic means and suppose without loss of generality that Y^n converges to some predictable process Y of bounded variation almost surely at each point $t \in [0,T]$. Using Lemma 3.6.11, we conclude that $Y \in \mathcal{Y}^0$. It remains to check that $\tilde{Z}Y$ dominates a martingale. By virtue of Lemma 3.6.10, the prelimit processes $\tilde{Z}Y^n$ are supermartingales. Since Y^n_T dominates $W_T \succeq_{G_T} -\kappa\mathbf{1}$, we have that $\tilde{Z}_T Y^n_T \geq -\kappa\tilde{Z}_T$. It follows that the supermartingale $\tilde{Z}Y^n$ dominates the martingale $-\kappa\tilde{Z}$, and so does the supermartingale $\tilde{Z}Y$. □

Lemma 3.6.22 *The set $A_{\mathbf{T}^m}(.) \cap L^\infty(P^{\nu^m})$ is Fatou-dense in $A_{\mathbf{T}^m}(.)$.*

Proof. Let $W \in A_{\mathbf{T}^m}(.)$ be dominated at the points of $\mathbf{T}^m$ by a portfolio $Y \in \mathcal{Y}^0_b(\tilde{Z})$. Let κ be a constant such that $W_t + \kappa\mathbf{1} \in G_t$ for $t \in \mathbf{T}^m$. Put

$$W^n := W I_{\{|W| \leq n\}} - \kappa I_{\{|W|>n\}}.$$

Then $W^n \in L^\infty(P \otimes \nu^m)$ and tends to W as $n \to \infty$. Since

$$Y_t - W^n_t = (Y_t - W_t) I_{\{|W_t| \leq n\}} + (Y_t + \kappa\mathbf{1}) I_{\{|W_t|>n\}} \in G_t, \quad t \in \mathbf{T}^m,$$

we have that $W^n \in A_{\mathbf{T}^m}(.)$. □

Let $L^0_b(P \otimes \nu^m)$ be the cone in $L^0(P \otimes \nu^m)$ formed by the elements W (interpreted as random vectors) which are adapted and bounded from below in the sense of partial ordering, i.e., such that $W_r + c\mathbf{1} \in L^0(G_r, \mathcal{F}_r)$ for all $r \in \mathbf{T}^m$. The notation $L^1(G^*, P \otimes \nu^m)$ has an obvious meaning.

The following assertion is nothing but Theorem 5.5.3 formulated in the notation adjusted to the considered situation (where one takes $W_0 = 0$).

Lemma 3.6.23 *Let A be a convex subset in $L^0_b(P\otimes\nu^m)$ which is Fatou-closed and such that the set $A^\infty := A\cap L^\infty(\mathbf{R}^d, P\otimes\nu^m)$ is Fatou-dense in A. Suppose that there is $W_0 \in A^\infty$ such that $W_0 - L^\infty(G, P\otimes\nu^m) \subseteq A^\infty$. Then*

$$A = \left\{W \in L^0_b(P\otimes\nu^m) : \; E^{\nu^m} ZW \le f(Z) \;\forall Z \in L^1(G^*, P\otimes\nu^m)\right\},$$

where $f(Z) = \sup_{Y\in A} E^{\nu^m} ZY$.

With the above preliminaries, we can complete the proof of Theorem 3.6.20 by establishing the remaining inclusion $D \subseteq \Gamma = \Gamma(\tilde{Z})$. Indeed, take a point $x \in D$. Suppose that $U - x \notin A_{\mathbf{T}^m}(.)$ for some m. By virtue of Lemma 3.6.23, there exists $Z \in L^1(G^*, P\otimes\nu^m)$ such that $E^{\nu^m} Z(U-x) > f(Z)$. But $f(Z) = 0$ as $A_{\mathbf{T}^m}(.)$ is a cone. We can identify Z with a right-continuous adapted process taking value Z_{t_k} at the points t_k. Since $E^{\nu^m} ZY \le 0$ for all $Y \in A_{\mathbf{T}^m}(.)$, the process $Z \in \mathcal{Z}(G^*, P, \nu^m)$. Thus, $x \notin D$, a contradiction. This means that $U - x \in A_{\mathbf{T}^m}(.)$ for all m, i.e., there exist admissible portfolio processes Y^n dominating $U - x$ at the points of $\mathbf{T}^n$. In particular, the sequence Y^n_T is bounded from below by a constant vector, and, by virtue of Lemma 3.6.13, the total variations $\|Y^n\|_T$ are bounded in a certain $L^1(Q)$ with $Q \sim P$. Using Lemma 3.6.14, we may assume without loss of generality that the sequence Y^n converges to some predictable process Y of bounded variation almost surely at each point $t \in [0, T]$. Recall that $U_t + \kappa\mathbf{1} \in G_t$. The limiting process Y dominates $U - x$ at all points from $\mathbf{T}$. Using the right continuity of the processes, we obtain that Y_+ dominates $U - x$ on the whole interval, and so does the "larger" process Y. So, $x \in \Gamma$. □

Remark. Theorem 3.6.20 implies as a corollary a hedging theorem for càdlàg portfolio processes under the assumption that the cone-valued process G is continuous. Indeed, let $\mathcal{X}^0$ be the set of all càdlàg processes X of bounded variation with $X_0 = 0$ and such that $dX/d\|X\| \in -G \; dP\, d\|X\|$-a.e. The notation $\mathcal{X}^x$ and $\mathcal{X}^x_b$ is obvious. Let

$$\Gamma_X := \left\{x \in \mathbf{R}^d : \; \exists X \in \mathcal{X}^x_b \text{ such that } X \ge_G U\right\}.$$

Arguments similar to (but simpler than) those used in the proof of Proposition 3.6.19 show that $\Gamma_X \subseteq D$.

Suppose that all generators ξ^k of G are continuous processes. It is easy to check that if the process $Y \in \mathcal{Y}^0$, then $Y_+ \in \mathcal{X}^0$. Thus, $\Gamma \subseteq \Gamma_X$, and Theorem 3.6.20 implies that if $\mathcal{D}(G) \ne \emptyset$, then $\Gamma_X = D$.

3.6.8 When Does a Consistent Price System Exist?

In this section we provide a sufficient condition for the existence of a martingale evolving in the interior of $\widehat{K}^*$. To simplify the notation, we consider the following setting:

Let C be a cone in $\mathbf{R}^d$ containing the vector $\mathbf{1} = (1,\dots,1)$ in its interior. Let $S = (S_t)_{t\in[0,1]}$ be an $\mathbf{R}^d$-valued continuous adapted process with strictly positive components defined on a stochastic basis $(\Omega, \mathcal{F}, \mathbf{F}, P)$. The random diagonal operators $\Sigma_t : (x^1,\dots,x^d) \mapsto (S_t^1 x^1,\dots,S_t^d x^d)$ define the cone-valued adapted process $\Sigma C = (\Sigma_t C)_{t\in[0,1]}$. The question is: *when is the set $\mathcal{M}_0^1(\Sigma C \setminus \{0\})$ nonempty?* That is, when does there exist an $\mathbf{R}^d$-valued martingale M with $M_t(\omega) \in \Sigma_t(\omega)C \setminus \{0\}$ for all ω and t?

In our usual language, $C = K^*$, the dual to the solvency cone in the model with constant transaction costs, $\Sigma_t = \phi_t^{-1}$ and $\Sigma_t C = \widehat{K}^*$.

To formulate the result we introduce two conditions.

Let τ and σ be two stopping times with values in $[0,1]$ such that $\sigma \ge \tau$. We denote by $A_{\tau,\sigma}$ the (random) topological support of the regular conditional distribution $P_{\tau,\sigma}(dx,\omega)$ of $S_\sigma - S_\tau$ with respect to $\mathcal{F}_\tau$.

(a) $0 \in \operatorname{ri}\operatorname{conv} A_{\tau,\sigma}$ a.s. on $\{\tau < 1\}$ for all stopping times τ and σ such that $\sigma \ge \tau$ (recall that ri means relative interior).
(b) $P(\sup_{\tau\le r\le 1} |S_r - S_\tau| \le \varepsilon | \mathcal{F}_\tau) > 0$ a.s. on $\{\tau < 1\}$ for all $\varepsilon > 0$ and all stopping times τ.

Theorem 3.1. *Assume that* (a) *and* (b) *hold. Then* $\mathcal{M}_0^1(\Sigma C \setminus \{0\}) \neq \emptyset$.

Proof. Fix $\theta > 1$. Define the sequence of stopping times, $\tau_0 = 0$,

$$\tau_n := \inf\Big\{t \ge \tau_{n-1} : \max_{i\le d}\big|\ln S_t^i - \ln S_{\tau_{n-1}}^i\big| \ge \ln\theta\Big\} \wedge 1, \quad n \ge 1,$$

and the stopping time $\tau_t := \min\{\tau_n : \tau_n > t\}$ for $t \in [0,1[$. We put also $\sigma_t := \max\{\tau_n : \tau_n \le t\}$ and $\nu := \max\{n : \tau_n < 1\}$. Since the ratios $S_t^i/S_{\sigma_t}^i$ and $S_{\tau_t}^i/S_{\sigma_t}^i$ take values in the interval $[\theta^{-1},\theta]$, we have the bounds

$$\theta^{-2} \le S_{\tau_t}^i/S_t^i \le \theta^2, \quad i \le d. \tag{3.6.11}$$

Set $X_n := S_{\tau_n} I_{\{\tau_n<1\}} + S_{\tau_\nu} I_{\{\tau_n=1\}}$, $\mathcal{G}_n := \mathcal{F}_{\tau_n}$. We apply, to the discrete-time process $X = (X_n)$, Proposition 2.2.14 conditions (i) and (iii) of which hold by virtue of (a) and (b), while (ii) is always fulfilled for continuous S. In virtue of this proposition, X is a uniformly integrable Q-martingale with respect to some probability measure $Q = Z_\infty P$ equivalent to P. Let us consider the continuous-time martingale $\tilde{S}_t := E_Q(X_\infty|\mathcal{F}_t)$, $t \in [0,1]$. Since $\tilde{S}_{\tau_n} = X_n$, we have the inequalities

$$\theta^{-1} \le \tilde{S}_{\tau_n}^i/S_{\tau_n}^i \le \theta,$$

where τ_n can be replaced by τ_t. Using this and the bounds (3.6.11), we get that

$$\theta^{-3} \le \tilde{S}_{\tau_t}^i/S_t^i \le \theta^3.$$

But $\tilde{S}_t^i/S_t^i = E_Q(\tilde{S}_{\tau_t}^i/S_t^i|\mathcal{F}_t)$, and, therefore, the ratios $\tilde{S}_t^i/S_t^i$ take values in the interval $[\theta^{-3},\theta^3]$. Thus, for θ sufficiently close to unit, the Q-martingale $\tilde{S}$ evolves in $\Sigma C \setminus \{0\}$, and so does also the P-martingale $M := Z\tilde{S}$. □

3.7 Asymptotic Arbitrage Opportunities of the Second Kind

In this section we shall work in a framework of a very general continuous-time model, with an arbitrary adapted cone-valued process satisfying the efficient friction condition and dominating the constant process $\mathbf{R}^d_+$. Regularity hypothesis is not imposed. We present a criterion of the NFL-property together with some results on the NAA2-property, No Asymptotic Arbitrage of the 2nd Kind, close to the NA2-property considered for the discrete-time model.

Let $(\Omega, \mathbf{F} = \mathcal{F}, (\mathcal{F}_t)_{t\le T}, P)$ be a continuous-time stochastic basis verifying the usual conditions. We are given a pair of set-valued adapted processes $G = (G_t)_{t\in[0,T]}$ and $G^* = (G^*_t)_{t\in[0,T]}$ whose values are closed cones in $\mathbf{R}^d$ in duality, i.e., $G^*_t(\omega) = \{y : \ yx \ge 0 \ \forall x \in G_t(\omega)\}$. "Adapted" means that the graphs

$$\{(\omega, x) \in \Omega \times \mathbf{R}^d : \ x \in G_t(w)\}$$

are $\mathcal{F}_t$-measurable.

We assume that all cones G_t are proper, i.e., $G_t \cap (-G_t) = \{0\}$ or, equivalently, $\operatorname{int} G^*_t \ne \emptyset$. We assume also that $G^*\setminus\{0\} \subset \operatorname{int} \mathbf{R}^d_+$.

Recall that in a more specific financial setting, the cones G_t are the solvency cones $\widehat{K}_t$, provided that the portfolio positions are expressed in physical units.

For each $s \in]0,T]$, we are given a convex cone $\mathcal{Y}^T_s$ of optional $\mathbf{R}^d$-valued processes $Y = (Y_t)_{t\in[s,T]}$ with $Y_s = 0$.

It is assumed that $\mathcal{Y}^T_s$ is stable under multiplication by the bounded $\mathcal{F}_s$-measurable random variables, i.e., by the elements of $L^\infty_+(\mathcal{F}_s) = L^\infty(\mathbf{R}_+, \mathcal{F}_s)$. Moreover, if sets $A^n \in \mathcal{F}_s$ form a countable partition of Ω and $Y^n \in \mathcal{Y}^T_s$, then $\sum_n Y^n I_{A^n} \in \mathcal{Y}^T_s$.

The following notation will be used in the sequel:

- for d-dimensional processes Y and Y', the relation $Y \ge_G Y'$ means that the difference $Y - Y'$ evolves in G, that is, $Y_t - Y'_t \in G_t$ a.s. for every t;
- $\mathcal{Y}^T_{s,b}$ denotes the subset of $\mathcal{Y}^T_s$ formed by the processes Y dominated from below in the sense of partial ordering generated by G, i.e., such that there is a constant κ such that the process $Y + \kappa\mathbf{1}$ evolves in G;
- $\mathcal{Y}^T_{s,b}(T)$ is the set of random variables Y_T where $Y \in \mathcal{Y}^T_{s,b}$;
- $\mathcal{A}^T_{s,b}(T) = (\mathcal{Y}^T_{s,b}(T) - L^0(G_T, \mathcal{F}_T)) \cap L^\infty(\mathbf{R}^d, \mathcal{F}_T)$, and $\overline{\mathcal{A}^T_{s,b}(T)}^w$ is the closure of this set in $\sigma\{L^\infty, L^1\}$;
- $\mathcal{M}^T_s(G^*)$ is the set of all d-dimensional martingales $Z = (Z_t)_{t\in[s,T]}$ "evolving" in G^*, i.e., such that $Z_t \in G^*_t$ almost surely for every $t \in [s,T]$.

Throughout the section we assume the following *standing hypotheses* on the sets $\mathcal{Y}^T_{s,b}$:

$\mathbf{S}_1$. $E\xi Z_T \le 0$ for all $\xi \in \mathcal{Y}^T_{s,b}(T)$, $Z \in \mathcal{M}^T_s(G^*)$, and $s \in [0,T[$.
$\mathbf{S}_2$. $\bigcup_{t\ge s} L^\infty(-G_t, \mathcal{F}_t) \subseteq \mathcal{Y}^T_{s,b}(T)$ for each $s \in [0,T]$.

Hypotheses $\mathbf{S}_1$ and $\mathbf{S}_2$ adopted in this section are fulfilled for both continuous-time models considered above. Hypothesis $\mathbf{S}_2$ means that an investor has the right to take any position less advantageous than zero and keeps it until the end of the planning horizon.

Now we introduce other properties of interest: No Free Lunch, No Asymptotic Arbitrage of the 2nd Kind, and Many Consistent Price Systems.

NFL. $\overline{\mathcal{A}^T_{s,b}(T)}^w \cap L^\infty(\mathbf{R}^d_+, \mathcal{F}_T) = \{0\}$ for each $s \in [0,T[$.

NAA2. For all $s \in [0,T[$ and $\xi \in L^\infty(\mathbf{R}^d, \mathcal{F}_s)$,

$$(\xi + \overline{\mathcal{A}^T_{s,b}(T)}^w) \cap L^\infty(\mathbf{R}^d_+, \mathcal{F}_T) \neq \emptyset$$

only if $\xi \in L^\infty(G_s, \mathcal{F}_s)$.

MCPS. For any $\eta \in L^1(\text{int } G^*_s, \mathcal{F}_s)$, there is $Z \in \mathcal{M}^T_s(G^* \setminus \{0\})$ with $Z_s = \eta$.

Finally, we recall one more condition:

B. If ξ is an $\mathcal{F}_s$-measurable $\mathbf{R}^d$-valued random variable such that $Z_s\xi \geq 0$ for any $Z \in \mathcal{M}^T_s(G^*)$, then $\xi \in G_s$ (a.s.).

The following assertion is a version of FTAP for the considered setting:

Theorem 3.7.1 **NFL** $\Leftrightarrow \mathcal{M}^T_0(G^*\setminus\{0\}) \neq \emptyset$.

Proof. ($\Leftarrow$) Let $Z \in \mathcal{M}^T_0(G^*\setminus\{0\})$. Then the components of Z_T are strictly positive, and $EZ_T\xi > 0$ for all $\xi \in L^\infty(\mathbf{R}^d_+, \mathcal{F}_T)$ except $\xi = 0$. On the other hand, $E\xi Z_T \leq 0$ for all $\xi \in \mathcal{Y}^T_{s,b}(T)$ and thus for all $\xi \in \overline{\mathcal{A}^T_{s,b}(T)}^w$.

($\Rightarrow$) The Kreps–Yan theorem on separation of closed cones in $L^\infty(\mathbf{R}^d, \mathcal{F}_T)$ implies the existence of $\eta \in L^1(\text{int}\,\mathbf{R}^d_+, \mathcal{F}_T)$ such that $E\xi\eta \leq 0$ for all $\xi \in \overline{\mathcal{A}^T_{s,b}(T)}^w$ and, hence, by hypothesis $\mathbf{S}_2$, for all $\xi \in L^\infty(-G_t, \mathcal{F}_t)$. Let us consider the martingale $Z_t = E(\eta|\mathcal{F}_t)$, $t \geq s$, with strictly positive components. Since $EZ_t\xi = E\xi\eta \geq 0$, $t \geq s$, for every $\xi \in L^\infty(G_t, \mathcal{F}_t)$, it follows that $Z_t \in L^1(G_t, \mathcal{F}_t)$, and, therefore, $Z \in \mathcal{M}^T_s(G^*\setminus\{0\})$. □

Now we formulate the main result of this section, giving, in particular, a financial meaning of condition **B**.

Theorem 3.7.2 *Assume that* **NFL** *holds. Then*

$$\mathbf{MCPS} \Rightarrow \mathbf{B} \Leftrightarrow \mathbf{NAA2}.$$

If, moreover, the sets $\mathcal{Y}^T_{s,b}(T)$ are Fatou-closed for any $s \in [0,T[$, then all three conditions are equivalent.

In the above formulation the Fatou-closedness means that the set $\mathcal{Y}^T_{s,b}(T)$ contains the limit on any a.s. convergent sequence of its elements, provided that the latter is bounded from below in the sense of partial ordering induced by G_T.

Proof. **MCPS** $\Rightarrow$ **B** Let ξ be $\mathcal{F}_s$-measurable random variable such that $Z_s\xi \geq 0$ for any martingale $Z \in \mathcal{M}_s^T(G^*)$. Since **MCPS** holds, we have that $\eta\xi \geq 0$ for all $\eta \in L^1(\text{int}\, G_s^*, \mathcal{F}_s)$ and, hence, for all $\eta \in L^0(G_s^*, \mathcal{F}_s)$. This implies that $\xi \in G_s$ (a.s.).

B $\Rightarrow$ **NAA2** Let $\xi \in L^\infty(\mathbf{R}^d, \mathcal{F}_s)$ and let $V \in \overline{\mathcal{A}_{s,b}^T(T)}^w$ be such that $\xi + V \in L^\infty(G_T, \mathcal{F}_T)$. For any $Z \in \mathcal{M}_s^T(G^*)$ and $\Gamma \in \mathcal{F}_s$ the process $ZI_\Gamma \in \mathcal{M}_s^T(G^*)$ and we have:

$$0 \leq EZ_T(\xi + V)I_\Gamma = EZ_s\xi I_\Gamma + EZ_T I_\Gamma V \leq EZ_s\xi I_\Gamma$$

because $EZ_T I_\Gamma V \leq 0$ due to the hypothesis $\mathbf{S}_1$. Thus, $EZ_s\xi I_\Gamma \geq 0$ for every $\Gamma \in \mathcal{F}_s$, i.e. $Z_s\xi \geq 0$. By virtue of **B** the random variable $\xi \in L^\infty(G_s, \mathcal{F}_s)$ and we conclude.

NAA2 $\Rightarrow$ **B** Let $\zeta \in L^\infty(\mathbf{R}_+^d)$. We define the convex set

$$\Gamma_\zeta := \left\{x \in \mathbf{R}^d : \ \zeta - x \in \overline{\mathcal{A}_{s,b}^T(T)}^w\right\}$$

and the closed convex set

$$D_\zeta := \left\{x \in \mathbf{R}^d : \ EZ_s x \geq EZ_T\zeta \ \ \forall Z \in \mathcal{M}_s^T(G^*)\right\}.$$

Lemma 3.7.3 *Suppose that* **NFL** *holds. Then* $\Gamma_\zeta = D_\zeta$.

Proof. The argument is standard, but we sketch it for the sake of completeness. The inclusion $\Gamma_\zeta \subseteq D_\zeta$ is obvious. For the converse, let us consider a point $x \in D_\zeta$ such that $\zeta - x \notin \overline{\mathcal{A}_{s,b}^T(T)}^w$. Using the Hahn–Banach theorem, we separate $\zeta - x$ and $\overline{\mathcal{A}_{s,b}^T(T)}^w$ by a hyperspace given by some $\eta \in L^1(\mathbf{R}^d)$ and define the martingale $Z_t^\zeta = E(\eta|\mathcal{F}_t)$ for which $EZ_T^\zeta\xi \leq 0$ for all ξ from $\mathcal{A}_{s,b}^T(T)$. By our hypothesis the latter set is rich enough to ensure that $Z^\zeta \in \mathcal{M}_s^T(G^*)$. The point $\zeta - x$ lays in the interior of the complementary subspace, i.e., the inequality $EZ_T^\zeta(\zeta - x) > 0$ holds. This contradicts to the definition of D_ζ. Thus, $\Gamma_\zeta = D_\zeta$. $\square$

Suppose that $\xi \in L^\infty(\mathbf{R}^d, \mathcal{F}_s)$ is such that $Z_t\xi \geq 0$ for any $Z \in \mathcal{M}_s^T(G^*)$. It follows that $0 \in D_{-\xi}$ and, by the above lemma, $-\xi \in \overline{\mathcal{A}_{s,b}^T(T)}^w$. The last property means that

$$0 = \xi - \xi \in \left(\xi + \overline{\mathcal{A}_{s,b}^T(T)}^w\right) \cap L^\infty(\mathbf{R}_+^d).$$

In virtue of the condition **NAA2**, this may happen only if $\xi \in G_s$ a.s. So, the condition **B** is fulfilled.

B $\Rightarrow$ **MCPS** Now we assume that the sets $\mathcal{Y}_{s,b}^T(T)$ are Fatou-closed.

Lemma 3.7.4 *Assume that* **B** *and* **NFL** *hold. Then, for any* $\eta \in L^1(\text{int}\, G_s^*, \mathcal{F}_s)$, *there exists a sequence* $Z^n \in \mathcal{M}_s^T(G^* \backslash \{0\})$ *such that* $Z_s^n \to \eta$ *in* L^1.

Proof. Suppose that $\eta \in L^1(\operatorname{int} G_s^*, \mathcal{F}_s)$ does not belong to the set $\bar{M}_s$, the closure in L^1 of the convex cone $M_s := \{Z_s : Z \in \mathcal{M}_s^T(G^*\backslash\{0\})\}$. By the Hahn–Banach theorem, there exists $\xi \in L^\infty(G_s, \mathcal{F}_s)$ such that

$$EZ_s\xi \le E\eta\xi \quad \forall Z \in \mathcal{M}_s^T(G^*\backslash\{0\}).$$

Since the set $\mathcal{M}_s^T(G^*\backslash\{0\})$ is a cone, the left-hand side of the above inequality is negative for Z from this set.

We take a martingale $\tilde{Z} \in \mathcal{M}_s^T(G^*\backslash\{0\})$ existing by virtue of Theorem 3.7.1. For any $Z \in \mathcal{M}_s^T(G^*)$, $\Gamma \in \mathcal{F}_s$, and $k > 0$, the process $ZI_\Gamma + k^{-1}\tilde{Z}$ belongs to $\mathcal{M}_s^T(G^*\backslash\{0\})$ and $E(Z_sI_\Gamma + k^{-1}\tilde{Z}_s)\xi \le 0$. We deduce from here that $Z_s\xi \le 0$ for every $Z \in \mathcal{M}_s^T(G^*)$. Condition **B** implies that $\xi \in -G_s$ a.s., leading to a contradiction since $E\eta\xi > 0$. Hence, $\eta \in \bar{M}_s$, i.e., there exists a sequence $Z^n \in \mathcal{M}_s^T(G^*\backslash\{0\})$ such that $Z_s^n \to \eta$ in L^1. □

Since the components of Z^n in the above are positive, the expectations of components of the vector Z_T^n coincide with the expectations of components of Z_s^n. It follows that the sequence Z_T^n is bounded in L^1 and the Komlós theorem can be applied. Replacing the original sequence by a sequence of the Césaro means, from the latter theorem we obtain a sequence in $\mathcal{M}_s^T(G^*\backslash\{0\})$ the terminal values of which converge a.s. to a random variable $Z_T \in L^1(G_T^*)$. The following lemma shows that we could do better.

Lemma 3.7.5 *Assume that* **B** *and* **NFL** *hold. Then, for any* $\eta \in L^1(\operatorname{int} G_s^*, \mathcal{F}_s)$*, there exists a sequence* $Z^n \in \mathcal{M}_s^T(G^*\backslash\{0\})$ *such that* $Z_s^n \to \eta$ *in* L^1 *and* $Z_T^n \to Z_T$ *a.s. where* $Z_T \in L^1(G_T^* \setminus \{0\})$.

Proof. Let $\eta \in L^1(\operatorname{int} G_s^*, \mathcal{F}_s)$. We may assume without loss of generality that $E|\eta| \le 1/2$. We start with an arbitrary $\widetilde{Z}^1 \in \mathcal{M}_s^T(G^*\backslash\{0\}) \ne \emptyset$. Using the measurable selection, we find $\alpha_s^1 \in L^0(]0,1[, \mathcal{F}_s)$ such that the difference $\eta - \alpha_s^1\widetilde{Z}_s^1 \in \operatorname{int} G_s^*$ a.s. The process $\alpha_s^1\widetilde{Z}^1 \in \mathcal{M}_s^T(G^*\backslash\{0\})$; we may assume that $E|\eta - \alpha_s^1\widetilde{Z}_s^1| \le 1$.

Now, we proceed by induction. Put $\bar{Z}^1 := \alpha_s^1\widetilde{Z}^1$. Since $\eta - \overline{Z}_s^1 \in \operatorname{int} G_s^*$ a.s., we apply Lemma 3.7.4 and find $\widetilde{Z}^2 \in \mathcal{M}_s^T(G^*\backslash\{0\})$ such that

$$E|\eta - \overline{Z}_s^1 - \widetilde{Z}_s^2| \le 1/2.$$

Using measurable selection, we find $\alpha_s^2 \in L^0(]0,1[, \mathcal{F}_s)$ such that

$$\eta - \bar{Z}_s^1 - \alpha_s^2\widetilde{Z}_t^2 \in \operatorname{int} G_s^*,$$

where

$$\alpha_s^2\widetilde{Z}^2 \in \mathcal{M}_s^T(G^*\backslash\{0\}).$$

We put $\bar{Z}^2 := \bar{Z}^1 + \alpha_s^2\widetilde{Z}^2$. Suppose that we have already defined the processes $\bar{Z}^{n-1} = \sum_{i=1}^{n-1} \alpha_s^i\widetilde{Z}^i$, $\widetilde{Z}^{n-1}$, where $\widetilde{Z}^i \in \mathcal{M}_s^T(G^*\backslash\{0\})$ and $\alpha_s^i \in L^0(]0,1[, \mathcal{F}_s)$, such that

$$\eta - \overline{Z}_s^{n-1} \in \operatorname{int} G_s^* \quad \text{a.s.}, \qquad E|\eta - \bar{Z}_s^{n-2} - \widetilde{Z}_s^{n-1}| \leq 2^{-(n-1)}.$$

By Lemma 3.7.4 there is $\widetilde{Z}^n \in \mathcal{M}_s^T(G^*\backslash\{0\})$ such that

$$E|\eta - \bar{Z}_s^{n-1} - \widetilde{Z}_s^n| \leq 2^{-n}$$

and, by virtue of measurable selection arguments, there is $\alpha_s^n \in L^0(]0,1[,\mathcal{F}_s)$ such that

$$\eta - \bar{Z}_s^{n-1} - \alpha_s^n \widetilde{Z}_s^n \in \operatorname{int} G_s^*, \quad \alpha_s^n \widetilde{Z}^n \in \mathcal{M}_s^T(G^*\backslash\{0\}).$$

We put $\bar{Z}^n := \bar{Z}^{n-1} + \alpha_s^n \widetilde{Z}^n$, and the induction step is done.

Due to our standing assumption, $G_T^*\backslash\{0\}$ lies in the interior of $\mathbf{R}_+^d$. It follows that $\bar{Z}_T^{n-1}$ is a componentwise increasing sequence bounded in L^1, and, therefore, this sequence converges a.s. and in L^1 to some random variable $\bar{Z}_T \in L^1(G_T^*\backslash\{0\},\mathcal{F}_T)$. Automatically, $\bar{Z}_t^{n-1}$ converges (increasingly) to $E(\bar{Z}_T|\mathcal{F}_t)$ a.s. and in L^1 for each $t \geq s$. By construction, $\bar{Z}_s^{n-1} + \widetilde{Z}_s^n$ converges to η in L^1. The sequence of terminal values of martingales $Z^n := \bar{Z}^{n-1} + \widetilde{Z}^n$ evolving in $\mathcal{M}_s^T(G^*\backslash\{0\})$ is bounded L^1, and the Komlós theorem can be applied. That is, passing to a sequence of Césaro, we may assume without loss of generality that $\widetilde{Z}_T^n \to \widetilde{Z}_T$ where $\widetilde{Z}_T \in L^1(G_T,\mathcal{F}_T)$. Hence, the properties claimed in the lemma hold for the sequence of Z^n. □

We need some further auxiliary results.

For $\eta \in L^1(\operatorname{int} G_s^*,\mathcal{F}_s)$, we define the random half-space $\widetilde{G}_s$ by putting $\widetilde{G}_s^* = \mathbf{R}_+\eta$. Note that $(-\widetilde{G}_s) \cap G_s = \{0\}$.

Let $L_b^0(\mathbf{R}^d) := \{\xi \in \mathbf{R}^d : \exists \kappa_\xi \text{ such that } \xi + \kappa_\xi \mathbf{1} \in G_T\}$, and let

$$\widetilde{A}_{s,b}^T(T) := \left(L^0(-\widetilde{G}_s,\mathcal{F}_s) + \mathcal{Y}_s^T(T)\right) \cap L_b^0(\mathbf{R}^d).$$

Lemma 3.7.6 *Assume that* **B** *and* **NFL** *hold. If the set* $\mathcal{Y}_{s,b}^T(T)$ *is Fatou-closed, then* $\widetilde{A}_{s,b}^T(T)$ *is also Fatou-closed.*

Proof. We consider a sequence $Y_T^n := \xi_s^n + \gamma_T^n$, where

$$\xi_s^n \in L^0(-\widetilde{G}_s,\mathcal{F}_s), \qquad \gamma_T^n \in \mathcal{Y}_s^T(T),$$

are such that $Y_T^n + k\mathbf{1} \in G_T$ a.s. for some constant k and $Y_T^n \to Y_T$ a.s. Let $\Gamma_s = \{\sup_n |\xi_s^n| = \infty\}$. According to the lemma on subsequences, there exists a strictly increasing sequence of integer-valued $\mathcal{F}_s$-measurable random variables θ_n such that $|\xi_s^{\theta_n}| \to \infty$ on Γ_s.

Put

$$\widetilde{\xi}_s^n := \frac{\xi_s^{\theta_n}}{|\xi_s^{\theta_n}| \vee 1} I_{\Gamma_s}, \qquad \widetilde{\gamma}_T^n := \frac{\gamma_T^{\theta_n}}{|\xi_s^{\theta_n}| \vee 1} I_{\Gamma_s}, \qquad \widetilde{Y}_T^n := \frac{Y_T^{\theta_n}}{|\xi_s^{\theta_n}| \vee 1} I_{\Gamma_s}.$$

The sequence $\widetilde{Y}_T^n$ is bounded from below (in the sense of partial ordering induced by G_T). Since $\widetilde{\xi}_s^n$ takes values in the unit ball, this implies that the sequence $\widetilde{\gamma}_T^n$ is bounded from below and its elements belong to $\mathcal{Y}_{s,b}^T(T)$ (note that $\gamma_T^{\theta_n} = \sum_k \gamma_T^k I_{\{\theta_n=k\}} \in \mathcal{Y}_s^T(T)$ due to our assumption). Applying again the lemma on subsequences (this time to $(\widetilde{\xi}_s^n)$) and taking into account that $\widetilde{Y}_T^n \to 0$, we may assume without loss of generality that

$$\widetilde{\xi}_s^n \to \widetilde{\xi}_s \in L^\infty(-\widetilde{G}_s, \mathcal{F}_s), \qquad \widetilde{\gamma}_T^n \to \widetilde{\gamma}_T = -\widetilde{\xi}_s.$$

Due to the Fatou-closedness of $\mathcal{Y}_{s,b}^T(T)$, we have that $\widetilde{\gamma}_T \in \mathcal{Y}_{s,b}^T(T)$.

Let $Z \in \mathcal{M}_s^T(G^*)$, and let $\Gamma \in \mathcal{F}_s$. It follows from hypothesis $\mathbf{S}_1$ that

$$0 = EZ_T I_\Gamma(\widetilde{\xi}_s + \widetilde{\gamma}_T) \le EZ_T I_\Gamma \widetilde{\xi}_s = EZ_s I_\Gamma \widetilde{\xi}_s.$$

Thus, $Z_s I \widetilde{\xi}_s \ge 0$, and, by virtue of condition $\mathbf{B}$, $\widetilde{\xi}_s \in L^\infty(G_s, \mathcal{F}_s)$. Hence, $\widetilde{\xi}_s \in L^\infty((-\widetilde{G}_s) \cap G_s, \mathcal{F}_s)$, i.e., $\widetilde{\xi}_s = 0$ a.s. But $|\widetilde{\xi}_s| = 1$ on Γ_s. Therefore, $P(\Gamma_s) = 0$.

We may assume, passing to a subsequence, that $\xi_s^n \to \xi_s$ and $\gamma_T^n \to \gamma_T$ a.s. In the same spirit as above, we define $\bar{Y}_T^n = \bar{\xi}_s^n + \bar{\gamma}_T^n$, where

$$\bar{\xi}_s^n := \frac{\xi_s^n}{|\xi_s^n| + 1} \in L^\infty(-\widetilde{G}_s, \mathcal{F}_s), \qquad \bar{\gamma}_T^n := \frac{\gamma_T^n}{|\xi_s^n| + 1} \in \mathcal{Y}_b^s(T).$$

By virtue of the Fatou-closedness of $\mathcal{Y}_b^s(T)$, we obtain that

$$\bar{\gamma}_T^n \to \bar{\gamma}_T = \frac{\gamma_T}{|\xi_s| + 1} \in \mathcal{Y}_{b,s}^T(T).$$

Thus, $Y_T = \xi_s + (1 + |\xi_s|)\bar{\gamma}_T$ is an element of $\widetilde{A}_{s,b}^T(T)$. □

Lemma 3.7.7 *Assume that* **B** *and* **NFL** *hold. If the set $\mathcal{Y}_{s,b}^T(T)$ is Fatou-closed, then $\widetilde{A}_{s,b}^T(T) \cap L^\infty$ is Fatou dense in $\widetilde{A}_{s,b}^T(T)$.*

Proof. Let $Y_T = \xi_s + \gamma_T \in \widetilde{A}_{s,b}^T(T)$ and $Y_T + \kappa\mathbf{1} \ge_{G_T} 0$. Put

$$Y_T^n := Y_T I_{\{|Y_T| \le n\}} - \kappa \mathbf{1} I_{\{|Y_T| > n\}}.$$

From the identity $Y_T - Y_T^n = (Y_T + \kappa\mathbf{1}) I_{\{|Y_T|>n\}} \in G_T$ we easily obtain that $Y_T^n \in \widetilde{A}_{s,b}^T(T)$. Clearly, Y_T^n form a sequence Fatou-convergent to Y_T. □

By virtue of the above lemmas, we obtain the following dual characterization of the Fatou-closed set $\widetilde{A}_{s,b}^T(T)$ (see Appendix 5.5):

$$\widetilde{A}_{s,b}^T(T) = \Big\{\xi \in L_b^0(\mathbf{R}^d) : \ E\xi\eta \le \sup_{X \in \widetilde{A}_{s,b}^T(T)} EX\eta, \ \forall \eta \in L^1(G_T^*)\Big\}. \quad (3.7.1)$$

Lemma 3.7.8 *Assume that* **B** *and* **NFL** *hold. If the set* $\mathcal{Y}^T_{s,b}(T)$ *is Fatou-closed, then*

$$\widetilde{A}^T_{s,b}(T) \cap L^\infty(\mathbf{R}^d_+) = \{0\}.$$

Proof. Let us consider

$$Y_T := \xi_s + \gamma_T \in \widetilde{A}^T_{s,b}(T) \cap L^\infty(\mathbf{R}^d_+).$$

Using the notation introduced above, we rewrite Y_T in the form

$$Y_T := \big(1 + |\xi_s|\big)(\overline{\xi}_s + \overline{\gamma}_T),$$

where $\overline{\gamma}_T \in \mathcal{Y}^T_{s,b}(T)$ and $\overline{\xi} \in L^\infty(-\widetilde{G}_s, \mathcal{F}_s)$. For the sequence Z^n from Lemma 3.7.5, we have by the Fatou lemma that

$$0 \le E(\overline{\xi}_s + \overline{\gamma}_T)Z_T \le \liminf_n \big(E\overline{\xi}_s Z^n_T + E\overline{\gamma}_T Z^n_T\big),$$

where

$$E\overline{\xi}_s Z^n_T = E\overline{\xi}_s Z^n_s \to E\overline{\xi}_s \eta \le 0,$$

and $E\overline{\gamma}_T Z^n_T \le 0$ under condition $\mathbf{S}_1$. This implies that $Y_T = 0$ a.s. □

With the above lemma, we get the implication $\mathbf{B} \Rightarrow$ **MCPS** by a standard argument. Indeed, the Kreps–Yan separation theorem ensures the existence of $Z_T \in L^1(\operatorname{int}\mathbf{R}^d_+, \mathcal{F}_T)$ such that $EZ_T\xi \le 0$ for all $\xi \in \widetilde{A}^T_{s,b}(T)$. Define the martingale $Z_t := E(Z_T|\mathcal{F}_t)$, $s \ge t$, whose components are strictly positive. Since $\widetilde{A}^T_{s,b}(T)$ contains $L^\infty(-G_t, \mathcal{F}_t)$ for $t \ge s$ and $L^\infty(-\widetilde{G}_s, \mathcal{F}_s)$, we infer that $Z_t \in L^1(G^*_t, \mathcal{F}_t)$ for $t \ge s$ and $Z_s \in L^1(\mathbf{R}_+\eta, \mathcal{F}_s)$. Since Z_T is defined up to a scalar strictly positive multiplier, we choose it to have the equality $Z_s = \eta$ and get a processes claimed in condition **MCPS**.

Theorem 3.7.2 is proven. □

4

Consumption–Investment Problems

4.1 Consumption–Investment without Friction

4.1.1 The Merton Problem

The study of consumption–investment problems in continuous time was initiated by Merton. He considered a model of frictionless market where the price processes are geometric Brownian motions and the investor's goal is to maximize the expected discounted utility of consumption on the infinite time interval. For the power utility function, he obtained an explicit solution of the optimal control problem. This solution has a clear financial meaning: the optimal investment is to keep the proportions of the total wealth held in risky securities equal to a constant vector. The latter is easily calculated from the model parameters. This work was extended by many authors in various directions including models with transaction costs, which are the main objects of our interest. Taking into account that the Merton problem is classical and exposed in a number of textbooks, we give here a rather sketchy presentation needed to understand basic ideas and methods as well as their evolution. The results of this section will be used at the end of this chapter, where we discuss an asymptotical behavior of the consumption–investment problem for small transaction cost coefficients.

We are given a stochastic basis with an m-dimensional standard Wiener process w. The market contains a nonrisky security, which is the *numéraire*, i.e., its price is identically equal to unit, and m risky securities with the price evolution

$$dS_t^i = S_t^i\big(\mu^i\,dt + dM_t^i\big), \quad i = 1, \dots, m, \tag{4.1.1}$$

where $M = \Sigma w$ is a (deterministic) linear transform of w. Thus, M is a Gaussian martingale with $\langle M\rangle_t = At$; the covariance matrix $A = \Sigma\Sigma^*$ is assumed to be nondegenerate.

The evolution of the value process corresponding to a self-financing strategy H is given as $dV_t = H_t\,dS_t$. Assuming withdrawal of the funds for the

Yu. Kabanov, M. Safarian, *Markets with Transaction Costs*,
Springer Finance, DOI 10.1007/978-3-540-68121-2_4,

consumption with rate $c_t \geq 0$, we arrive at the dynamics

$$dV_t = H_t\,dS_t - c_t\,dt. \tag{4.1.2}$$

Of course, we can substitute dS_t by its expression given in (4.1.1). Since H_t^i is a number of units of the ith asset in the portfolio, the quantity $\alpha_t^i := H_t^i S_t^i / V_t$ can be interpreted as the proportion of the wealth invested in this asset. It is convenient to choose α together with c as the control parameters. With these considerations, the problem with infinite time horizon can be formulated in usual terms of stochastic optimal control theory in the following way.

The system dynamics is given by the controlled stochastic differential equation

$$dV_t = V_t\alpha_t(\mu\,dt + dM_t) - c_t\,dt, \qquad V_0 = x, \tag{4.1.3}$$

with the initial condition $x > 0$ and the control $\pi = (\alpha, c)$, which is a predictable process. We suppose that the consumption intensity process c has trajectories integrable on every finite interval, while the trajectories of α are uniformly bounded by a constant which may depend on the strategy.[1]

More substantially, we require from π to be in the class of *admissible controls* $\mathcal{A}(x)$ for which the process $V = V^{x,\pi}$ is positive. We assume also that after the bankruptcy time (which is the first instant when V hits zero), the control π is equal to zero, and the process V stops.

The investor's goal is the following:

$$EJ_\infty^\pi \to \max, \tag{4.1.4}$$

i.e., to maximize the expectation of the limiting expected value of the utility process J^π defined as

$$J_t^\pi := \int_0^t e^{-\beta s} u(c_s)\,ds. \tag{4.1.5}$$

The standard economically meaningful assumptions on the utility function are that u is increasing and concave. For the sake of simplicity, we add to this that u is positive and $u(0) = 0$. The parameter $\beta > 0$ shows to which extent the agent prefers to consume today rather than in the future.

A typical example is the power utility function $u(c) = c^\gamma/\gamma$, $\gamma \in]0,1[$.

Define the Bellman function

$$W(x) := \sup_{\pi \in \mathcal{A}(x)} EJ_\infty^\pi, \quad x > 0. \tag{4.1.6}$$

By convention, $\mathcal{A}(0) := \{0\}$ and $W(0) := 0$.

Notice that the Bellman function W inherits the properties of u. Namely, it is increasing (as $\mathcal{A}(\tilde{x}) \supseteq \mathcal{A}(x)$ when $\tilde{x} \geq x$). With the chosen α-parameterization, its concavity appears not to be so obvious, but we get it immediately

[1] This assumption is not very wise but allows us to avoid discussions of integrability. It is done because, in the Merton problem, the optimal strategy in a wider class possesses this property.

by turning back to the initial H-parameterization. Indeed, suppose that the strategies $\pi_j = (\alpha_j, c_j)$, $\pi_j \in \mathcal{A}(x_j)$, $j = 1, 2$, generate the value processes V_j. The convex combination of these processes, $V = \lambda V_1 + (1-\lambda)V_2$, is of the form (4.1.2), where H and c are the convex combinations with the same coefficients of the corresponding controls. The process H admits the representation via the process α with the components

$$\alpha^i = H^i S^i / V = \frac{\lambda V_1}{\lambda V_1 + (1-\lambda)V_2} \alpha_1^i + \frac{(1-\lambda)V_2}{\lambda V_1 + (1-\lambda)V_2} \alpha_2^i;$$

α is bounded because both α_j are bounded. Thus, $\pi = (\alpha, \lambda c_1 + (1-\lambda)c_2)$ belongs to $\mathcal{A}(x)$ with $x = \lambda x_1 + (1-\lambda)x_2$, and, therefore,

$$W\big(\lambda x_1 + (1-\lambda)x_2\big) \geq EJ_\infty^\pi \geq \lambda EJ_\infty^{\pi_1} + (1-\lambda)EJ_\infty^{\pi_2}$$

due to the concavity of u. With this, we obtain the concavity of W by taking supremum over π_1 and π_2.

Notice that we cannot guarantee without additional assumptions that W is finite. If the latter property holds, then, due to the concavity, $W(x)$ is continuous for $x > 0$, but the question whether it is continuous at zero remains open.

At last, when the utility u is a power function, the Bellman function W, if finite, is proportional to u. Indeed, the linear dynamics of the control system implies that $W(\nu x) = \nu^\gamma W(x)$ for all $\nu > 0$, i.e., the Bellman function is positive homogeneous of the same order as the utility function. In a scalar case this homotheticity property defines, up to a multiplicative constant, a unique finite function, namely x^γ.

Now we formulate the Merton theorem.

Theorem 4.1.1 *Let u be the power utility function. Assume that the parameters of the model are such that the constant*

$$\kappa_M := \frac{1}{1-\gamma}\left(\beta - \frac{1}{2}\frac{\gamma}{1-\gamma}\left|A^{-1/2}\mu\right|^2\right) > 0. \qquad (4.1.7)$$

Then the optimal strategy $\pi^o = (\alpha^o, c^o)$ is given by the formulae

$$\alpha^o = \theta := \frac{1}{1-\gamma} A^{-1}\mu, \qquad (4.1.8)$$

$$c_t^o = \kappa_M V_t^o, \qquad (4.1.9)$$

where V^o is the solution of the linear stochastic equation

$$dV^o = V_t^o \theta(\mu\, dt + dM_t) - \kappa_M V_t^o\, dt, \qquad V_0^o = x. \qquad (4.1.10)$$

The process V^o is optimal, and the Bellman function is

$$W(x) = \kappa_M^{\gamma-1} x^\gamma / \gamma = \mathbf{m} x^\gamma. \qquad (4.1.11)$$

Note that W is proportional to x^γ, and, therefore, the last assertion is about the exact value of the coefficient $\mathbf{m}$, which happens to be finite and equal to $\kappa_M^{\gamma-1}/\gamma$ with κ_M given by (4.1.7).

4.1.2 The HJB Equation and a Verification Theorem

The most powerful and efficient method to solve stochastic control problems is the method of dynamic programming based on the analysis of the Hamilton–Jacobi–Bellman equation (HJB, in short). For our infinite-horizon problem, the latter is

$$\sup_{(\alpha,c)}\left[\frac{1}{2}|A^{1/2}\alpha|^2x^2f''(x)+\alpha\mu xf'(x)-\beta f(x)-f'(x)c+u(c)\right]=0, \quad (4.1.12)$$

where $x>0$, and the supremum is taken over all $\alpha\in\mathbf{R}^d$ and $c\in\mathbf{R}_+$.

To solve the consumption–investment problem with the power utility function, we use a very elementary tool, namely, the so-called verification theorem for the HJB equation. It is based on the following considerations.

Let $f:\mathbf{R}_+\to\mathbf{R}_+$ and $\pi\in\mathcal{A}(x)$. We consider the nonnegative process $X^f=X^{f,x,\pi}$ with

$$X_t^f:=e^{-\beta t}f(V_t)+J_t^\pi, \quad (4.1.13)$$

where $V=V^{x,\pi}$. If f is smooth, the Itô formula for the process V given by (4.1.3) implies the following important representation, which is the key point to explain how the HJB equation arises:

$$X_t^f=f(x)+D_t+N_s, \quad (4.1.14)$$

where

$$D_t:=\int_0^t e^{-\beta s}L(V_s,\alpha_s,c_s)\,ds \quad (4.1.15)$$

with $L(x,\alpha,c)$ standing for the expression in square brackets of the formula (4.1.12), and

$$N_t:=\int_0^t e^{-\beta s}f'(V_s)V_s\alpha_s\,dM_s. \quad (4.1.16)$$

The process N is a continuous local martingale up to the bankruptcy time σ. That is, there exist stopping times $\sigma_n\uparrow\sigma$ such that the stopped processes N^{σ_n} are uniformly integrable martingales. In the case where $\sigma=\infty$ and N is a martingale, we shall take $\sigma_n=n$.

Suppose now that a smooth function f is a supersolution of (4.1.12), i.e.,

$$\sup_{(\alpha,c)}\left[\frac{1}{2}|A^{1/2}\alpha|^2x^2f''(x)+\alpha\mu xf'(x)-\beta f(x)-f'(x)c+u(c)\right]\le 0. \quad (4.1.17)$$

Then the integrand in the definition of D does not exceed zero, and, therefore, the process D is decreasing with $D_0=0$. This implies, in particular, the inequality $N\ge -f(x)$. It follows (as usual, by applying the Fatou lemma) that N, being bounded from below, is a supermartingale. Due to the inequality $-D_t\le f(x)+N_t$, the (negative) random variable D_t is integrable: we obtain that the process X_t^f is a supermartingale, and, hence,

$$EJ_t = EX_t^f - Ee^{-\beta t} f(V_t) \le EX_t^f \le f(x). \tag{4.1.18}$$

Since $EJ_t \to EJ_\infty$ as $t \to \infty$, we infer that $W(x) \le f(x)$, i.e., f provides a "cap" for the Bellman function, implying, in particular, that the latter is finite. If, moreover, the supersolution f vanishes at zero, the function W (being positive) is necessarily continuous at zero. Summarizing, we formulate the outcome of this reasoning in the following statement.

Proposition 4.1.2 *If f is a supersolution of (4.1.12), then $W \le f$, and, hence, $W \in C(\mathbf{R}_+ \setminus \{0\})$. If, moreover, $f(0+) = 0$, then $W \in C(\mathbf{R}_+)$.*

An inspection of the above reasoning shows that if, in addition, it happened that the process D (depending on the control) vanishes and

$$\lim_n Ee^{-\beta\sigma_n} f(V_{\sigma_n}) = 0, \tag{4.1.19}$$

then $W = f$, and the corresponding control is optimal. With these observations, we arrive at the promised verification theorem, which can be obtained, of course, in a much more general context.

Theorem 4.1.3 *Let $f \in C(\mathbf{R}_+) \cap C^2(\mathbf{R}_+ \setminus \{0\})$ be a positive concave function solving the HJB equation (4.1.12) and vanishing at zero. Suppose that the supremum in (4.1.12) is attained on $\alpha(x)$ and $c(x)$ such that α is a bounded measurable function, c is a positive measurable function, and the equation*

$$dV_t^o = V_t^o \alpha(V_t^o)(\mu\, dt + dM_t) - c(V_t^o)\, dt, \qquad V_0^o = x, \tag{4.1.20}$$

admits a strong solution V_t^o. If condition (4.1.19) holds for the process V^o, then $W = f$, and the optimal control $\pi^o = (\alpha(V^o), c(V^o))$.

4.1.3 Proof of the Merton Theorem

With the above provision, we return to the HJB equation (4.1.12) and calculate the supremum.

Put

$$u^*(p) := \sup_{c \ge 0} [u(c) - cp];$$

the function u^* is the Fenchel transform of the function $-u(-.)$. In particular, for the power utility $u(c) = c^\gamma/\gamma$, we have that

$$u^*(p) = \frac{1-\gamma}{\gamma} p^{\gamma/(\gamma-1)} \tag{4.1.21}$$

because the supremum in the definition of u^* is attained at the point $p^{1/(\gamma-1)}$. Expecting that $f'' < 0$, we find easily that the maximum of the quadratic form over α is attained at the point

$$\alpha^o(x) = -A^{-1}\mu \frac{f'(x)}{x f''(x)}.$$

Thus, the HJB equation can be transformed to the following one:

$$-\frac{1}{2}|A^{-1/2}\mu|^2 \frac{(f'(x))^2}{f''(x)} - \beta f(x) + \frac{1-\gamma}{\gamma}(f'(x))^{\frac{\gamma}{\gamma-1}} = 0.$$

We find easily that its solution of the form $f(x) = \mathbf{m}x^\gamma$ should have the coefficient $\mathbf{m} = \kappa_M^{\gamma-1}/\gamma$ with $\kappa_M > 0$ given in (4.1.7).

Now the function $\alpha^o(x) = A^{-1}\mu/(1-\gamma)$ is constant, $c^o(x) = \kappa_M x$, and (4.1.20), pretending to describe the optimal dynamics, is linear:

$$\frac{dV_t^o}{V_t^o} = \left(\frac{1}{1-\gamma}|A^{-1/2}\mu|^2 - \kappa_M\right) dt + \frac{A^{-1}\mu}{1-\gamma}\, dM, \qquad V_0^o = x,$$

and its solution is the geometric Brownian motion, which never hits zero. Noticing that $\langle A^{-1}\mu M\rangle_t = |A^{-1/2}\mu|^2 t$, it can be given by the following explicit formula:

$$V_t^o = x\exp\left\{\left(\frac{1}{1-\gamma} - \frac{1}{2}\frac{1}{(1-\gamma)^2}\right)|A^{-1/2}\mu|^2 t - \kappa_M t + \frac{A^{-1}\mu}{1-\gamma}M_t\right\}.$$

Since $E(V_t^o)^p = x^p e^{\kappa_p t}$ where κ_p is a constant, the process N for this control is a true martingale, and we may take the localizing sequence σ_n deterministic.

In the particular case where $p = \gamma$, the corresponding constant

$$\kappa_\gamma = \frac{1}{2}\frac{\gamma}{1-\gamma} - \gamma\kappa_M = \beta - \kappa_M$$

in virtue of (4.1.7). Thus,

$$e^{-\beta t}E\big(V_t^o\big)^\gamma = x^\gamma e^{-\kappa_M t},$$

and, therefore, (4.1.19) holds. The Merton theorem is proven.

4.1.4 Discussion

1. The optimal strategy in the Merton problem with the power utility functions prescribes to keep constant proportions of wealth in each position. Let us consider the special case $m = 1$, i.e., the model with a single risky asset. Then the quantities $V_t^{2o} := \alpha^o V_t^o$ and $V_t^{1o} = (1-\alpha^o)V_t^o$ are, respectively, the optimal holdings in the risky and nonrisky assets,

$$\alpha^o = \theta = \frac{1}{1-\gamma}\frac{\mu}{\sigma^2}.$$

Thus,

$$V_t^{2o} := \frac{\alpha^o}{1-\alpha^o} V_t^{1o} = \frac{\theta}{1-\theta} V_t^{1o}.$$

This means that the two-dimensional process (V_t^{1o}, V_t^{2o}) on the plain (v^1, v^2) evolves along the straight line with the slope $\theta/(1-\theta)$, called in the literature the *Merton line.* The parameter θ is referred to as the *Merton proportion.*

2. In our presentation we consider the case where the price of the nonrisky asset is constant over time as it would pay the interest $r = 0$. The reader may be accustomed with the tradition to treat the model with an arbitrary $r \geq 0$. However, it is easy to see that, for the power utility function, considering the model with zero interest rate does not lead to any loss in generality. Indeed, due to the identity

$$u(e^{rs} c_s) = e^{\gamma rs} u(c_s),$$

the maximization problem where the consumption is measured in "money" is the same as that where the consumption is measured in "bonds" but with the coefficient β replaced by $\tilde{\beta} := \beta - \gamma r$. Thus, there is no real reason to retain r in calculations.

3. An analysis of the proof of Theorem 4.1.1 shows that, after minor changes, it works well also for the power utility function with $\gamma < 0$, and, hence, the same explicit formulae represent the optimal solution also in this case. The HJB approach can be extended to the model with the logarithmic utility function $u(c) = \ln c$ (corresponding to the value $\gamma = 0$). Of course, one needs to impose an additional constraint to the consumption process ensuring the integrability of J_∞^π.

4. Turning back to the multi-asset case, let us define the scalar process $\tilde{M}$ with $d\tilde{M} = \theta(\mu\, dt + dM_t)$. Let us consider the same consumption–investment problem imposing the restriction that the investments should be shared between money and the risky asset the price evolution of which follows the process $\tilde{M}$. Any value process and consumption process in this two-asset model are those of the original one. One can imagine a financial institution (a mutual fund) which offers such an artificial asset, called the *market portfolio.* This allows the agent to allocate his wealth only in the nonrisky asset and the market portfolio. Due to this economical interpretation, the Merton theorem sometimes is referred to as the *mutual fund theorem.*

5. Formula (4.1.11) shows that, for a positive initial capital, the value $W(x) \to \infty$ as $\kappa_M \downarrow 0$. It follows that, for small values of the discount parameter β, namely, when

$$\beta \leq \frac{1}{2} \frac{\gamma}{1-\gamma} |A^{-1/2}\mu|^2,$$

the Bellman function $W(x) = \infty$, $x > 0$.

4.1.5 Robustness of the Merton Solution

There is an interesting question about the sensitivity of the Merton solution with respect to errors in determining the optimal proportion. It happens that it is quite robust: a deviation of order ε from the Merton proportion leads to losses in the expected utility only of order ε^2. To see this, suppose that in the two-asset model the investor's strategy is to maintain the proportion $\alpha^o+\varepsilon$ and consume a constant part $(1+\delta)\kappa_M$ of the current wealth optimizing the expected utility with respect to δ. Assume, for simplicity, that the initial endowment $x = 1$. For such a strategy, the dynamics is given by the linear equation

$$\frac{dV_t}{V_t} = (\alpha^o + \varepsilon)(\mu\, dt + \sigma\, dw_t) - (1+\delta)\kappa_M\, dt$$

the solution of which is the geometric Brownian motion

$$V_t = \exp\Big\{ (\alpha^o + \varepsilon)\mu t - \frac{1}{2}(\alpha^o + \varepsilon)^2\sigma^2 t - (1+\delta)\kappa_M t + (\alpha^o + \varepsilon)\sigma w_t \Big\}.$$

We have that

$$EV_t^\gamma = e^{\kappa_\gamma(\varepsilon,\delta)t},$$

where

$$\kappa_\gamma(\varepsilon,\delta) = \beta - \kappa_M - \frac{1}{2}\gamma(1-\gamma)\sigma^2\varepsilon^2 - \gamma\kappa_M\delta,$$

and, in particular, $\kappa_\gamma(0,0) = \kappa_\gamma = \beta - \kappa_M$.

Notice that the coefficient at ε is zero, and this is a crucial fact. It follows that

$$EJ_\infty = \frac{1}{\gamma}\kappa_M^\gamma(1+\delta)^\gamma \int_0^\infty e^{-\beta t} EV_t^\gamma\, dt = \frac{1}{\gamma}\kappa_M^{\gamma-1}\frac{(1+\delta)^\gamma}{1+\frac{1}{2\kappa_M}\gamma(1-\gamma)\sigma^2\varepsilon^2 + \gamma\delta}.$$

Maximization over δ gives us the optimal value $\delta^o = \frac{1}{2\kappa_M}\gamma\sigma^2\varepsilon^2$, for which

$$EJ_\infty = \frac{1}{\gamma}\kappa_M^{\gamma-1}(1+\delta^o)^{\gamma-1} = \mathbf{m} - \frac{1}{2}(1-\gamma)\kappa_M^{\gamma-2}\sigma^2\varepsilon^2 + O(\varepsilon^4),$$

and we get the claimed asymptotic.

Of course, the robustness of the Merton solution is of great practical importance.

4.2 Consumption–Investment under Transaction Costs

4.2.1 The Model

The setting described in this section is, in some aspects, slightly more general than that of the standard model of financial market under constant proportional transaction costs. In particular, the cone K is not supposed to be polyhedral. On the other hand, it is more restrictive with respect to the price processes: they are assumed to be geometric Brownian motions. Our framework appeals to a well-developed theory of viscosity solutions (in fact, only to basic elements of the latter) and allows us to catch essential properties of the Bellman function before going to the specific case of the two-asset model with the power utility function the detailed analysis of which is our ultimate goal.

Let $Y = (Y_t)$ be an $\mathbf{R}^d$-valued semimartingale on a stochastic basis $(\Omega, \mathcal{F}, \mathbf{F}, P)$ with trivial initial σ-algebra. Let K and $\mathcal{C}$ be *proper* cones in $\mathbf{R}^d$ such that $\mathcal{C} \subseteq \operatorname{int} K \neq \emptyset$. Define the set $\mathcal{A}$ of controls $\pi = (B, C)$ as the set of adapted càdlàg processes of bounded variation such that, up to an evanescent set,

$$\dot{B} \in -K, \qquad \dot{C} \in \mathcal{C}. \tag{4.2.1}$$

Let $\mathcal{A}_a$ be the set of controls with absolutely continuous C and $\Delta C_0 = 0$. For the elements of $\mathcal{A}_a$, we have $c := dC/dt \in \mathcal{C}$.

The controlled process $V = V^{x,\pi}$ is the solution of the linear system

$$dV_t^i = V_{t-}^i \, dY_t^i + dB_t^i - dC_t^i, \quad V_{0-}^i = x^i, \quad i = 1, \ldots, d. \tag{4.2.2}$$

For $x \in \operatorname{int} K$, we consider the subsets $\mathcal{A}^x$ and $\mathcal{A}_a^x$ of "admissible" controls for which the processes $V^{x,\pi}$ never leave the set $\operatorname{int} K \cup \{0\}$ and have the origin as an absorbing point. Thus, if $V_{s-}(\omega) \in \partial K$, then $\Delta B_s(\omega) = -V_{s-}(\omega)$.

The important hypothesis that the cone K is proper, i.e., $K \cap (-K) = \{0\}$, or equivalently, $\operatorname{int} K^* \neq \emptyset$, corresponds to the model of financial market with *efficient friction*. In a financial context, K (usually containing $\mathbf{R}_+^d$) is interpreted as the solvency region and $C = (C_t)$ as the consumption process; the process $B = (B_t)$ describes the accumulated fund transfers.

Let $G := (-K) \cap \partial\mathcal{O}_1(0)$, where $\partial\mathcal{O}_1(0) = \{x \in \mathbf{R}^d : |x| = 1\}$ in accordance with the notation for the open ball $\mathcal{O}_r(y) := \{x \in \mathbf{R}^d : |x - y| < r\}$ The set G is a compact, and $-K = \operatorname{cone} G$. We denote by Σ_G the *support function* of G, given by the relation $\Sigma_G(p) = \sup_{x \in G} px$.

We shall work using the following assumption:

H₁. The process Y is a continuous process with independent increments with mean $EY_t = \mu t$, $\mu \in \mathbf{R}^d$, and covariance $DY_t = At$.

To facilitate references, we formulate also a more specific hypothesis (frequent in the literature), where the matrix A is diagonal with $a^{ii} = (\sigma^i)^2$, i.e., the components of the driving noise are independent.

H₂. The components of Y are of the form $dY_t^i = \mu^i \, dt + \sigma^i \, dw_t^i$, where w is a standard Wiener process in $\mathbf{R}^d$.

In our proof of the dynamic programming principle (needed to derive the HJB equation) we shall assume that the stochastic basis is a canonical one, that is, the space of continuous functions with the Wiener measure.

The efficient friction assumption, together with the hypothesis $\mathbf{H_1}$, ensures that the L^2-norm of the "maximal function" of the portfolio trajectories admits an exponential bound which is uniform with respect to strategies. This result will be used in the sequel to claim that certain stochastic integrals are not just local martingales but true martingales. For future references, we immediately give a precise formulation and proof.

Proposition 4.2.1 *There is a constant $\kappa > 0$ such that*

$$E \sup_{t\le T} |V_t|^2 \le \kappa |x|^2 e^{\kappa T^2} \tag{4.2.3}$$

for any value process $V = V^{x,\pi}$, $x \in \operatorname{int} K$, and $T \ge 0$.

Proof. As usual, κ denotes a "generic" positive constant which may be different in different formulae. Let us take an arbitrary vector $p \in \operatorname{int} K^*$ with $|p| = 1$. Making use that $p\,dB \le 0$ and $p\,dC \ge 0$ (in the sense of densities), we obtain from (4.2.2) that

$$pV_s \le px + \int_0^s \tilde{p} V_r \, dr + \int_0^s V_r \, d\tilde{M}_r,$$

where $\tilde{p}^i := p^i \mu^i$, and $\tilde{M}^i = p^i M^i$ with M denoting the martingale part of Y. The crucial observation is that there is $\kappa > 0$ such that $\kappa^{-1}|y| \le py$ for any $y \in K$. Since $|py| \le |y|$ for any $y \in \mathbf{R}^d$, we easily obtain the estimate

$$|V_s| \le \kappa|x| + \kappa \int_0^s |V_r| \, dr + \kappa \left| \int_0^s V_r \, d\tilde{M}_r \right|.$$

Notice that the right-hand side of this inequality is a continuous process, and, hence, V is locally bounded, i.e., there exists a sequence of stopping times $\tau_n \uparrow \infty$ such that each stopped process $V^{\tau_n} = (V_{t\wedge\tau_n})$ is bounded. With this observation, the proof is completed by a fairly standard argument, which we only sketch on. Squaring the above inequality, we get, by elementary estimates combined with the Cauchy–Schwarz and Doob inequalities, that the (bounded) function $\varphi_t^{(n)} := E \sup_{s\le t\wedge\tau_n} |V_s|^2$ satisfies the inequality

$$\varphi_t^{(n)} \le \kappa|x|^2 + \kappa(T+1) \int_0^t \varphi_s^{(n)} \, ds.$$

The Gronwall–Bellman lemma implies that

$$\varphi_T^{(n)} \le \kappa|x|^2 e^{\kappa(T+1)T}.$$

Taking here the limit in n and enlarging the constant, we arrive at the required bound. □

4.2.2 Goal Functionals

Let $U : \mathcal{C} \to \mathbf{R}_+$ be a concave function such that $U(0) = 0$ and $U(x)/|x| \to 0$ as $|x| \to \infty$. With every $\pi = (B, C) \in \mathcal{A}_a^x$, we associate the "utility process"

$$J_t^\pi := \int_0^t e^{-\beta s} U(c_s)\, ds, \quad t \geq 0,$$

where $\beta > 0$. We consider the infinite-horizon maximization problem with the *goal functional* EJ_∞^π and define its *Bellman function* W by

$$W(x) := \sup_{\pi \in \mathcal{A}_a^x} EJ_\infty^\pi, \quad x \in \operatorname{int} K. \tag{4.2.4}$$

If π_i, $i = 1, 2$, are admissible strategies for the initial points x_i, then the strategy $\lambda\pi_1 + (1 - \lambda)\pi_2$ is an admissible strategy for the initial point $\lambda x_1 + (1 - \lambda)x_2$ for any $\lambda \in [0, 1]$, and the corresponding absorbing time is the maximum of the absorbing times for both π_i. It follows that the function W is *concave* on $\operatorname{int} K$. Since $A_a^{x_1} \subseteq A_a^{x_2}$ when $x_2 - x_1 \in K$, the function W is *increasing* with respect to the partial ordering $\geq_K$ generated by the cone K. It is convenient to put W equal to zero on the boundary of K and extend it to the whole space $\mathbf{R}^d$ as a concave function just by putting $W := -\infty$ outside K.

Remark 1. In financial models, usually, $\mathcal{C} = \mathbf{R}_+ e_1$ and $\sigma^0 = 0$, i.e., the only first (nonrisky) asset is consumed. Our presentation in this section is oriented to the scalar power utility function $u(c) = c^\gamma/\gamma$, $\gamma \in]0, 1[$. As we already mentioned in the previous section, in this case there is no need to consider a nonzero interest rate for the nonrisky asset, which can be chosen as the numéraire. Of course, for other types of utility functions, adding to the model an interest rate may have sense.

Remark 2. We consider here a model with mixed "regular–singular" controls. In fact, the assumption that the consumption process has an intensity $c = (c_t)$ and the agent's utility depends on this intensity is not very satisfactory from the economical point of view. One can consider models with an intertemporal substitution and the consumption by "gulps," i.e., dealing with "singular" controls of the class $\mathcal{A}^x$ and the goal functionals like

$$J_t^\pi := \int_0^t e^{-\beta s} U(\bar{C}_s)\, ds,$$

where

$$\bar{C}_s = \int_0^s K(s, r)\, dC_r$$

with a suitable kernel $K(s, r)$ (the exponential kernel $e^{-\gamma(s-r)}$ is the common choice).

4.2.3 The Hamilton–Jacobi–Bellman Equation

Assume that hypothesis $\mathbf{H}_1$ on the structure of driving noise holds. In the sequel we denote by U^* the convex function $U^*(p) := \sup_{x\in\mathcal{C}}(U(x) - px)$. We introduce a continuous function of four variables by putting

$$F(X, p, W, x) := \max\{F_0(X, p, W, x) + U^*(p), \Sigma_G(p)\},$$

where X belongs to $\mathcal{S}_d$, the set of $d\times d$ symmetric matrices, $p, x \in \mathbf{R}^d$, $W \in \mathbf{R}$, and the function F_0 is given by

$$F_0(X, p, W, x) := \frac{1}{2}\operatorname{tr} A(x)X + \mu(x)p - \beta W,$$

where $A^{ij}(x) := a^{ij}x^ix^j$, $\mu^i(x) := \mu^i x^i$, $1 \le i, j \le d$. In the detailed form we have that

$$F_0(X, p, W, x) = \frac{1}{2}\sum_{i,j=1}^{d} a^{ij}x^ix^jX^{ij} + \sum_{i=1}^{d}\mu^ix^ip^i - \beta W.$$

If ϕ is a smooth function, we put

$$\mathcal{L}\phi(x) := F\big(\phi''(x), \phi'(x), \phi(x), x\big).$$

In a similar way, $\mathcal{L}_0$ corresponds to the function F_0.

We show, under mild hypotheses, that W is the unique viscosity solution of the Dirichlet problem for the HJB equation

$$F\big(W''(x), W'(x), W(x), x\big) = 0, \quad x \in \operatorname{int} K, \tag{4.2.5}$$

$$W(x) = 0, \quad x \in \partial K, \tag{4.2.6}$$

with the boundary condition understood in the usual classical sense.

We do not suppose that the reader is acquainted with the theory of viscosity solutions. Necessary prerequisites, adapted to our needs, are given in the next sections.

4.2.4 Viscosity Solutions

Since, in general, W may have no derivatives at some points $x \in \operatorname{int}K$ (and this is, indeed, the case for the model considered here), the notation (4.2.5) needs to be interpreted. The idea of viscosity solutions is to plug into F the derivatives and Hessians of quadratic functions touching W from above and below. Formal definitions (adapted to the case we are interested in) are as follows.

Let f and g be functions defined in a neighborhood of zero. We shall write $f(.) \lessapprox g(.)$ if $f(h) \le g(h) + o(|h|^2)$ as $|h| \to 0$. The notation $f(.) \gtrapprox g(.)$ and $f(.) \approx g(.)$ has an obvious meaning.

For $p \in \mathbf{R}^d$ and $X \in \mathcal{S}_d$, we consider the quadratic function

$$Q_{p,X}(z) := pz + (1/2)\langle Xz, z\rangle, \quad z \in \mathbf{R}^d,$$

and define the *superjets* and *subjets* of a function v at the point x:

$$J^+v(x) := \big\{(p,X):\ v(x+.) \lessapprox v(x) + Q_{p,X}(.)\big\},$$
$$J^-v(x) := \big\{(p,X):\ v(x+.) \gtrapprox v(x) + Q_{p,X}(.)\big\}.$$

In other words, $J^+v(x)$ (resp. $J^-v(x)$) is the family of coefficients of quadratic functions $v(x) + Q_{p,X}(y - .)$ dominating the function $v(.)$ (resp., dominated by this function) in a neighborhood of the point x with precision up to the second order included and coinciding with $v(.)$ at this point.

A function $v \in C(K)$ is called a *viscosity supersolution* of (4.2.5) if

$$F\big(X, p, v(x), x\big) \le 0 \quad \forall (p,X) \in J^-v(x),\ x \in \operatorname{int} K.$$

A function $v \in C(K)$ is called a *viscosity subsolution* of (4.2.5) if

$$F\big(X, p, v(x), x\big) \ge 0 \quad \forall (p,X) \in J^+v(x),\ x \in \operatorname{int} K.$$

A function $v \in C(K)$ is a *viscosity solution* of (4.2.5) if v is simultaneously a viscosity supersolution and subsolution of (4.2.5).

At last, a function $v \in C(K)$ is called a *classical supersolution* of (4.2.5) if $v \in C^2(\operatorname{int} K)$ and $\mathcal{L}v \le 0$ on $\operatorname{int} K$. We add the adjective *strict* when $\mathcal{L}v < 0$ on the set $\operatorname{int} K$.

Of course, the above notions[2] can be formulated also for open subsets of K.

If v is smooth at a point x, then

$$J^+v(x) := \big\{(p,X):\ p = v'(x), X \ge v''(x)\big\},$$
$$J^-v(x) := \big\{(p,X):\ p = v'(x), X \le v''(x)\big\},$$

where the inequality between matrices is understood in the sense of partial ordering induced by the cone of positive semidefinite matrices. The pair $(v'(x), v''(x))$ is the unique element belonging to the intersection of $J^-v(x)$ and $J^+v(x)$. Thus, any viscosity solution v which is in $C^2(\operatorname{int} K)$ is a classical solution of (4.2.5). It is not difficult to check that a classical solution solves (4.2.5) in the viscosity sense: the needed property that F is increasing in X with respect to the partial ordering holds in our case.

Remark on a mnemonic rule. The monotonicity allows us to memorize easily the signs of the inequalities for F. In the smooth case for the second-order Taylor approximation, i.e., for the quadratic function $(v'(x), v''(x))$, we

[2] The reader may notice that the introduced concepts are related only with the operator and, therefore, could be called viscosity super-, sub-, and median functions, which seems to be a more natural terminology. We have no courage to deviate from the tradition already established in the theory of viscosity solutions.

have the equality. Thus, if $X \geq v''(x)$ for the pair $(v'(x), X)$ which is an element of $J^+v(x)$, we have obviously the inequality ≥ 0. Note that in the literature, the equation is quite often written with the opposite sign, and so its left-hand side is decreasing in X...

For the sake of simplicity and having in mind the specific case we shall work on, we incorporated in the definitions the requirement that the viscosity super- and subsolutions are continuous on K including the boundary. For other cases, this might be too restrictive, and more general and flexible formulations can be used.

The next criterion gives a flexibility to manipulate with the above concepts. It allows us to use smooth local majorants/minorants of a function, which is the supposed viscosity solution, as test functions (to be inserted with their derivatives into the operator).

Lemma 4.2.2 *Let $v \in C(K)$. Then the following conditions are equivalent:*

(a) *the function v is a viscosity supersolution of (4.2.5);*
(b) *for any ball $\mathcal{O}_r(x) \subseteq K$ and any $f \in C^2(\mathcal{O}_r(x))$ such that $v(x) = f(x)$ and $v \geq f$ on $\mathcal{O}_r(x)$, the inequality $\mathcal{L}f(x) \leq 0$ holds.*

Proof. (a) $\Rightarrow$ (b). Obvious: the pair $(f'(x), f''(x))$ is in $J^-v(x)$ according to the Taylor formula.

(b) $\Rightarrow$ (a). Take (p, X) in $J^-v(x)$. To conclude, we construct a smooth function f with $f'(x) = p$ and $f''(x) = X$ satisfying the requirements of (b).

By definition,

$$v(x+h) - v(x) - Q_{p,X}(h) \geq |h|^2\varphi(|h|),$$

where $\varphi(u) \to 0$ as $u \downarrow 0$. We consider on $]0, r[$ the function

$$\delta(u) := \sup_{\{h:\ |h| \leq u\}} \frac{1}{|h|^2}\big(v(x+h) - v(x) - Q_{p,X}(h)\big)^- \leq \sup_{\{y:\ 0 \leq y \leq u\}} \varphi^-(y).$$

Obviously, δ is continuous, increasing, and $\delta(u) \to 0$ as $u \downarrow 0$. The function

$$\Delta(u) := \frac{2}{3}\int_u^{2u}\int_\eta^{2\eta} \delta(\xi)\, d\xi\, d\eta$$

vanishes at zero with its two right derivatives, and $u^2\delta(u) \leq \Delta(u) \leq u^2\delta(4u)$. It follows that the function $x \mapsto \Delta(|x|)$ belongs to $C^2(\mathcal{O}_r(0))$, its Hessian vanishes at zero, and

$$v(x+h) - v(x) - Q_{p,X}(h) \geq -|h|^2\delta(|h|) \geq -\Delta(|h|).$$

Thus, $f(y) := v(x) + Q_{p,X}(y - x) - \Delta(|y - x|)$ is the needed function. □

For subsolutions, we have a similar result with the inverse inequalities. Using the alternative definition, we can easily establish the following:

Lemma 4.2.3 *Suppose that the function v is a viscosity solution of (4.2.5). If v is twice differentiable at x_0, then it satisfies (4.2.5) at this point in the classical sense.*

Proof. One needs to be more precise with definitions since it is not assumed that v' is defined at every point of a neighborhood of x_0. "Twice differentiable" means here that the Taylor formula at x_0 holds:

$$v(x) = v(x_0) + \langle v'(x_0), x - x_0\rangle + \frac{1}{2}\langle v''(x_0)(x - x_0), x - x_0\rangle + o(|x - x_0|^2).$$

Let us consider the C^2-function

$$f_\varepsilon(x) = v(x_0) + \langle v'(x_0), x - x_0\rangle + \frac{1}{2}\langle v''(x_0)(x - x_0), x - x_0\rangle + \varepsilon|x - x_0|^2$$

with $f_\varepsilon(x_0) = v(x_0)$. If $\varepsilon < 0$, then $v \geq f_\varepsilon$ in a sufficiently small neighborhood of x_0. Thus, by virtue of the previous lemma $\mathcal{L}f_\varepsilon(x_0) \leq 0$. Letting ε tend to zero, we obtain that $\mathcal{L}v(x_0) \leq 0$. Taking in the above definition $\varepsilon > 0$, we get the opposite inequality. □

Obviously, one can give a slightly different formulation saying that v is a viscosity supersolution of the second-order differential equation if and only if, for every $x \in \operatorname{int} K$, the inequality

$$F(\phi''(x), \phi'(x), v(x), x) \leq 0 \tag{4.2.7}$$

holds for any C^2-function ϕ such that, at the point x, the difference $v - \phi$ attains its local minimum equal to zero. The reader may ask why we replace in the inequality $\phi(x)$ by $v(x)$, which is the same number. This has sense! We can skip in the suggested reformulation the words "equal to zero" due to the following assertion, which happens to be useful in the sequel.

Lemma 4.2.4 *A function $v \in C(K)$ is a viscosity supersolution of (4.2.5) if and only if, for every point $x \in \operatorname{int} K$, inequality (4.2.7) holds for any C^2-function ϕ defined in a neighborhood of the point x and such that the difference $v - \phi$ attains its local minimum at x.*

Proof. In one direction the claim is trivial, and we need to check only that, for a supersolution, the mentioned inequality (4.2.7) holds when $v - \phi$ has a local minimum at x, i.e., when, for all y from a certain neighborhood $\mathcal{O}_\varepsilon(x)$, we have the bound

$$v(y) - \phi(y) > v(x) - \phi(x), \quad y \neq x.$$

Let $\bar{v}$ be a C^2-function dominated by v, and let g be a smooth function on $\mathbf{R}_+$ taking values in the interval $[0,1]$ and such that $g(t) = 1$ for $t \leq \varepsilon/2$ and

$g(t) = 0$ for $t \geq \varepsilon$. Let us consider the C^2-function $\tilde{\phi} = \tilde{\phi}(y)$ with

$$\tilde{\phi}(y) = \big[\phi(y) + v(x) - \phi(x)\big]g\big(|x-y|\big) + \big(1 - g\big(|x-y|\big)\big)\bar{v}(y).$$

The difference $v - \tilde{\phi}$ attains its minimal value equal to zero at point x, and, therefore, by the supersolution property, (4.2.7) holds for $\tilde{\phi}$ and, hence, for ϕ because the two derivatives of both functions coincide at x. □

Again, a corresponding result holds for subsolutions. Notice also that specific features of the set K (with nonempty interior) were not used in the above discussions.

Now we give an application of the last characterization of the viscosity solution to prove an assertion claiming that, for a "regular" ordinary differential equation, a C^1-function known to be the viscosity solution is, in fact, a smooth one satisfying the equation in the classical sense. In the present context, the "regular" means, roughly speaking, that the equation can be solved with respect to the second derivative and the resulting right-hand side is continuous in all variables. More precisely, we have the following:

Lemma 4.2.5 *Let $\psi \in C^1(a,b)$ be a viscosity solution of the equation*

$$\psi''(z) = G\big(\psi'(z), \psi(z), z\big).$$

Suppose that the right-hand side here is a continuous function. Then the function $\psi \in C^2(a,b)$, and the equation holds in the classical sense.

Proof. Take a subinterval $[z_1, z_2]$ of $]a,b[$ and consider on it the C^2-function $\psi_\varepsilon(z)$ such that

$$\psi_\varepsilon''(z) = G\big(\psi'(z), \psi(z), z\big) + \varepsilon, \qquad \psi_\varepsilon(z_i) = \psi(z_i), \quad i = 1, 2.$$

Of course, this function could be expressed by an explicit formula, but we need not it. The parameter ε here is an arbitrary real number. We first argue with $\varepsilon > 0$. Suppose that $\psi - \psi_\varepsilon$ attains a local minimum at an interior point z of $[z_1, z_2]$. Then, necessarily, $\psi_\varepsilon'(z) = \psi'(z)$. According to the above criterion for the supersolution,

$$\psi_\varepsilon''(z) \leq G\big(\psi_\varepsilon'(z), \psi(z), z\big) = G\big(\psi'(z), \psi(z), z\big),$$

in contradiction with the definition of ψ_ε. Thus, the difference $\psi - \psi_\varepsilon$ is minimal at the extremities of $[z_1, z_2]$, where it is equal to zero. This means that $\psi(z) \geq \psi_\varepsilon(z)$ for all $z \in [z_1, z_2]$. Letting $\varepsilon \downarrow 0$ and noting that $\psi_\varepsilon(z) \to \psi_0(z)$ (even uniformly), we obtain the inequality $\psi(z) \geq \psi_0(z)$. Arguing in the same way with $\varepsilon < 0$ and using the subsolution property, we obtain the reverse inequality. So, $\psi = \psi_0$ on $[z_1, z_2]$. This means that ψ_0 is a classical solution on this interval, and it coincides with ψ. It is easily seen that such a property implies the claim of the lemma. □

4.2.5 Ishii's Lemma

The only result we need from the theory of viscosity solutions (or, better to say, from convex analysis) is the following simplified version of Ishii's lemma, see Crandall et al. [42] or Fleming and Soner [72].

Lemma 4.2.6 *Let v and $\tilde{v}$ be two continuous functions on an open subset $\mathcal{O} \subseteq \mathbf{R}^d$. Consider the function $\Delta(x,y) := v(x) - \tilde{v}(y) - \frac{1}{2}n|x-y|^2$ with $n > 0$. Suppose that Δ attains a local maximum at $(\widehat{x}, \widehat{y})$. Then there are symmetric matrices X and Y such that*

$$\big(n(\widehat{x}-\widehat{y}), X\big) \in \bar{J}^+ v(\widehat{x}), \qquad \big(n(\widehat{x}-\widehat{y}), Y\big) \in \bar{J}^- \tilde{v}(\widehat{y}),$$

and

$$\begin{pmatrix} X & 0 \\ 0 & -Y \end{pmatrix} \le 3n \begin{pmatrix} I & -I \\ -I & I \end{pmatrix}. \tag{4.2.8}$$

In this statement, I is the identity matrix, and $\bar{J}^+ v(x)$ and $\bar{J}^- v(x)$ are values of the set-valued mappings whose graphs are closures of graphs of the set-valued mappings $J^+ v$ and $J^- v$, respectively.

Of course, if v is smooth, the claim follows directly from the necessary conditions of a local maximum (with $X = v''(\widehat{x})$, $Y = \tilde{v}''(\widehat{y})$ and the constant 1 instead of 3 in inequality (4.2.8)).

The following assertion is an easy exercise from linear algebra.

Lemma 4.2.7 *The inequality (4.2.8) implies that, for any $d \times m$ matrices B and C,*

$$\operatorname{tr}(BB'X - CC'Y) \le 3n|B-C|^2. \tag{4.2.9}$$

Proof. For a symmetric matrix $S \ge O$ and any matrix G of appropriate dimension, $\operatorname{tr} GG'S = \operatorname{tr} G'S^{1/2}S^{1/2}G \ge 0$. Manipulating with block matrices and using this observation, we have

$$\begin{aligned} \operatorname{tr}(BB'X - CC'Y) &= \operatorname{tr}\begin{pmatrix} BB' & BC' \\ CB' & CC' \end{pmatrix}\begin{pmatrix} X & 0 \\ 0 & -Y \end{pmatrix} \\ &\le 3n \operatorname{tr}\begin{pmatrix} BB' & BC' \\ CB' & CC' \end{pmatrix}\begin{pmatrix} I & -I \\ -I & I \end{pmatrix} \\ &= 3n \operatorname{tr}(BB' - BC' - CB' + CC') \\ &= 3n \operatorname{tr}(B-C)(B-C)' = 3n|B-C|^2, \end{aligned}$$

and the result is proven. □

Notice that $A(x) = \operatorname{diag} x A \operatorname{diag} x$. We denote by $\operatorname{diag} x$ the diagonal matrix whose entries on the diagonal are the coordinates of the vector x. Applying the above lemma with the matrices $B = \operatorname{diag} x A^{1/2}$ and $C = \operatorname{diag} y A^{1/2}$, we obtain the following inequality which we need in the sequel:

$$\operatorname{tr}\big(A(x)X - A(y)Y\big) \le 3n\big|A^{1/2}\big|^2|x-y|^2. \tag{4.2.10}$$

Remark. We can obtain the similar inequality

$$\operatorname{tr}\big(A(x)X - A(y)Y\big) \le 3n \operatorname{tr} A|x-y|^2$$

by "probabilistic" considerations using the above lemma in its simpler version with $m = 1$. Indeed, let ξ be a standard Gaussian vector-column, and let $\eta = A^{1/2}\xi$.

Applying the lemma with $B = \operatorname{diag} x\,\eta$ and $C = \operatorname{diag} y\,\eta$, we get the inequality

$$\operatorname{tr}\,(BB'X - CC'Y) \le 3n\big|\operatorname{diag}(x-y)\big|^2|\eta|^2.$$

It remains to take the expectation and note that $EBB' = A(x)$, $ECC' = A(y)$, and $E|\eta|^2 = \operatorname{tr} A$.

4.3 Uniqueness of the Solution and Lyapunov Functions

4.3.1 Uniqueness Theorem

The following concept plays a crucial role in the proof of a purely analytic result on the uniqueness of the viscosity solution, which we establish by a classical method of doubling variables using the Ishii lemma.

Definition. We say that a positive function $\ell \in C(K) \cap C^2(\operatorname{int} K)$ is the *Lyapunov function* if the following properties are satisfied:

(1) $\ell'(x) \in \operatorname{int} K^*$ and $\mathcal{L}_0\ell(x) \le 0$ for all $x \in \operatorname{int} K$,
(2) $\ell(x) \to \infty$ as $|x| \to \infty$.

Theorem 4.3.1 *Suppose that there exists a Lyapunov function ℓ. Then the Dirichlet problem (4.2.5)–(4.2.6) has at most one viscosity solution in the class of continuous functions satisfying the growth condition*

$$W(x)/\ell(x) \to 0, \quad |x| \to \infty. \tag{4.3.1}$$

Proof. Let W and $\tilde{W}$ be two viscosity solutions of (4.2.5) coinciding on the boundary ∂K. Suppose that $W(z) > \tilde{W}(z)$ for some $z \in K$. Take $\varepsilon > 0$ such that

$$W(z) - \tilde{W}(z) - 2\varepsilon\ell(z) > 0.$$

We introduce the family of continuous functions $\Delta_n : K \times K \to \mathbf{R}$ by putting

$$\Delta_n(x,y) := W(x) - \tilde{W}(y) - \frac{1}{2}n|x-y|^2 - \varepsilon\big[\ell(x) + \ell(y)\big], \quad n \ge 0.$$

Note that $\Delta_n(x,x) = \Delta_0(x,x)$ for all $x \in K$ and $\Delta_0(x,x) \le 0$ for $x \in \partial K$. From the assumption that the function l has a higher growth rate than W we deduce that $\Delta_n(x,y) \to -\infty$ as $|x|+|y| \to \infty$. It follows that the level sets $\{\Delta_n \ge a\}$ are compacts and the function Δ_n attains its maximum. That is, there exist $(x_n,y_n) \in K \times K$ such that

$$\Delta_n(x_n,y_n) = \bar{\Delta}_n := \sup_{(x,y)\in K\times K} \Delta_n(x,y) \ge \bar{\Delta} := \sup_{x\in K} \Delta_0(x,x) > 0.$$

All (x_n,y_n) belong to the compact set $\{(x,y):\ \Delta_0(x,y) \ge 0\}$. It follows that the sequence $n|x_n-y_n|^2$ is bounded. We continue to argue (without introducing new notation) with a subsequence along which (x_n,y_n) converge to some limit $(\widehat{x},\widehat{x})$. Necessarily, $n|x_n-y_n|^2 \to 0$ (otherwise we would have $\Delta_0(\widehat{x},\widehat{x}) > \bar{\Delta}$). It is easily seen that $\bar{\Delta}_n \to \Delta_0(\widehat{x},\widehat{x}) = \bar{\Delta}$. Thus, $\widehat{x}$ is an interior point of K, and so are x_n and y_n for sufficiently large n.

By virtue of the Ishii lemma applied to the functions $v := W - \varepsilon\ell$ and $\tilde{v} := \tilde{W} + \varepsilon\ell$ at the point (x_n,y_n), there exist matrices $X^n = (X^n_{ij})$ and $Y^n = (Y^n_{ij})$ satisfying (4.2.8) and such that

$$\big(n(x_n-y_n), X^n\big) \in \bar{J}^+ v(x_n), \qquad \big(n(x_n-y_n), Y^n\big) \in \bar{J}^- \tilde{v}(y_n).$$

Using the notation $p_n := n(x_n-y_n) + \varepsilon\ell'(x_n)$, $q_n := n(x_n-y_n) - \varepsilon\ell'(y_n)$, $X_n := X^n + \varepsilon\ell''(x_n)$, $Y_n := Y^n - \varepsilon\ell''(y_n)$, we may rewrite the last relations in the following equivalent form:

$$(p_n, X_n) \in \bar{J}^+ W(x_n), \qquad (q_n, Y_n) \in \bar{J}^- \tilde{W}(y_n). \tag{4.3.2}$$

Since W and $\tilde{W}$ are viscosity sub- and supersolutions,

$$F\big(X_n, p_n, W(x_n), x_n\big) \ge 0 \ge F\big(Y_n, q_n, \tilde{W}(y_n), y_n\big).$$

The second inequality implies that $mq_n \le 0$ for each $m \in G = (-K)\cap\partial\mathcal{O}_1(0)$. But for the Lyapunov function, $\ell'(x) \in \operatorname{int} K^*$ for $x \in \operatorname{int} K$, and, therefore,

$$mp_n = mq_n + \varepsilon m\big(\ell'(x_n) + \ell'(y_n)\big) < 0.$$

Since G is a compact, $\Sigma_G(p_n) < 0$. It follows that

$$F_0\big(X_n, p_n, W(x_n), x_n\big) + U^*(p_n) \ge 0 \ge F_0\big(Y_n, q_n, \tilde{W}(y_n), y_n\big) + U^*(q_n).$$

Recall that U^* is decreasing with respect to the partial ordering generated by $\mathcal{C}^*$ and, hence, also by K^*. Thus, $U^*(p_n) \le U^*(q_n)$, and we obtain the inequality

$$b_n := F_0\big(X_n, p_n, W(x_n), x_n\big) - F_0\big(Y_n, q_n, \tilde{W}(y_n), y_n\big) \ge 0.$$

Clearly,

$$b_n = \frac{1}{2}\sum_{i,j=1}^{d}\left(a^{ij}x_n^i x_n^j X_{ij}^n - a^{ij}y_n^i y_n^j Y_{ij}^n\right) + n\sum_{i=1}^{d}\mu^i\left(x_n^i - y_n^i\right)^2$$
$$- \frac{1}{2}\beta n|x_n - y_n|^2 - \beta\Delta_n(x_n, y_n) + \varepsilon\big(\mathcal{L}_0\ell(x_n) + \mathcal{L}_0\ell(y_n)\big).$$

By virtue of (4.2.10), the first sum is dominated by const$\times n|x_n - y_n|^2$; a similar bound for the second sum is obvious; the last term is negative according to the definition of a Lyapunov function. It follows that $\limsup b_n \le -\beta\bar{\Delta} < 0$, and we get a contradiction arising from the assumption $W(z) > \tilde{W}(z)$. □

An inspection of the arguments shows that they lead to the following slightly more general and useful comparison result.

Theorem 4.3.2 *Assume that there exists a Lyapunov function ℓ. Let W and $\tilde{W}$ be, respectively, viscosity sub- and supersolution of the equation in an open set $\mathcal{O} \subseteq K$ coinciding on $\partial\mathcal{O}$ and such that*

$$W(x) = o\big(\ell(x)\big), \quad \tilde{W}(x) = o\big(\ell(x)\big), \quad |x| \to \infty.$$

Then $W(x) \le \tilde{W}(x)$ for all $x \in \bar{\mathcal{O}}$.

Remark. The definition of a Lyapunov function does not depend on U (it is a property of the operator with $U^* = 0$), and we have the uniqueness for any utility function U for which U^* is decreasing with respect to the partial ordering induced by K^*. However, to apply the uniqueness theorem, we should know that W is not growing faster than a certain Lyapunov function.

4.3.2 Existence of Lyapunov Functions and Classical Supersolutions

Results on the uniqueness of a solution to the HJB equation are all based on work with specific Lyapunov functions. The following general considerations explain how the latter can be constructed.

Let $u \in C(\mathbf{R}_+) \cap C^2(\mathbf{R}_+ \setminus \{0\})$ be an increasing strictly concave function with $u(0) = 0$ and $u(\infty) = \infty$. Introduce the function $R := -u'^2/(u''u)$. Assume that $\bar{R} := \sup_{z>0} R(z) < \infty$.

For $p \in K^* \setminus \{0\}$, we define the function $f(x) = f_p(x) := u(px)$ on K. If $y \in K$ and $x \neq 0$, then $yf'(x) = (py)u'(px) \le 0$. The inequality is strict when $p \in \operatorname{int} K^*$.

Recall that $A(x)$ is the matrix with $A^{ij}(x) = A^{ij}x^i x^j$ and the vector $\mu(x)$ has the components $\mu^i x^i$. Suppose that $\langle A(x)p, p\rangle \neq 0$. Putting $z := px$ for brevity, we obtain by obvious transformations intended to isolate full square that

$$\mathcal{L}_0 f(x) = \frac{1}{2}\left[\langle A(x)p,p\rangle u''(z) + 2\langle \mu(x),p\rangle u'(z) + \frac{\langle \mu(x),p\rangle^2}{\langle A(x)p,p\rangle}\frac{u'^2(z)}{u''(z)}\right]$$
$$+\frac{1}{2}\frac{\langle \mu(x),p\rangle^2}{\langle A(x)p,p\rangle}R(z)u(z) - \beta u(z). \tag{4.3.3}$$

Since $u'' \le 0$, the expression in the square brackets is negative, and so is the whole right-hand side of the above formula if $\beta \ge \eta(p)\bar{R}$, where

$$\eta(p) := \frac{1}{2}\sup_{x\in K}\frac{\langle \mu(x),p\rangle^2}{\langle A(x)p,p\rangle}.$$

Of course, if $\langle A(x)p,p\rangle = 0$ we cannot argue in this way, but if in such a case also $\langle \mu(x),p\rangle = 0$, then $\mathcal{L}_0 f(x) = -\beta u(z) \le 0$ for any $\beta \ge 0$.

These simple observations lead us to the following existence result for Lyapunov functions:

Proposition 4.3.3 *Let $p \in \operatorname{int} K^*$. Suppose that $\langle \mu(x),p\rangle$ vanishes on the set $\{x \in \operatorname{int} K : \ \langle A(x)p,p\rangle = 0\}$. If $\beta \ge \eta(p)\bar{R}$, then f_p is a Lyapunov function.*

Let $\bar{\eta} := \sup_{p\in K^*}\eta(p)$. Note that $\eta(p) = \eta(p/|p|)$. Continuity considerations show that $\bar{\eta}$ is finite if $\langle A(x)p,p\rangle \ne 0$ for all $x \in K\setminus\{0\}$ and $p \in K^*\setminus\{0\}$. Obviously, if $\beta \ge \bar{\eta}\bar{R}$, then f_p is a Lyapunov function for $p \in \operatorname{int} K^*$.

The representation (4.3.3) is useful also in the search of classical supersolutions for the operator $\mathcal{L}$. Since $\mathcal{L}f = \mathcal{L}_0 f + U^*(f')$, it is natural to choose u related to U. For a particular case where $\mathcal{C} = \mathbf{R}^d_+$ and $U(c) = u(e_1 c)$, with u satisfying the postulated properties (except, maybe, unboundedness) and assuming, moreover, that the inequality

$$u^*\big(au'(z)\big) \le g(a)u(z) \tag{4.3.4}$$

holds, we get, using the homogeneity of $\mathcal{L}_0$, the following result.

Proposition 4.3.4 *Assume $\langle A(x)p,p\rangle \ne 0$ for all $x \in \operatorname{int} K$ and $p \in K^*\setminus\{0\}$. Suppose that (4.3.4) holds for all $a, z > 0$ with $g(a) = o(a)$ as $a \to \infty$. If $\beta > \bar{\eta}\bar{R}$, then there exists a_0 such that, for every $a \ge a_0$, the function af_p is a classical supersolution of (4.2.5) whatever is $p \in K^*$ with $p^1 \ne 0$. Moreover, if $p \in \operatorname{int} K^*$, then af_p is a strict supersolution on any compact subset of* $\operatorname{int} K$.

For the power utility function $u(z) = z^\gamma/\gamma$, $\gamma \in]0,1[$, we have

$$R(z) = \gamma/(1-\gamma) = \bar{R}$$

and $u^*(au'(z)) = (1-\gamma)a^{\gamma/(\gamma-1)}u(z)$. Therefore, inequality (4.3.4) holds with $g(a) = o(a)$, $a \to 0$.

If Y satisfies $\mathbf{H}_2$ with $\sigma^1 = 0$, $\mu^1 = 0$ (i.e., the first asset is the *numéraire*), and $\sigma^i \ne 0$ for $i \ne 1$, then, by the Cauchy–Schwarz inequality applied to

$\langle \mu(x), p \rangle$,

$$\eta(p) \le \frac{1}{2} \sum_{i=2}^{d} \left(\frac{\mu^i}{\sigma^i} \right)^2 .$$

The inequality

$$\beta > \frac{\gamma}{1-\gamma} \frac{1}{2} \sum_{i=2}^{d} \left(\frac{\mu^i}{\sigma^i} \right)^2 \tag{4.3.5}$$

(implying the relation $\beta > \bar{\eta}\bar{R}$) is a standing assumption in many studies on the consumption–investment problem under transaction costs, see Akian et al. [3] and Davis and Norman [47].

In particular, for the model with only one risky asset and the power utility function, by virtue of the above computations, we have, for the function $f(x) = au(px)$ given by $p \in K^*$ with $p^1 = 1$, that

$$\mathcal{L}_0 f(x) + U^*(f'(x)) = [\ldots] + \left(\frac{1}{2} \frac{\gamma}{1-\gamma} \frac{\mu^2}{\sigma^2} - \beta + (1-\gamma) a^{1/(\gamma-1)} \right) f(x),$$

where $[\ldots] \le 0$. This implies the following conclusion.

Proposition 4.3.5 *Suppose that, in the two-asset model with the power utility function, the Merton parameter*

$$\kappa_M := \frac{1}{1-\gamma} \left(\beta - \frac{1}{2} \frac{\gamma}{1-\gamma} \frac{\mu^2}{\sigma^2} \right) > 0.$$

Then the function

$$f(x) = \frac{1}{\gamma} \kappa_M^{\gamma-1} (px)^\gamma = \mathbf{m}(px)^\gamma \tag{4.3.6}$$

is a classical supersolution of the HJB equation whatever is $p \in K^*$ with $p^1 = 1$.

As we shall see in the next section, the existence of supersolutions has important implications for the Bellman function, ensuring, in particular, the finiteness of the latter.

4.4 Supersolutions and Properties of the Bellman Function

4.4.1 When is W Finite on K?

First, we present sufficient conditions ensuring that the Bellman function W of the considered maximization problem is finite.

Let Φ be the set of continuous functions $f : K \to \mathbf{R}_+$ increasing with respect to the partial ordering $\ge_K$ and such that, for every $x \in \operatorname{int} K$ and every $\pi \in \mathcal{A}_a^x$ the positive process $X^f = X^{f,x,\pi}$ given by the formula

$$X_t^f := e^{-\beta t} f(V_t) + J_t^\pi, \tag{4.4.1}$$

where $V = V^{x,\pi}$, is a supermartingale.

The set Φ of f with this property is convex and stable under the operation $\wedge$ (recall that the minimum of two supermartingales is a supermartingale). Any continuous function which is a monotone limit (increasing or decreasing) of functions from Φ also belongs to Φ.

Lemma 4.4.1 (a) *If $f \in \Phi$, then $W \le f$;*

(b) *if for any $y \in \partial K$, there exists $f \in \Phi$ such that $f(y) = 0$, then W is continuous on K.*

Proof. (a) Using the positivity of f, the supermartingale property of X^f, and, finally, the monotonicity of f, we get the following chain of inequalities leading to the required property:

$$EJ_t^\pi \le EX_t^f \le f(V_0) \le f(V_{0-}) = f(x).$$

(b) Recall that a concave function is locally Lipschitz continuous on the interior of its domain, i.e., on the interior of the set where it is finite. Hence, if Φ is not empty, then W is continuous (and even locally Lipschitz continuous) on $\operatorname{int} K$. The continuity at a point $y \in \partial K$ follows from the assumed property because $0 \le W \le f$. $\square$

Lemma 4.4.2 *Let $f : K \to \mathbf{R}_+$ be a function in $C(K) \cap C^2(\operatorname{int} K)$. If f is a classical supersolution of (4.2.5), then $f \in \Phi$, i.e., X^f is a supermartingale.*

Proof. First, notice that a classical supersolution is increasing with respect to the partial ordering $\ge_K$. Indeed, by the finite increments formula we have that, for any $x, h \in \operatorname{int} K$,

$$f(x+h) - f(x) = f'(x + \vartheta h)h$$

for some $\vartheta \in [0,1]$. The right-hand side is greater or equal to zero because, for the supersolution f, we have the inequality $\Sigma_G(f'(y)) \le 0$ whatever is $y \in \operatorname{int} K$, or, equivalently, $f'(y)h \ge 0$ for every $h \in K$, just by the definition of the support function Σ_G and the choice of G as a generator of the cone $-K$. By continuity, $f(x+h) - f(x) \ge 0$ for every $x, h \in K$.

In order to be able to apply the Itô formula in a comfortable way, we introduce the process $\tilde{V} = V^{\sigma-} = VI_{[0,\sigma[} + V_{\sigma-}I_{[\sigma,\infty[}$, where σ is the first hitting time of zero by the process V. This process coincides with V on $[0,\sigma[$ but, in contrast to the latter, either always remains in $\operatorname{int} K$ (due to the stopping at σ if $V_{\sigma-} \in \operatorname{int} K$) or exits to the boundary in a continuous way and stops there. Let $\tilde{X}^f$ be defined by (4.4.1) with V replaced by $\tilde{V}$. Since

$$X^f = \tilde{X}^f + e^{-\beta\sigma}\big(f(V_{\sigma-} + \Delta B_\sigma) - f(V_{\sigma-})\big)I_{[\sigma,\infty[},$$

by the monotonicity of f it is sufficient to verify that $\tilde{X}^f$ is a supermartingale.

Applying Itô's formula to $e^{-\beta t} f(\tilde{V}_t)$, we obtain on $[0,\sigma[$ the representation

$$\tilde{X}_t^f = f(x) + \int_0^t e^{-\beta s}\left[\mathcal{L}_0 f(V_s) - c_s f'(V_s) + U(c_s)\right] ds + R_t + m_t, \quad (4.4.2)$$

where m is a process such that $m^{\sigma_n} = (m_{t\wedge\sigma_n})$ are continuous martingales for some stopping times σ_n increasing to σ, and

$$R_t := \int_0^t e^{-\beta s} f'(\tilde{V}_{s-})\, dB_s^c + \sum_{s\le t} e^{-\beta s}\left[f(\tilde{V}_{s-} + \Delta B_s) - f(\tilde{V}_{s-})\right]. \quad (4.4.3)$$

By the definition of a supersolution, for any $x \in \operatorname{int} K$,

$$\mathcal{L}_0 f(x) \le -U^*\left(f'(x)\right) \le cf'(x) - U(c) \quad \forall c \in K.$$

Thus, the integral in (4.4.2) is a decreasing process. The process R is also decreasing because the terms of the sum in (4.4.3) are less or equal to zero by monotonicity of f, while the integral is negative since

$$f'(\tilde{V}_{s-})\, dB_s^c = I_{\{\Delta B_s = 0\}} f'(\tilde{V}_{s-}) \dot{B}_s\, d\|B\|_s,$$

where $f'(\tilde{V}_{s-})\dot{B}_s \le 0$ since $\dot{B}$ takes values in K. Taking into account that $\tilde{X}^f \ge 0$, we obtain from (4.4.2) that for each n, the negative decreasing process $R_{t\wedge\sigma_n}$ dominates an integrable process, and so it is integrable. The same conclusion holds for the stopped integral. Being a sum of integrable decreasing process and a martingale, the process $\tilde{X}^f_{t\wedge\sigma_n}$ is a positive supermartingale and, hence, by the Fatou lemma, $\tilde{X}^f$ is a supermartingale as well. □

Lemma 4.4.2 implies that the existence of a smooth positive supersolution f of (4.2.5) ensures the finiteness of W on K. Sometimes, e.g., in the case of power utility function, it is possible to find such a function in a rather explicit form.

Remark. Let $\bar{\mathcal{O}}$ be the closure of an open subset $\mathcal{O}$ of K, and let $f : \bar{\mathcal{O}} \to \mathbf{R}_+$ be a classical supersolution in $\bar{\mathcal{O}}$. Let $x \in \mathcal{O}$, and let τ be the exit time of the process $V^{x,\pi}$ from $\bar{\mathcal{O}}$. The above arguments imply that the process $X^f_{t\wedge\tau}$ is a supermartingale, and, therefore,

$$E\left[e^{-\beta(t\wedge\tau)} f(V_{t\wedge\tau}) + J^\pi_{t\wedge\tau}\right] \le f(x). \quad (4.4.4)$$

4.4.2 Strict Local Supersolutions

The next, slightly more technical result, the proof of which is also based on the analysis of (4.4.2), is of great importance. It will play a crucial role in deducing from the Dynamic Programming Principle that W is a subsolution of the HJB equation.

We fix a ball $\bar{\mathcal{O}}_r(x) \subseteq \operatorname{int} K$ and define τ^π as the exit time of $V^{\pi,x}$ from $\mathcal{O}_r(x)$, i.e.,

$$\tau^\pi := \inf\{t \geq 0 : |V_t^{\pi,x} - x| \geq r\}.$$

For simplicity, we assume that f is smooth in a neighborhood of $\bar{\mathcal{O}}_r(x)$.

Lemma 4.4.3 *Let $f \in C^2(\bar{\mathcal{O}}_r(x))$ be such that $\mathcal{L}f \leq -\varepsilon < 0$ on $\bar{\mathcal{O}}_r(x)$. Then there exist a constant $\eta > 0$ and an interval $]0, t_0]$ such that*

$$\sup_{\pi \in \mathcal{A}_a^x} E X_{t\wedge\tau^\pi}^{f,x,\pi} \leq f(x) - \eta t \quad \forall t \in]0, t_0].$$

Proof. We fix a strategy π and omit its symbol in the notation below. In what follows, only the behavior of the processes on $[0, \tau]$ does matter. Taking into account the monotonicity of f and modifying, if necessary, the strategy at the date τ by reducing the size of the jump ΔB_τ, we may assume without loss of generality that $|V_\tau - x| = r$ on the set $\{\tau < \infty\}$. As in the proof of Lemma 4.4.2, we apply the Itô formula. By assumption, for y from the ball $\bar{\mathcal{O}}_r(x)$, we have the bounds $\mathcal{L}_0 f(y) \leq -\varepsilon - U^*(y)$ and $\Sigma_G(f'(y)) \leq -\varepsilon$; the latter means that $kf'(y) \leq -\varepsilon|k|$ for $k \in -K$ (hence, $f'(\bar{\mathcal{O}}_r(x)) \subset \operatorname{int} K^*$). This implies the inequality

$$E X_{t\wedge\tau}^{f,x} \leq f(x) - e^{-\beta t} E N_t,$$

where

$$N_t := \varepsilon(t \wedge \tau) + \int_0^{t\wedge\tau} H(c_s, f'(V_s))\, ds + \varepsilon \int_0^{t\wedge\tau} |\dot{B}_s|\, d\|B\|_s$$

with $H(c, p) := U^*(p) + pc - U(c) \geq 0$. It remains to verify that EN_t dominates, on a certain interval $]0, t_0]$, a strictly increasing linear function which is independent of π.

Being the image of a closed ball under continuous mapping, the set $f'(\bar{\mathcal{O}}_r(x))$ is a compact in $\operatorname{int} K^*$. The lower bound of U^* on $f'(\bar{\mathcal{O}}_r(x))$ is finite. For any p from $f'(\bar{\mathcal{O}}_r(x))$ and $c \in \mathcal{C} \subseteq K$, we have the inequality $(c/|c|)p \geq \varepsilon$. At last, $U(c)/|c| \to 0$ as $c \to \infty$. Combining these facts, we infer that there is a constant κ ("large"; for convenience, $\kappa \geq 1$) such that

$$\inf_{p \in \bar{f}'(\bar{\mathcal{O}}_r(x))} H(c, p) \geq \kappa^{-1}|c|, \quad \forall c \in \mathcal{C},\ |c| \geq \kappa.$$

Thus, for the first integral in the definition of N_t, we have

$$\int_0^{t\wedge\tau} H(c_s, f'(V_s))\, ds \geq \kappa^{-1} \int_0^{t\wedge\tau} I_{\{|c_s|\geq\kappa\}} |c_s|\, ds.$$

Notice that the second integral dominates $\tilde{\kappa}\|B\|_{t\wedge\tau}$ for some $\tilde{\kappa} > 0$. To see this consider the absolute norm $|.|_1$ in $\mathbf{R}^d$. Then the total variation of B with respect to this norm is $\sum_i \operatorname{Var} B^i$, and

$$|\dot{B}|_1 = \sum_i |\dot{B}^i| = \sum_i \left|\frac{dB^i}{d\|B\|}\right| = \sum_i \left|\frac{dB^i}{d\operatorname{Var} B^i}\right| \frac{d\operatorname{Var} B^i}{d\|B\|} = \frac{d\sum_i \operatorname{Var} B^i}{d\|B\|}.$$

But all the norms in $\mathbf{R}^d$ are equivalent, i.e., $\tilde{\kappa}^{-1}|\cdot| \le |\cdot|_1 \le \tilde{\kappa}|\cdot|$ for some strictly positive constant $\tilde{\kappa}$, and the same inequalities relate the corresponding total-variation processes.

Summarizing, we conclude that it is sufficient to check the domination property for $E\tilde{N}_t$ with the simpler processes

$$\tilde{N}_t := t \wedge \tau + \int_0^{t\wedge\tau} I_{\{|c_s| \ge \kappa\}} |c_s|\, ds + \|B\|_{t\wedge\tau}. \tag{4.4.5}$$

The idea of the concluding reasoning is very simple: on a certain set of strictly positive probability, where one may neglect the random fluctuations, either τ is "large," or the total variation of the control is "large."

The formal arguments are as follows. Take $\delta > 1$. By the stochastic Cauchy formula the solution of the linear equation (4.2.2) can be written as

$$V_t^i = \mathcal{E}_t(Y^i)x^i + \mathcal{E}_t(Y^i) \int_0^t \mathcal{E}_s^{-1}(Y^i)\, d(B_s^i - C_s^i), \quad i = 1, \dots, d,$$

with the Girsanov exponential

$$\mathcal{E}(Y^i) := e^{Y^i - (1/2)\langle Y^i \rangle}.$$

Using only the fact that $\mathcal{E}_{0+}(Y^i) = \mathcal{E}_0(Y^i) = 1$, we get immediately from this representation that there exist a number $t_0 > 0$ and a measurable set Γ with $P(\Gamma) > 0$ on which

$$|V^{x,\pi} - x| \le r/2 + \delta(\|B\| + \|C\|) \quad \text{on } [0, t_0]$$

whatever is the control $\pi = (B, C)$. Of course, diminishing t_0, we may assume without loss of generality that $\kappa t_0 \le r/(4\delta)$. For any $t \le t_0$, we have on the set $\Gamma \cap \{\tau \le t\}$ the inequality $\|B\|_\tau + \|C\|_\tau \ge r/(2\delta)$, and, hence,

$$\tilde{N}_t \ge \|B\|_\tau + \|C\|_\tau - \int_0^\tau I_{\{|c_s| < \kappa\}} |c_s|\, ds \ge \frac{r}{2\delta} - \kappa t_0 \ge \kappa t_0 \ge t_0 \ge t.$$

On the set $\Gamma \cap \{\tau > t\}$, obviously, $\tilde{N}_t \ge t$. Thus, $E\tilde{N}_t \ge tP(\Gamma)$ on $[0, t_0]$, and the result is proven. □

4.5 Dynamic Programming Principle

The following property of the Bellman function is usually referred to as the (weak) "dynamic programming principle":

Theorem 4.5.1 *Assume that $W(x) < \infty$ for $x \in \operatorname{int} K$. Then for any finite stopping time τ,*

$$W(x) = \sup_{\pi \in \mathcal{A}_a^x} E\big(J_\tau^\pi + e^{-\beta\tau} W(V_{\tau-}^{x,\pi})\big). \tag{4.5.1}$$

It is corollary of two more precise results given in Lemmas 4.5.2 and 4.5.3, which will be our tools to derive the HJB equation for the Bellman function (though nicely looking, the above formulation does not suit this purpose).

We work on the canonical filtered space of continuous functions equipped with the Wiener measure. The generic point $\omega = \omega_{.}$ of this space is a continuous function on $\mathbf{R}_+$, zero at the origin. Let $\mathcal{F}_t^\circ := \sigma\{\omega_s,\ s \le t\}$ and $\mathcal{F}_t := \bigcap_{\varepsilon>0} \mathcal{F}_{t+\varepsilon}$. We add the superscript P to denote σ-algebras augmented by all P-null sets from Ω. Recall that $\mathcal{F}_t^{\circ,P}$ coincides with $\mathcal{F}_t^P$ (this assertion follows easily from the predictable representation theorem).

A particular structure of Ω allows us to consider such operators as the stopping $\omega_{.} \mapsto \omega_{.}^s$, $s \ge 0$, where $\omega_{.}^s = \omega_{s\wedge .}$, and the translation $\omega_{.} \mapsto \omega_{s+.} - \omega_s$. Taking Doob's theorem into account, one can describe $\mathcal{F}_s^\circ$-measurable random variables as those of the form $g(w_{.}) = g(w_{.}^s)$ where g is a measurable function on Ω.

We define also the "concatenation" operator as the measurable mapping

$$g : \mathbf{R}_+ \times \Omega \times \Omega \to \Omega$$

with $g_t(s, \omega_{.}, \tilde\omega_{.}) = \omega_t I_{[0,s[}(t) + (\tilde\omega_{t-s} + \omega_s) I_{[s,\infty[}(t)$.

Notice that

$$g_t\big(s, \omega_{.}^s, \omega_{.+s} - \omega_s\big) = \omega_t.$$

Thus, $\pi(\omega) = \pi(g(s, \omega_{.}^s, \omega_{.+s} - \omega_s))$.

Let π be a fixed strategy from $\mathcal{A}_a^x$, and let $\vartheta = \vartheta^{x,\pi}$ be a hitting time of zero for the process $V^{x,\pi}$.

We need the following general fact on conditional distributions.

Let ξ and η be two random variables taking values in Polish spaces X and Y equipped with their Borel σ-algebras $\mathcal{X}$ and $\mathcal{Y}$. Then ξ admits a regular conditional distribution given $\eta = y$, which we shall denote by $p_{\xi|\eta}(\Gamma, y)$, and

$$E\big(f(\xi,\eta)|\eta\big) = \int f(x,y) p_{\xi|\eta}(dx, y)\Big|_{y=\eta} \qquad \text{(a.s.)}$$

for any measurable function $f(x,y) \ge 0$.

We shall apply the above relation to the random variables $\xi = (\omega_{.+\tau} - \omega_\tau)$ and $\eta = (\tau, \omega^\tau)$. In this case, according to the Dynkin–Hunt theorem, the conditional distribution $p_{\xi|\eta}(\Gamma, y)$ admits a version which is independent of y and coincides with the Wiener measure P.

At last, for fixed s and w^s, the shifted control $\pi(g(s, \omega_{.}^s, \tilde\omega.), s + dr)$ is admissible for the initial condition $V_{s-}^{x,\pi}(\omega)$. Here we denote by $\tilde\omega.$ a generic point of the canonical space.

Lemma 4.5.2 *Let $\mathcal{T}_f$ and $\mathcal{T}_b$ be, respectively, the sets of all finite and bounded stopping times. Then*

$$W(x) \le \sup_{\pi \in \mathcal{A}_a^x} \inf_{\tau \in \mathcal{T}_f} E\big(J_\tau^\pi + e^{-\beta\tau} W\big(V_{\tau-}^{x,\pi}\big)\big). \tag{4.5.2}$$

If $W(x) < \infty$ for all $x \in \operatorname{int} K$, then

$$W(x) \le \sup_{\pi \in \mathcal{A}_a^x} \inf_{\tau \in \mathcal{T}_b} E\big(J_\tau^\pi + e^{-\beta\tau} W\big(V_{\tau}^{x,\pi}\big)\big). \tag{4.5.3}$$

Proof. For arbitrary $\pi \in \mathcal{A}_a^x$ and $\mathcal{T}_f$, we have that

$$EJ_\infty^\pi = EJ_\tau^\pi + Ee^{-\beta\tau} \int_0^\infty e^{-\beta r} u(c_{r+\tau})\, dr.$$

According to the above discussion, we can rewrite the second term of the right-hand side as

$$Ee^{-\beta\tau} \int \Big(\int_0^\infty e^{-\beta r} u\big(c_{r+\tau}\big(g(\tau, \omega^\tau, \tilde{\omega})\big)\big)\, dr \Big) P(d\tilde{\omega})$$

and dominate it by $Ee^{-\beta\tau} W(V_{\tau-}^{x,\pi})$. Thus,

$$EJ_\infty^\pi \le EJ_\tau^\pi + Ee^{-\beta\tau} W\big(V_{\tau-}^{x,\pi}\big).$$

This bound leads directly to the first announced inequality. To obtain the second, we note that W is dominated by a linear function and consider, for a bounded stopping time τ, the sequence $\tau_n := \tau + 1/n$; for τ_n, the above bound holds. Clearly, $V_{\tau_n-}^{x,\pi} \to V_\tau^{x,\pi}$. Since W is continuous in $\operatorname{int} K$ and zero is an adsorbing point, $W(V_{\tau_n-}^{x,\pi}) \to W(V_\tau^{x,\pi})$. At last, Proposition 4.2.1 allows us to apply the dominated convergence theorem and remove the annoying minus in the bound, which leads, in this modified form, to (4.5.3). □

The proof of the opposite inequality is based on different ideas.

Lemma 4.5.3 *Assume that $W(x) < \infty$ for all $x \in \operatorname{int} K$. Then for any finite stopping time τ,*

$$W(x) \ge \sup_{\pi \in \mathcal{A}_a^x} E\big(J_\tau^\pi + e^{-\beta\tau} W\big(V_{\tau-}^{x,\pi}\big)\big). \tag{4.5.4}$$

Proof. Fix $\varepsilon > 0$. Being concave, the function W is continuous on $\operatorname{int} K$. For each $x \in \operatorname{int} K$, we can find an open ball $\mathcal{O}_r(x) = x + \mathcal{O}_r(0)$ with $r = r(\varepsilon, x) < \varepsilon$ contained in the open set $\{y \in \operatorname{int} K : \; |W(y) - W(x)| < \varepsilon\}$. Moreover, we can find a smaller ball $\mathcal{O}_{\tilde{r}}(x)$ contained in the set $y(x) + K$ with $y(x) \in \mathcal{O}_r(x)$. Indeed, take a ball $x_0 + \mathcal{O}_\delta(0) \subseteq K$. Since K is a cone,

$$x + \mathcal{O}_{\lambda\delta}(0) \subseteq x - \lambda x_0 + K$$

for every $\lambda > 0$. Clearly, the requirement is met for $y(x) = x - \lambda x_0$ and $\tilde{r} = \lambda\delta$ when $\lambda|x_0| < \varepsilon$ and $\lambda\delta < r$. The family of sets $\mathcal{O}_{\tilde{r}(x)}(x)$, $x \in \operatorname{int} K$, is an open covering of $\operatorname{int} K$. But any open covering of a separable metric space contains a countable subcovering (this is the Lindelöf property; in our case, where $\operatorname{int} K$ is a countable union of compacts, it is obvious). Take a countable subcovering indexed by points x_n. For simplicity, we shall denote its elements by $\mathcal{O}_n$ and $y(x_n)$ by y_n. Put $A_1 := \mathcal{O}_1$ and $A_n = \mathcal{O}_n \setminus \bigcap_{k<n} \mathcal{O}_k$. The sets A_n are disjoint, and their union is $\operatorname{int} K$.

Let $\pi^n = (B^n, C^n) \in \mathcal{A}_a^{y_n}$ be an ε-optimal strategy for the initial point y_n, i.e., such that

$$EJ^{\pi_n} \geq W(y_n) - \varepsilon.$$

Let $\pi \in \mathcal{A}_a^x$ be an arbitrary strategy. We consider the strategy $\tilde{\pi} \in \mathcal{A}_a^x$ defined by the relation

$$\tilde{\pi} = \pi I_{[0,\tau[} + \sum_{n=1}^{\infty} \big[\big(y_n - V_{\tau-}^{x,\pi}, 0\big) + \bar{\pi}^n \big] I_{[\tau,\infty[} I_{A_n}\big(V_{\tau-}^{x,\pi}\big) I_{\{\tau<\vartheta\}},$$

where $\bar{\pi}^n$ is the translation of the strategy π^n: namely, for a point $\omega_.$ with $\tau(\omega) = s < \infty$, we have

$$\bar{\pi}_t^n(\omega_.) := \pi_{t-s}^n(\omega_{.+s} - \omega_s).$$

In other words, the measure $d\tilde{\pi}$ coincides with $d\pi$ on $[0,\tau[$ and with the shift of $d\pi^n$ on $]\tau,\infty[$ when $V_{\tau-}^{x,\pi}$ is a subset of A_n; the correction term guarantees that in the latter case the trajectory of the control system corresponding to the control $\tilde{\pi}$ passes at time τ through the point y_n.

Now, using the same considerations as in the previous lemma, we have

$$\begin{aligned} W(x) \geq EJ_\infty^{\tilde{\pi}} &= EJ_\tau^\pi + \sum_{n=1}^{\infty} EI_{A_n}\big(V_{\tau-}^{x,\pi}\big) I_{\{\tau<\vartheta\}} \int_\tau^\infty e^{-\beta s} u(\bar{c}_s^n)\, ds \\ &\geq EJ_\tau^\pi + \sum_{n=1}^{\infty} EI_{A_n}\big(V_{\tau-}^{x,\pi}\big) I_{\{\tau<\vartheta\}} e^{-\beta\tau} \big(W(y_n) - \varepsilon\big) \\ &\geq EJ_\tau^\pi + Ee^{-\beta\tau} W\big(V_{\tau-}^{x,\pi}\big) - 2\varepsilon. \end{aligned}$$

Since π and ε are arbitrary, the result follows. □

Remark. The previous lemmas imply the identity

$$W(x) = \sup_{\pi \in \mathcal{A}_a^x} \inf_{\tau \in \mathcal{T}_f} E\big(J_\tau^\pi + e^{-\beta\tau} W\big(V_{\tau-}^{x,\pi}\big)\big).$$

It can be considered as another form of the dynamic programming principle.

4.6 The Bellman Function and the HJB Equation

Theorem 4.6.1 *Assume that the Bellman function W is in $C(K)$. Then W is a viscosity solution of (4.2.5).*

Proof. The claim follows from the two lemmas below. □

Lemma 4.6.2 *If (4.5.4) holds, then W is a viscosity supersolution of (4.2.5).*

Proof. Let $x \in \mathcal{O} \subseteq \operatorname{int} K$. We choose a test function $\phi \in C^2(\mathcal{O})$ such that $\phi(x) = W(x)$ and $W \geq \phi$ in $\mathcal{O}$.

At first, we fix $m \in K$ and argue with $\varepsilon > 0$ small enough to ensure that $x - \varepsilon m \in \mathcal{O}$. The function W is increasing with respect to the partial ordering generated by K. Thus,

$$\phi(x) = W(x) \geq W(x - \varepsilon m) \geq \phi(x - \varepsilon m).$$

It follows that $-m\phi'(x) \leq 0$, and, therefore, $\Sigma_G(\phi'(x)) \leq 0$.

Take now π with $B_t = 0$ and $c_t = c \in \mathcal{C}$. Let τ_r be the exit time of the continuous process $V = V^{x,\pi}$ from the ball $\bar{\mathcal{O}}_r(x) \subseteq \operatorname{int} K$. The identity (4.5.4) implies that

$$W(x) \geq E\big(J^{\pi}_{t\wedge\tau_r} + e^{-\beta(t\wedge\tau_r)} W(V_{t\wedge\tau_r})\big),$$

and this inequality holds true if replace W by ϕ. Writing all terms of the latter in the right-hand side and applying the Itô formula (4.4.2), we get that

$$\begin{aligned} 0 &\geq E\bigg(\int_0^{t\wedge\tau_r} e^{-\beta s} U(c_s)\, ds + e^{-\beta(t\wedge\tau_r)} \phi(V_{t\wedge\tau_r})\bigg) - \phi(x) \\ &\geq E \int_0^{t\wedge\tau_r} e^{-\beta s} \big[\mathcal{L}_0 \phi(V_s) - c\phi'(V_s) + U(c)\big]\, ds \\ &\geq \min_{y \in \bar{\mathcal{O}}_r(x)} \big[\mathcal{L}_0\phi(y) - c\phi'(y) + U(c)\big] E\bigg[\frac{1}{\beta}\big(1 - e^{-\beta(t\wedge\tau_r)}\big)\bigg]. \end{aligned}$$

Dividing the resulting inequality by t and taking successively the limits as t and r converge to zero, we infer that $\mathcal{L}_0\phi(x) - c\phi'(x) + U(c) \leq 0$. Maximizing over $c \in \mathcal{C}$ yields the bound $\mathcal{L}_0\phi(x) + U^*(\phi'(x)) \leq 0$, and, therefore, W is a supersolution of the HJB equation. □

Lemma 4.6.3 *If (4.5.2) holds, then W is a viscosity subsolution of (4.2.5).*

Proof. Let $x \in \mathcal{O} \subseteq \operatorname{int} K$. Let $\phi \in C^2(\mathcal{O})$ be a function such that $\phi(x) = W(x)$ and $W \leq \phi$ on $\mathcal{O}$. Assume that the subsolution inequality for ϕ fails at x. Thus, there exists $\varepsilon > 0$ such that $\mathcal{L}\phi \leq -\varepsilon$ on some ball $\bar{\mathcal{O}}_r(x) \subseteq \mathcal{O}$. By virtue of Lemma 4.4.3 (applied to the function ϕ), there are $t_0 > 0$ and $\eta > 0$ such that on the interval $]0, t_0]$, for any strategy $\pi \in \mathcal{A}^x_a$,

$$E\big(J^{\pi}_{t\wedge\tau^\pi} + e^{-\beta\tau^\pi} \phi(V^{x,\pi}_{t\wedge\tau^\pi})\big) \leq \phi(x) - \eta t,$$

where τ^π is the exit time of the process $V^{x,\pi}$ from the ball $\bar{\mathcal{O}}_r(x)$. Fix $t \in]0, t_0]$. By the second claim of Lemma 4.5.2), there exists $\pi \in \mathcal{A}_a^x$ such that

$$W(x) \le E\big(J^\pi_{t\wedge\tau} + e^{-\beta\tau} W\big(V^{x,\pi}_{t\wedge\tau}\big)\big) + \frac{1}{2}\eta t$$

for every stopping time τ, in particular, for τ^π.

Using the inequality $W \le \phi$ and applying Lemma 4.4.3, we obtain from the above relations that $W(x) \le \phi(x) - (1/2)\eta t$. This is a contradiction because at the point x the values of W and ϕ are the same. □

4.7 Properties of the Bellman Function

4.7.1 The Subdifferential: Generalities

The subdifferential of the function W at a point $x \in \operatorname{int} K$ is defined as the set

$$\partial W(x) := \big\{ w \in \mathbf{R}^d : \ W(y) \le W(x) + w(y - x) \ \forall y \in K \big\}.$$

Since W is concave, this set is nonempty; obviously, it is closed and bounded. If W is unbounded, zero does not belong to $\partial W(x)$.

Recall that, for a concave function f of scalar argument, the subdifferential $\partial f(x) = [D^+ f(x), D^- f(x)]$, the interval between the values of the right and left derivatives at x.

Lemma 4.7.1 *Let x_1, x_2 be two points in* $\operatorname{int} K$. *Then*

$$\big(\partial W(x_1) - \partial W(x_2)\big)(x_1 - x_2) \le 0. \tag{4.7.1}$$

Proof. Let $w_i \in \partial W(x_i)$, $i = 1, 2$. From the definition we have the inequalities

$$W(x_2) \le W(x_1) + w_1(x_2 - x_1), \qquad W(x_1) \le W(x_2) + w_2(x_1 - x_2).$$

Adding them, we obtain that $(w_1 - w_2)(x_1 - x_2) \le 0$, the relation we need. □

Lemma 4.7.2 *The set $\partial W(x)$ is a singleton if and only if W is differentiable at x; in this case the unique element of $\partial W(x)$ is $W'(x)$.*

Lemma 4.7.3 *Let $\mathcal{O}$ be an open subset of K. The function W is of class $C^1(\mathcal{O})$ if $\partial W(x)$ is a singleton at any point $x \in \mathcal{O}$.*

Now we exploit some specific properties of the Bellman function.

The following lemma follows from the monotonicity of W with respect to the partal ordering induced by the cone K.

Lemma 4.7.4 *For every $x \in \operatorname{int} K$, we have the inclusion $\partial W(x) \subseteq K^*$.*

Proof. If $w \in \partial W(x)$, the linear function $\varphi(.) := W(x) + w(. - x)$ dominates $W(.)$ on K. Then, for any $y \in K$,

$$\varphi(x) = W(x) \leq W(x+y) \leq \varphi(x+y) = W(x) + wy.$$

Thus, $wy \geq 0$ for all $y \in K$, and the result follows. □

The Bellman function in the model with the power utility inherits the homotheticity property of the latter. Namely,

$$W(\nu x) = \nu^\gamma W(x) \quad \forall \nu > 0. \tag{4.7.2}$$

This implies a homotheticity property for the subdifferential.

Lemma 4.7.5 *If W satisfies (4.7.2), then*

$$\partial W(\nu x) = \nu^{\gamma-1} \partial W(x) \quad \forall \nu > 0. \tag{4.7.3}$$

Proof. Taking into account that K is a cone, we have

$$\begin{aligned}
\partial W(\nu x) &= \big\{w \in \mathbf{R}^d : \ W(\nu y) \leq W(\nu x) + w(\nu y - \nu x) \ \forall y \in K\big\} \\
&= \big\{w \in \mathbf{R}^d : \ \nu^\gamma W(y) \leq \nu^\gamma W(x) + w(\nu y - \nu x) \ \forall y \in K\big\} \\
&= \big\{w \in \mathbf{R}^d : \ W(y) \leq W(x) + \nu^{1-\gamma} w(y - x) \ \forall y \in K\big\}.
\end{aligned}$$

Since the right-hand side is $\nu^{\gamma-1}\partial W(x)$, we get the claim. □

Corollary 4.7.6 *If $W \neq 0$ satisfies (4.7.2), then $0 \notin \partial W(x)$.*

Proof. If $0 \in \partial W(x)$, then $0 \in \partial W(\nu x)$, $\nu > 0$. Thus, W attains its maximum at every point νx. In virtue of (4.7.2), this is possible only if $W = 0$. □

Lemma 4.7.7 *If W satisfies (4.7.2), then the projection of $\partial W(x)$ on* Lin x *is the singleton $\gamma W(x)|x|^{-2}x$. In particular, for $d = 2$, if not a singleton, the subdifferential $\partial W(x)$ is a closed interval orthogonal to x.*

Proof. Let $w \in \partial W(x)$, and let $w = \kappa x + w^\perp$ where $w^\perp x = 0$. Then, for any real $t > 0$, we have, in virtue of the definition of a subdifferential, that

$$W(tx) \leq W(x) + wx(t-1) = W(x) + \kappa |x|^2 (t-1).$$

On the other hand, for the smooth scalar function $\psi(t) := W(tx) = t^\gamma W(x)$, the subdifferential $\partial\psi(1)$ is the singleton $\gamma W(x)$. Thus, $\gamma W(x) = \kappa|x|^2$. □

4.7.2 The Bellman Function of the Two-Asset Model

Now we investigate the structure of the Bellman function for the case $d = 2$ assuming its homotheticity. Let g_1, g_2 be the generators of K. First, let us consider the ray

$$r_1 := g_2 + \mathbf{R}_+ g_1 = \{x \in \mathbf{R}^2 : \ x = g_2 + tg_1, \ t \geq 0\},$$

parallel to g_1 and starting from the point g_2. Relation (4.7.2) allows us to recover the whole function W from its values on r_1, i.e., from the values of the concave increasing function $W_1(t) := W(g_2 + tg_1)$, $t > 0$, with $W(0+) = 0$. Its subdifferential $\partial W_1(t)$ is the interval $[D^+W_1(t), D^-W_1(t)] \subseteq \mathbf{R}_+$; if $\tilde{t} > t$, then the interval $\partial W_1(\tilde{t})$ lays leftwards with respect to the interval $\partial W_1(t)$. Put $t_1 := \inf\{t \geq 0 : \ D^+W_1(t) = 0\}$. Necessarily, $\partial W_1(t) = \{0\}$ for $t > t_1$.

Define the cone $K_1 := \operatorname{cone}\{g_1, g_2 + t_1 g_1\}$ contained in K; by convention, $g_2 + \infty\, g_1 = g_1$, i.e., $K_1 := \operatorname{cone}\{g_1\}$ when $t_1 = \infty$.

Notice that $g_1 \partial W(g_2 + tg_1) \subseteq \partial W_1(t)$. Indeed, if $w \in \partial W(g_2 + tg_1)$, then, for all $s > 0$,

$$W(g_2 + sg_1) \leq W(g_2 + tg_1) + wg_1(t - s),$$

i.e., $wg_1 \in \partial W_1(t)$.

In particular, $g_1 \partial W(g_2 + tg_1) = \{0\}$ for $t > t_1$. On the other hand, by Lemma 4.7.7 the projection of $\partial W(g_2 + tg_1)$ on the direction $g_2 + tg_1$ is a singleton. Thus, $\partial W(g_2 + tg_1)$ i also a singleton. Using Lemma 4.7.3 and Lemma 4.7.3, we arrive at the following conclusion: W is C^1 on $\operatorname{int} K_1$, and $g_1 W' = 0$ on this set.

Changing the role of indices, we may introduce also the function W_2, the value t_2, and the cone $K_2 := \operatorname{cone}\{g_2, g_1 + t_2 g_2\}$ degenerating to the ray $\operatorname{cone}\{g_2\}$ when $t_2 = \infty$. Similarly, W is C^1 on K_2, and $g_2 W' = 0$ on this set.

Notice that $\operatorname{int} K_1 \cap \operatorname{int} K_2 = \emptyset$. Indeed, at a common point one would have the identities $g_i W'(x) = 0$, possible only if $W'(x) = 0$. This contradicts to Corollary 4.7.6. Therefore, $K_0 := \operatorname{cone}\{g_2 + t_1 g_1, g_1 + t_2 g_2\}$ is a cone lying in between K_1 and K_2; the interiors of these three cones are disjoint.

Lemma 4.7.8 *For every $x \in \operatorname{int} K_0$, we have the inclusion $\partial W(x) \subseteq \operatorname{intK}^*$ or, equivalently, $wg_i > 0$ for all $w \in \partial W(x)$, $i = 1, 2$.*

Proof. As we just proved, $g_1 \partial W(g_2 + tg_1) \subseteq \partial W_1(t)$. But for $t < t_1$, the set $\partial W_1(t)$ lies in $]0, \infty[$. It follows that $g_1 \partial W(x) > 0$ for x belonging to the intersection of the ray $g_2 + \mathbf{R}_+ g_1$ with $\operatorname{int} K_0$ and, hence, by Lemma 4.7.3, for all $x \in \operatorname{int} K_0$. The arguments for the generator g_2 are similar. □

To check that $\operatorname{int} K_1$ and $\operatorname{int} K_2$ are nonempty as well as $\operatorname{int} K_0$, we use more particular properties of the HJB equation. This will be done in the next section.

4.7.3 Lower Bounds for the Bellman Function

For the model where the functional depends only on the first asset which is the numéraire, one can get easily tractable lower bounds for the Bellman function which will be used later.

Let $l(x)$ be the *liquidation function*, i.e.,

$$l(x) := \sup\{z \in \mathbf{R}_+ : \ x - ze_1 \in K\}.$$

We consider the subset of admissible strategies with $\Delta B_0 = e_1 l(x) - x$ and $B_t = B_0$ for $t > 0$. This means that the agent liquidates his position in the risky asset entering the market and remains afterwards only with money. For a strategy π of this type,

$$J^\pi_\infty = \int_0^\tau e^{-\beta t} u(c_t)\, dt,$$

where τ is the instant when the process $X_t := l(x) - \int_0^t c_s\, ds$ hits zero. In particular, if the consumption is proportional to the wealth, i.e., $c_t = \kappa X_t$ with some constant $\kappa > 0$, we have the dynamics $X_t = l(x)e^{-\kappa t}$ with $\tau = \infty$ and

$$J^\pi_\infty = \int_0^\infty e^{-\beta t} u\big(\kappa l(x) e^{-\kappa t}\big)\, dt.$$

Thus,

$$W(x) \ge \sup_{\kappa > 0} \int_0^\infty e^{-\beta t} u\big(\kappa l(x) e^{-\kappa t}\big)\, dt. \tag{4.7.4}$$

In the specific case of the power utility function,

$$J^\pi_\infty = \frac{\kappa^\gamma}{\gamma(\beta + \kappa\gamma)} l^\gamma(x).$$

The maximum of the right-hand side over κ is attained at $\kappa_* = \beta/(1-\gamma)$. This gives us a useful lower bound for the Bellman function, which we formulate as follows:

Lemma 4.7.9 *In the problem with the power utility function,*

$$W(x) \ge \frac{1}{\gamma}\kappa_*^{\gamma-1} l^\gamma(x) = \frac{1}{\gamma}\left(\frac{\beta}{1-\gamma}\right)^{\gamma-1} l^\gamma(x). \tag{4.7.5}$$

In particular,

$$W(e_1) \ge \frac{1}{\gamma}\kappa_*^{\gamma-1} = \frac{1}{\gamma}\left(\frac{\beta}{1-\gamma}\right)^{\gamma-1}. \tag{4.7.6}$$

This result will be used in the sequel for the two-asset model with the transaction cost coefficients $\lambda^{12} = \lambda^{21} = \lambda$. For such a case, at any point

$x = (\xi, \eta)$ which lies in the intersection of the solvency region K with the upper half-plane, the value of liquidation function is

$$l(x) = \xi + \frac{\eta}{1+\lambda} = \frac{p_2 x}{1+\lambda}$$

(the stock holding η is converted into $\eta/(1+\lambda)$ units of money). Therefore, we can write more explicitly that

$$W(x) \geq \frac{1}{\gamma}\kappa_*^{\gamma-1}\left(\xi + \frac{\eta}{1+\lambda}\right)^{\gamma}. \tag{4.7.7}$$

In particular, for $x = (1-z, z)$ with $z \in [0, 1+1/\lambda]$, we have the lower bound

$$W(1-z, z) \geq \frac{1}{\gamma}\left(\frac{\beta}{1-\gamma}\right)^{\gamma-1}\frac{1}{(1+\lambda)^{\gamma}}(1+\lambda-\lambda z)^{\gamma}. \tag{4.7.8}$$

Remark. Another lower bound for the Bellman function can be obtained by considering the strategy π which prescribes to convert immediately the portfolio into a single-asset one with holdings in a fixed risky asset and to consume proportionally to the current portfolio value (this means that shares permanently should be sold paying the transaction costs, which can be also interpreted as a consumption tax). Since the wealth in this case evolves accordingly to a stochastic linear equation, EJ_∞^π can be easily calculated.

4.8 The Davis–Norman Solution

4.8.1 Two-Asset Model: The Result

Let us consider the two-asset model with the price dynamics given by

$$\begin{aligned} dS_t^1 &- 0, \\ dS_t^2 &= S_t^2(\mu\, dt + \sigma\, dw_t), \end{aligned}$$

where w is a Wiener process, and $\sigma > 0$. That is, the first asset ("bond", "money", or "bank account") is the *numéraire*. The price of the risky asset follows a geometric Brownian motion. The portfolio values evolve as

$$\begin{aligned} dV_t^1 &= dL_t^{21} - (1+\lambda^{12})\, dL_t^{12} - c_t\, dt, \\ dV_t^2 &= V_t^2(\mu\, dt + \sigma\, dw_t) + dL_t^{12} - (1+\lambda^{21})\, dL_t^{21}, \end{aligned}$$

where L^{12} and L^{21} are adapted right-continuous increasing processes.

The optimization problem is of the form

$$E\int_0^\infty e^{-\beta s}u(c_s)\, ds \to \max, \tag{4.8.1}$$

where $u : \mathbf{R}_+ \to \mathbf{R}_+$ is a utility function. The maximum is taken over the set of strategies for which the value process evolves in the solvency cone K generated by the vectors $g_1 := (1 + \lambda^{12}, -1)$ and $g_2 := (-1, 1 + \lambda^{21})$ lying respectively in the forth and second quadrants. So, K and K^* are simply sectors; the generators of the dual cone are the vectors $p_1 := (1, 1 + \lambda^{12})$ and $p_2 := (1 + \lambda^{21}, 1)$ lying in $\mathbf{R}^2_+$ and orthogonal, respectively, to g_1 and g_2.

For the power utility function, the structure of the solution was found by Norman and Davis in 1990, though it was conjectured already in the pioneering paper of Magill and Constantinides of 1976. It was thoroughly analyzed using methods of viscosity solutions in the paper by Shreve and Soner of 1994. In our presentation we follow the latter.

We consider here the model with $u(c) = c^\gamma/\gamma$, $\gamma \in]0,1[$, supposing always that the Bellman function W is finite. As we already know, such a property is guaranteed if $\kappa_M > 0$, i.e.,

$$\beta > \frac{1}{2}\frac{\gamma}{1-\gamma}\frac{\mu^2}{\sigma^2}. \tag{4.8.2}$$

Note that the above inequality ensures the finiteness of W even in the classical Merton problem without friction. In the model with friction one can find other sufficient conditions for the finiteness of the Bellman function. We discuss this issue later.

It follows from the general theory that the Bellman function W is the viscosity solution of the HJB equation with zero boundary condition. It is unique in the class of functions with growth rate $\gamma' \geq \gamma$ such that the above bound still holds with γ'.

We assume, moreover, that the instantaneous interest rate of the risky asset $\mu > 0$, and this hypothesis will be used immediately in proving (4.8.5).

The HJB equation can be written as follows:

$$\max\left\{\frac{1}{2}\sigma^2\eta^2 W_{\eta\eta} + \mu\eta W_\eta - \beta W + u^*(W_\xi), -g_1 W', -g_2 W'\right\} = 0. \tag{4.8.3}$$

Of course, at the moment, we have no information about the existence of the involved derivatives of W, and the above relation has to be understood in the viscosity sense. Since this section is a case study and our intention is to obtain an explicit solution, we abandon the standard notation: to improve the perception of the formulae, we use the notation (ξ, η) and (t, z) for generic points in $\mathbf{R}^2_+$. Moreover, for the sake of simplicity, we suppose that the transaction costs for buying and selling are the same, i.e., $\lambda^{12} = \lambda^{21} = \lambda > 0$.

The principal result is the following:

Theorem 4.8.1 *There are vectors* $\tilde{g}_1, \tilde{g}_2$ *such that the solvency cone* K *is the union of* $K_1 = \operatorname{cone}\{g_1, \tilde{g}_1\}$, $K_2 = \operatorname{cone}\{g_2, \tilde{g}_2\}$, *and* $K_0 = \operatorname{cone}\{\tilde{g}_1, \tilde{g}_2\}$, *three sectors with disjoint nonempty interiors. The Bellman function* W *is a concave positive homogeneous function of order* γ *and belongs to the class*

$C^1(\text{int}\, K)$. *On* K_i, $i = 1, 2$, *it is given by the formulae* $a_i u(p_i x)$. *On* $\text{int}\, K_0$, *it is a classical* C^2*-solution of the equation*

$$\frac{1}{2}\sigma^2\eta^2 W_{\eta\eta} + \mu\eta W_\eta - \beta W + u^*(W_\xi) = 0, \tag{4.8.4}$$

where $u^*(p) = (1-\gamma)\gamma^{-1}p^{\gamma/(\gamma-1)}$.

4.8.2 Structure of Bellman Function

First of all, we recall that the functions $au(p_i x)$ for large a are classical supersolutions, and, hence, they dominate $W(x)$ "globally," i.e., on the whole K. We refine this result by showing that, on certain sectors with nonempty interiors and adjacent to the boundaries of the solvency region, the Bellman function W coincides with functions of this particular type. Specifically, we have:

Proposition 4.8.2 (a) *There exists* $a_1 > 0$ *such that*

$$W(x) = a_1 u(p_1 x) \quad \text{on cone}\{g_1, e_1\}. \tag{4.8.5}$$

(b) *There exists* $a_2 > 0$ *such that*

$$W(x) = a_2 u(p_2 x) \quad \text{on cone}\big\{g_2, 2\theta(1+\lambda)^{-1}g_2 + e_2\big\}. \tag{4.8.6}$$

Proof. (a) Take $a_1 := W(e_1)/u(p_1 e_1) = \gamma W(e_1)$. Due to the homotheticity property, this choice implies that the function $\varphi(x) := a_1 u(p_1 x)$ coincides with $W(x)$ on the whole ray $\mathbf{R}_+ e_1$. This immediately implies that $W(x) \geq \varphi(x)$ on the sector cone$\{g_1, e_1\}$ because, along each ray $\xi e_1 + \mathbf{R}_+ g_1$, $\xi > 0$, the function W is increasing while φ remains constant. On the other hand, both functions are zero on the boundary ray $\mathbf{R}_+ g_1$ (which is a part of ∂K). Let us check that φ is a classical supersolution of our HJB equation on the sector cone$\{g_1, e_1\}$, and, hence, we have the reverse inequality $W(x) \leq \varphi(x)$ on this set due to Lemmas 4.4.1 and 4.4.2. The only problem is to check, in the interior of the sector, the inequality $\mathcal{L}_0\varphi + u^*(\varphi_\xi) \leq 0$, which, in the detailed notation, is simply

$$\frac{1}{2}\sigma^2\eta^2\varphi_{\eta\eta} + \mu\eta\varphi_\eta - \beta\varphi + u^*(\varphi_\xi) \leq 0.$$

The first term in the left-hand side is always negative (due to the concavity). The second one is negative because, for $x = (\xi, \eta)$ in the considered set, the coordinate $\eta < 0$, the parameter $\mu > 0$ by assumption, and φ is increasing in η. At last, by virtue of the bound (4.7.6),

$$a_1 := \gamma W(e_1) \geq \kappa_*^{\gamma-1} = \left(\frac{\beta}{1-\gamma}\right)^{\gamma-1}, \tag{4.8.7}$$

and we easily get that

$$u^*(\varphi_\xi) = (1-\gamma)a_1^{\frac{1}{\gamma-1}}\varphi \le \beta\varphi. \tag{4.8.8}$$

(b) We put $N := 2\theta(1+\lambda)^{-1}$ and note that

$$\frac{1}{2}\sigma^2(1+\lambda)(\gamma-1)N + \mu = 0.$$

Take now $a_2 := W(Ng_2+e_2)/u(p_2e_2) = \gamma W(Ng_2+e_2)$. By the same general arguments as above, we obtain that the function $\psi(x) := a_2u(p_2x)$ coincides with $W(x)$ on the whole ray generated by the vector Ng_2+e_2 and satisfies the inequality $W(x) \ge \psi(x)$ on the sector cone$\{g_2, Ng_2+e_2\}$. To obtain the reverse inequality, it remains to verify that $\mathcal{L}_0\psi + u^*(\psi_\xi) \le 0$ at any point $x = (\xi, \eta)$ from the interior of this sector. Such a point admits a unique representation

$$x = \bar{\xi}g_2 + \bar{\eta}(Ng_2+e_2)$$

with some reals $\bar{\xi}, \bar{\eta} > 0$. We have: $p_2x = \bar{\eta}p_2e_2 = \bar{\eta}$ and

$$\eta = xe_2 = \bar{\xi}g_2e_2 + \bar{\eta}(Ng_2+e_2)e_2 \ge N\bar{\eta}g_2e_2 = N\bar{\eta}(1+\lambda).$$

Thus,

$$\begin{aligned}\frac{1}{2}\sigma^2\eta^2\psi_{\eta\eta}(x) + \mu\eta\psi_\eta(x) &= a_2(p_2x)^{\gamma-1}\eta\left(\frac{1}{2}\sigma^2(\gamma-1)\frac{\eta}{p_2x} + \mu\right)\\ &\le a_2(p_2x)^{\gamma-1}\eta\left(\frac{1}{2}\sigma^2(\gamma-1)N(1+\lambda) + \mu\right) = 0\end{aligned}$$

due to our choice of N.

On the other hand, the value of liquidation function is $l(Ng_2+e_2) = 1/(1+\lambda)$ (the intersection of the ray $Ng_2+e_2-\mathbf{R}_+g_2$ and the axis of abscises is the point $(1/(1+\lambda), 0)$), and, therefore, due to the bound (4.7.6), we have that

$$a_2 \ge \kappa_*^{\gamma-1}\frac{1}{(1+\lambda)^\gamma} = \left(\frac{\beta}{1-\gamma}\right)^{\gamma-1}\frac{1}{(1+\lambda)^\gamma}. \tag{4.8.9}$$

It follows that

$$u^*\big(\psi_\xi(x)\big) = (1-\gamma)a_2^{\frac{1}{\gamma-1}}(1+\lambda)^{\frac{\gamma}{\gamma-1}}\psi(x) \le \beta\psi(x), \tag{4.8.10}$$

and we get the result. □

We denote by K_i, $i = 1, 2$, the largest sectors on which the Bellman function is given by the formulae $W(x) = a_i(p_ix)^\gamma$. By the above we have:

Corollary 4.8.3 *The sectors K_i, $i = 1, 2$, are nonempty, and*

$$a_1 \ge \kappa_*^{\gamma-1}, \qquad a_2 \ge \kappa_*^{\gamma-1}\frac{1}{(1+\lambda)^\gamma}. \tag{4.8.11}$$

Our next aim is to show that $\operatorname{int} K_0 \neq \emptyset$ or, in other words, that K_1 and K_2 have no common boundary points (except zero). Though the crucial information will be obtained by a reduction to a one-dimensional problem, we make the first step in this direction immediately by inspecting the above formulae and establishing the following simple assertion:

Lemma 4.8.4 *If*

$$W(e_1) = \frac{1}{\gamma}\left(\frac{\beta}{1-\gamma}\right)^{\gamma-1}, \tag{4.8.12}$$

then the axis of abscises is not the common boundary of K_1 and K_2.

Proof. Suppose the opposite. Then the function $\psi(x) = a_2(p_2 x)^\gamma$ coincides with $W(x)$ on the sector cone$\{g_2, e_1\}$, and we can determine the value of a_2 using (4.8.12). It corresponds to the equalities in (4.8.9) and (4.8.10). Hence, in the considered case,

$$\mathcal{L}_0\psi(x) + u^*\left(\psi_\xi(x)\right) = a_2(p_2x)^{\gamma-1}\eta\left(\frac{1}{2}\sigma^2(\gamma-1)\frac{\eta}{p_2x} + \mu\right).$$

The right-hand side is strictly positive for points $x = (\xi, \eta)$ with sufficiently small coordinate $\eta > 0$. This means that, for such points, ψ cannot be the solution of the HJB equation. □

Usually, to find the constants a_i, one has to solve a free-boundary problem. However, in the special case where $\theta = 1$, the value a_2 can be calculated easily. The optimal strategy is just to sell a constant proportion of the stock to generate a flow of money for consumption. The precise result is here.

Proposition 4.8.5 *Suppose that the Merton parameter*

$$\kappa_M := \frac{1}{1-\gamma}\left(\beta - \frac{1}{2}\frac{\gamma}{1-\gamma}\frac{\mu^2}{\sigma^2}\right) > 0$$

and $\theta := (1-\gamma)^{-1}\mu\sigma^{-2} = 1$ (i.e., $\mu = (1-\gamma)\sigma^2$).
Then

$$W(x) = \mathbf{m}\frac{1}{(1+\lambda)^\gamma}(p_2x)^\gamma \quad \text{on } K \cap \{x = (\xi,\eta) :\ \xi \le 0\}. \tag{4.8.13}$$

Proof. Let us consider the process $V = (V^1, V^2)$ with $V^1 = 0$ and

$$dV_t^2 = V_t^2(\mu\,dt + \sigma\,dw_t) - \kappa_M V_t^2\,dt, \qquad V_0^2 = 1.$$

It corresponds to the strategy when the agent, having as the initial endowment a unit of stock, converts instantaneously a constant proportion of his wealth into cash and uses it immediately for the consumption with intensity given by the formula $c_t = (1+\lambda)^{-1}\kappa_M V_t^2$.

Apparently,

$$E(V_t^2)^\gamma = e^{\gamma(\mu-\sigma^2/2-\kappa_M)t+(1/2)\gamma^2\sigma^2 t} = e^{(\beta-\kappa_M)t},$$

the second equality holding because of the assumed identity $\mu = (1-\gamma)\sigma^2$. Thus, for the considered strategy,

$$J_\infty^\pi = \int_0^\infty e^{-\beta t} Eu(c_t)\,dt = \frac{1}{\gamma}\kappa_M^{\gamma-1}\frac{1}{(1+\lambda)^\gamma} = \mathbf{m}\frac{1}{(1+\lambda)^\gamma} = f(e_2),$$

where we denote by f the right-hand side of (4.8.13). It follows that $W(e_2) \geq f(e_2)$. In fact, we have the equality here since we know already that f is a classical supersolution (see Proposition 4.3.5), and, hence, $W \leq f$ in K. By the homotheticity, $W = f$ on $\mathbf{R}_+ e_2$. Since along each ray $\eta e_2 - \mathbf{R}_+ g_2$, $\eta > 0$, the function f is constant while W is decreasing, we have, on the sector $K \cap \{x = (\xi,\eta) :\ \xi \leq 0\}$, the inequality $W \leq f$ and, hence, the equality. □

4.8.3 Study of the Scalar Problem

The utility function is homogeneous of degree γ, and this property, due to the linearity of the dynamics, is inherited by the Bellman function, i.e.,

$$W(x) = \nu^\gamma W(x/\nu) \quad \forall \nu > 0. \tag{4.8.14}$$

Thus, knowing W on the intersection of the line $\{(\xi,\eta) : \xi+\eta = 1\}$ with the interior of K, that is, on the interval with the extremities $(-1/\lambda, 1+1/\lambda)$ and $(1+1/\lambda, -1/\lambda)$, one can reconstruct this function, using the homotheticity property, on the whole domain by the formula

$$W(\xi,y) = (\xi+\eta)^\gamma W\left(\frac{\xi}{\xi+\eta}, \frac{\eta}{\xi+\eta}\right), \qquad (\xi,\eta) \in \operatorname{int} K. \tag{4.8.15}$$

Let us consider the bijection mapping $T : (\xi,\eta) \mapsto (\xi = \eta, \eta/(\xi+\eta))$ of $\operatorname{int} K$ onto the rectangular $]0,\infty[\times]-1/\lambda, 1+1/\lambda[$; clearly, $T \in C^\infty$. It follows that the function $\Phi(t,z) = t^\gamma\psi(z)$ with $\psi(z) := W(1-z,z)$ is a viscosity solution of the equation obtained by the change of variables.

Specifically, let $t = t(\xi,\eta) = \xi+\eta$, $z = z(\xi,\eta) = \eta/(\xi+\eta)$ (and, hence, $\xi = t(1-z)$, $\eta = tz$). Differentiating the identity

$$W(\xi,\eta) = \big[t(\xi,\eta)\big]^\gamma \psi(z(\xi,\eta)),$$

we obtain the following formulae for derivatives:

$$\begin{aligned}
W_\xi(\xi,\eta) &= t^{\gamma-1}\big[\gamma\psi(z) - z\psi'(z)\big],\\
W_\eta(\xi,\eta) &= t^{\gamma-1}\big[\gamma\psi(z) + (1-z)\psi'(z)\big],\\
W_{\eta\eta}(\xi,\eta) &= t^{\gamma-2}\big[\gamma(\gamma-1)\psi(z) + 2(\gamma-1)(1-z)\psi'(z) + (1-z)^2\psi''(z)\big].
\end{aligned}$$

Therefore, we obtain that

$$g_1 W'(\xi,\eta) = t^{\gamma-1}\big[\lambda\gamma\psi(z) - (1+\lambda z)\psi'(z)\big],$$
$$g_2 W'(\xi,\eta) = t^{\gamma-1}\big[\lambda\gamma\psi(z) + (1+\lambda-\lambda z)\psi'(z)\big].$$

The formal substitution into (4.8.3) yields the following equation in the viscosity sense on the interval $[-1/\lambda, 1+1/\lambda]$ for the continuous function ψ vanishing at the extremities:

$$\max_{0\le i\le 2} \ell_i\psi = 0 \qquad (4.8.16)$$

with two first-order operators

$$\ell_1\psi := -\lambda\gamma\psi + (1+\lambda z)\psi', \qquad \ell_2\psi := -\lambda\gamma\psi - (1+\lambda-\lambda z)\psi',$$

and the second-order operator

$$\ell_0\psi = f_2\psi'' + f_1\psi' + f_0\psi + \frac{1-\gamma}{\gamma}[\gamma\psi - z\psi']^{\frac{\gamma}{\gamma-1}},$$

where

$$f_2(z) := \frac{1}{2}\sigma^2 z^2(1-z)^2,$$
$$f_1(z) := -\sigma^2(1-\gamma)z(1-z)(z-\theta),$$
$$f_0(z) := \frac{1}{2}\sigma^2\gamma(\gamma-1)z^2 + \gamma\mu z - \beta,$$

and $\theta := (1-\gamma)^{-1}\mu\sigma^{-2}$ is the Merton proportion.

The function ψ, being concave, has left and right derivatives continuous from the left and right, respectively, and satisfying the inequality $D^+\psi \le D^-\psi$, which can be strict only on a countably set. Outside this set, the derivative ψ' exists and is continuous.

Moreover, ψ is twice differentiable (in the sense of the Taylor formula) almost everywhere, and, therefore, (4.8.16) holds in the classical sense almost everywhere, see Lemma 4.2.3. This means that, at each point outside an exceptional null-set, we have three inequalities

$$\ell_1\psi(z) \le 0, \qquad \ell_0\psi(z) \le 0, \qquad \ell_2\psi(z) \le 0, \qquad (4.8.17)$$

and at least one of them is "active," that is, holds with the equality. By continuity, on the whole interval,

$$-\lambda\gamma\psi + (1+\lambda z)D^{\pm}\psi \le 0, \qquad -\lambda\gamma\psi - (1+\lambda-\lambda z)D^{\pm}\psi \le 0. \qquad (4.8.18)$$

Lemma 4.8.6 *The function ψ is continuously differentiable on the interval $I :=]-1/\lambda, 1+1/\lambda[$ except, maybe, zero. If ψ' has a discontinuity at zero, then*

$$\psi(0) = \frac{1}{\gamma}\left(\frac{1-\gamma}{\beta}\right)^{1-\gamma} = \frac{1}{\gamma}\kappa_*^{1-\gamma}. \qquad (4.8.19)$$

Proof. Since ψ is concave, it has left and right derivatives satisfying the inequality $D^+\psi \le D^-\psi$. Suppose that at a point z the inequality is strict. Let $p \in]D^+\psi(z), D^-\psi(z)[$. It follows that $(p, X) \in J^+\psi(z)$ whatever is $X \in \mathbf{R}$. By the definition of viscosity subsolution, we should have at least one of the following three inequalities: $\ell_i Q_{p,X}(z) \ge 0$, $i = 0, 1, 2$. This leads to an immediate contradiction if $z \ne 0$ or $z \ne 1$. Indeed, the coefficient at the second derivative being strictly positive, $\ell_0 Q_{p,X}(z) \to -\infty$ as $X \to -\infty$. A nonconstant linear function negative at the extremities of an interval is strictly negative in its interior, and, therefore, (4.8.18) implies that $\ell_i Q_{p,X}(z) < 0$ for $i = 1, 2$.

For the point $z = 1$, we can say only that

$$\ell_0 Q_{p,n}(z) = f_0(1)\psi(1) + u^*\big(\gamma\psi(1) - p\big) \ge 0. \tag{4.8.20}$$

To obtain a contradiction, we recall the classical fact that, for the monotone function ψ', the derivative ψ'' exists almost everywhere. Another fact (less known) is that ψ'' is locally integrable. On the other hand, the function $1/f_2$ has a nonintegrable singularity at 1. With this, we can find a sequence $z_n \uparrow 1$ (or $z_n \downarrow 1$) such that $\psi''(z_n)$ does exist, $\lim_n f_2(z_n)\psi''(z_n) = 0$, and inequalities (4.8.17) hold at z_n. The passage to the limit in the central one yields the inequalities

$$f_0(1)\psi(1) + u^*\big(\gamma\psi(1) - D^\pm\psi(1)\big) \le 0.$$

The function u^* being strictly monotone,

$$f_0(1)\psi(1) + u^*\big(\gamma\psi(1) - p\big) < 0,$$

contradicting to (4.8.20).

The above arguments for $z = 0$ do not work. An attempt to repeat them leads to a conclusion that if ψ' has a discontinuity at zero, then, necessarily,

$$f_0(0)\psi(0) + u^*\big(\gamma\psi(0)\big) = 0.$$

Solving this equation, we get the formula (4.8.19). □

Remark. On $\operatorname{int} K$, the Bellman function W always has a continuous derivative in the radial direction. Thus, the second claim of the lemma means that if W has no derivative in transversal directions at the ray $\mathbf{R}_+ e_1$, then, necessarily,

$$W(\xi, 0) = \frac{1}{\gamma}\left(\frac{1-\gamma}{\beta}\right)^{1-\gamma} \xi^\gamma.$$

Due to the continuity of the derivative, we may guess that the regions where inequalities in (4.8.16) are active are intervals. The concavity of ψ makes plausible a more specific structure of ψ, namely, that this function satisfies the above three differential inequalities, the first is a differential equation on the interval $]-1/\lambda, z_1[$, the second on $]z_1, z_2[$, and the third on $]z_2, 1+1/\lambda[$. The first-order differential equations $\ell_i\psi = 0$ with zero boundary conditions can be readily solved, and we have the following explicit formulae for the external intervals:

$$\psi(z) = \kappa_1(1+\lambda z)^\gamma, \quad z \in [-1/\lambda, z_1],$$
$$\psi(z) = \kappa_2(1+\lambda-\lambda z)^\gamma, \quad z \in [z_2, 1+1/\lambda],$$

where the constants $\kappa_i > 0$ are to be specified.

Lemma 4.8.7 *The interior of K_0 is nonempty.*

Proof. Suppose the opposite. This means that there exists a point x different from the origin which belongs to the common boundary of K_1 and K_2. If the function W is differentiable at x, we obtain that $W^{1/\gamma}$ has vanishing partial derivatives in the directions g_1 and g_2. It follows that $W'(x) = 0$, which is impossible. If W is not differentiable at x, then x is on the axis of abscises, and we refer simply to Lemmas 4.8.4 and 4.8.6. □

The most difficult part of the analysis is to show the following result claiming that the axis of abscises is not on the boundary of K_0.

Proposition 4.8.8 *The point e_1 belongs to* $\operatorname{int} K_1$.

Proof. Suppose that the assertion is note true, that is, the value $z_1 = 0$. If $\gamma W(e_1) > \kappa_*^{\gamma-1}$, then, as we know, $W \in C^1(\operatorname{int} K)$. In a right neighborhood of zero, $\psi(z) = W(1-z, z)$ is the solution of the second-order differential equation $\ell_0\psi = 0$. Since $f_1(0) = 0$ and $f_0(0) = -\beta$, we have that

$$\lim_{z\downarrow 0} \frac{1}{2}\sigma^2 z^2 \psi''(z) = \beta\psi(0) - \frac{1-\gamma}{\gamma}\left[\gamma\psi(0)\right]^{\frac{\gamma}{\gamma-1}}.$$

Noticing that the derivative of the function

$$H(y) := \beta y - \frac{1-\gamma}{\gamma}(\gamma y)^{\frac{\gamma}{\gamma-1}}, \quad y > 0,$$

is strictly positive and $H(\kappa_*^{\gamma-1}/\gamma) = 0$, we conclude that the limit above is also strictly positive. But this is impossible because ψ is concave and $\psi'' \le 0$.

Consider the "critical" case where $\gamma W(e_1) = \kappa_*^{\gamma-1}$ and the first derivative of ψ has a jump downwards at point zero.

Now we know the function $\psi(z)$ on the interval $[-1/\lambda, 0]$ explicitly. On the other hand, the lower bound for W implies the inequality

$$\psi(z) \ge h(z) := \frac{1}{\gamma}\kappa_*^{\gamma-1}\frac{1}{(1+\lambda)^\gamma}(1+\lambda-\lambda z)^\gamma.$$

In a right neighborhood of zero, $\psi(z)$ is the solution of the second-order differential equation. The difference $\tilde{\psi}(z) = \psi(z) - h(z) \ge 0$, and $\tilde{\psi}(0) = 0$. It is clear that

$$0 \le D^+\tilde{\psi}(0) = D^+\psi(0) - h'(0) \le D^-\psi(0) - h'(0) < \infty.$$

Substitution of $\psi(z) = h(z) + \tilde{\psi}(z)$ into the equation $\ell_0\psi(z) = 0$ yields the identity

$$g_1(z) + g_2(z) + g_3(z) = 0,$$

where

$$\begin{aligned} g_1(z) &:= \ell_0 h(z), \\ g_2(z) &:= \ell_0\tilde{\psi}(z) - u^*\big(\gamma\tilde{\psi}(z) - z\tilde{\psi}'(z)\big), \\ g_3(z) &:= u^*\big(\gamma h(z) - zh'(z) + \gamma\tilde{\psi}(z) - z\tilde{\psi}'(z)\big) - u^*\big(\gamma h(z) - zh'(z)\big). \end{aligned}$$

Note that

$$g_1(z) = f_2(z)h''(z) + f_1(z)h'(z) + \big(f_0(z) + \beta\big)h(z).$$

It follows that $g_1(0) = 0$ and

$$g_1'(0) = f_1'(0)h'(0) + f_0'(0)h(0) = \frac{\mu}{1+\lambda}\kappa_*^{\gamma-1}.$$

Observing that also $g_3(0+) = 0$, we infer that $g_2(0+) = 0$ as well, and, hence,

$$\lim_{z\downarrow 0} z^2\tilde{\psi}''(z) = 0. \tag{4.8.21}$$

The existence of the derivatives of g_1 and g_3 implies the existence of the derivatives of g_2. We have the identity for the derivatives

$$g_1'(z) + g_2'(z) + g_3'(z) = 0, \tag{4.8.22}$$

implying that

$$\lim_{z\downarrow 0}\left[g_2'(z) + g_3'(z)\right] = -\lim_{z\downarrow 0} g_1'(z) = -\frac{\mu}{1+\lambda}\kappa_*^{\gamma-1}. \tag{4.8.23}$$

The differentiability of g_2 implies that the derivative $\tilde{\psi}'''(z)$ does exist and is continuous for $z > 0$. Differentiating the expression

$$g_2(z) = f_2(z)\tilde{\psi}''(z) + f_1(z)\tilde{\psi}'(z) + f_0(z)\tilde{\psi}(z),$$

we obtain that

$$\lim_{z\downarrow 0} g_2'(z) = (\mu - \beta)D^+\tilde{\psi}(0) + \lim_{z\downarrow 0}\big[(1/2)\sigma^2 z^2\tilde{\psi}'''(z) + \big(\sigma^2 + \mu\big)z\tilde{\psi}''(z)\big].$$

Differentiating the formula

$$g_3(z) = \frac{1-\gamma}{\gamma}\left[\kappa_*^{\gamma-1}\left(\frac{1+\lambda-\lambda z}{1+\lambda}\right)^{\gamma-1} + \gamma\tilde{\psi}(z) - z\tilde{\psi}'(z)\right]^{\frac{\gamma}{\gamma-1}}$$
$$- \frac{1-\gamma}{\gamma}\kappa_*^{\gamma}\left(\frac{1+\lambda-\lambda z}{1+\lambda}\right)^{\gamma},$$

we infer that

$$\lim_{z\downarrow 0} g_3'(z) = -\kappa_*(\gamma-1)D^+\tilde{\psi}(0) + \kappa_* \lim_{z\downarrow 0} z\tilde{\psi}''(z).$$

Adding this identity with that for the limit of $g_2'(z)$, we arrive at the formula

$$\lim_{z\downarrow 0}\left[g_2'(z)+g_3'(z)\right] = \mu D^+\tilde{\psi}(0)+\lim_{z\downarrow 0}\left[(1/2)\sigma^2 z^2\tilde{\psi}'''(z)+\left(\sigma^2+\mu+\kappa_*\right)z\tilde{\psi}''(z)\right].$$

In virtue of the lemma below, the right-hand side is positive, in contradiction with identity (4.8.23). □

Lemma 4.8.9 *Let f be a bounded C^2-function on the interval $]0,\varepsilon[$, and let $\kappa \in \mathbf{R}$. Then*

$$\limsup_{z\downarrow 0}\left[z^2 f''(z) + \kappa z f'(z)\right] \geq 0.$$

Proof. Put $z := e^{-t}$ and consider the bounded function $\tilde{f}(t) = f(e^{-t})$. The claimed property means that

$$\limsup_{t\to\infty}\left[\tilde{f}''(t) + \kappa\tilde{f}'(t)\right] \geq 0$$

whatever is the constant κ. Suppose that the assertion fails. Then there exists $\kappa_1 > 0$ such that

$$\tilde{f}''(t) + \kappa\tilde{f}'(t) \leq -\kappa_1$$

for all sufficiently large t. The integration yields the inequality

$$\tilde{f}'(t) + \kappa\tilde{f}(t) \leq \kappa_2 - \kappa_1 t,$$

leading to an obvious contradiction: a function cannot be bounded while its derivative converges to $-\infty$. □

Proposition 4.8.8 completes the proof of Theorem 4.8.1. It provides the information that the suspicious point $z = 0$ belongs to the interval where the function ψ, given by the explicit formula, is smooth. Thus, ψ is C^1.

4.8.4 Skorohod Problem

In the classical Merton problem there are no difficulties to construct the optimal pair: the optimal wealth process is a solution of a simple linear stochastic equation, the optimal control is a linear function of the solution, and the same linear equation describes the optimal dynamics. In the model with transaction costs the situation is much more complicated. The optimal pair (i.e., the portfolio process and the control) is a solution of the stochastic Skorokhod problem (called also a stochastic differential equation with reflection). Moreover, the needed particular case of this problem has rather unpleasant features: the domain is a sector (so the boundary is not smooth), reflection is oblique, and the explicit form of the drift coefficient is not available. Since differential equations with reflections are rarely treated in the monographic literature, we provide in the Appendix a brief introduction with an elementary result which well serves our purpose here. In this subsection we use this result to check the existence and uniqueness of the optimal pair in the considered optimal control problem.

We have established that the solvency cone K can be decomposed into the union of three convex cones K_i (sectors, in fact) with disjoint nonempty interiors. The sectors K_i, $i = 1, 2$, share their "external" boundaries $\mathbf{R}_+ g_i$ with the solvency cone K, while the "internal" boundaries form the boundaries of K_0. The function $W^{1/\gamma}$ is linear in K_1 and K_2. Moreover, the axis of abscises is in the interior of K_1. The Bellman function W is C^1 in $\operatorname{int} K$.

Let $g : \partial K_0 \to \mathbf{R}^2$ be a vector-valued function with $g(x) = -g_i$ on the set $(\partial K_0 \cap \partial K_i) \setminus \{0\}$ and $g(0) = 0$. We consider on K_0 the Skorokhod problem formulated as follows: find a pair of adapted continuous processes, V, starting from $x \in K_0$, evolving in K_0, and trapped at zero, and k, scalar, starting at zero, and increasing, such that

$$dV_t^1 = -\big(W_\xi(V_t)\big)^{1/(\gamma-1)}\, dt + g_1(V_t)\, dk_t, \tag{4.8.24}$$

$$dV_t^2 = V_t^1(\mu\, dt + \sigma\, dw_t) + g_2(V_t)\, dk_t, \tag{4.8.25}$$

and

$$dk_t = I_{\{V_t \in \partial K_0\}}\, dk_t. \tag{4.8.26}$$

Proposition 4.8.10 *The Skorokhod problem has a solution.*

Proof. Let $\tilde{W}$ be the Bellman function of our optimal control problem but with the utility function $u_\gamma = u^\gamma/\gamma$. Let us introduce the polygons

$$K_0^n := K_0 \cap \{n^{-2/\gamma} \le x^1 + x^2 \le n^{2/\gamma}\}$$

and ice-cream-shaped closed regions $\tilde{K}_0^n$ having smooth boundaries and such that $K_0^n \subseteq \tilde{K}_0^n \subseteq K_0^{n+1}$. We define on the boundary $\partial\tilde{K}_0^n$ a smooth non-tangent reflection vector field coinciding on the lateral parts of the boundary with $g(x)$. According to Theorem 5.6.3 the Skorokhod problem in each region $\tilde{K}_0^n$ admits a unique solution (V^n, k^n). Let

$$\tau^n := \inf\{t : \left|V_t^n\right|_1 = n^{-2/\gamma}\}, \qquad \tau := \lim \tau^n,$$
$$\rho^n := \inf\{t : \left|V_t^n\right|_1 = n^{2/\gamma}\}, \qquad \rho := \lim \rho^n.$$

The uniqueness of solutions allows us to assert the existence of a pair of processes (V, k) defined on the time interval $[0, \tau \wedge \rho[$ and such that (V, k) coincides with (V^n, k^n) on $[0, \tau^n \wedge \rho^n]$. From the homotheticity property it follows that the Bellman function admits the upper bound of the form

$$W(x) \leq \bar{\kappa}|x|_1^\gamma, \quad x \in K,$$

and the lower bound

$$W(x) \geq \kappa|x|_1^\gamma, \quad x \in K_0.$$

Omitting in the dynamic programming inequality (4.5.4) the integral term and using afterwards the above lower bound, we obtain that

$$W(x) \geq Ee^{-\beta(\tau_n \wedge \rho_n \wedge t)} W(V_{\tau_n \wedge \rho_n \wedge t}) \geq \kappa e^{-t} E|V_{\tau_n \wedge \rho_n \wedge t}|_1^\gamma, \quad x \in K_0.$$

Since

$$E|V_{\tau_n \wedge \rho_n \wedge t}|_1^\gamma \geq EI_{\rho_n < \tau_n \wedge t}|V_{\tau_n \wedge \rho_n \wedge t}|_1^\gamma \geq n^2 P(\rho_n < \tau_n \wedge t),$$

this implies that $\sum_n P(\rho_n < \tau_n \wedge t) < \infty$. By virtue of the Borel–Cantelli lemma, $\rho_n \geq \tau_n \wedge t$ for sufficiently large n on the set of full measure. Thus, $\rho \geq \tau \wedge t$, and, because t is arbitrary, $\rho \geq \tau$.

So, we know that the processes V and k are defined on the stochastic interval $[0, \tau[$ and $\lim_n V_{\tau_n} = 0$.

One can show that $\lim_{t \uparrow \tau} V_{\tau_n} = 0$ (a.s.). In other words, the process V is absorbed at the origin. □

4.8.5 Optimal Strategy

Now we formulate the Davis–Norman theorem on the structure of the optimal solution.

Theorem 4.8.11 *Suppose that the initial endowment $x \in K_0$. Then the process V participating in the solution of the Skorokhod problem (4.8.24)–(4.8.26) defines the dynamics of the optimal portfolio, and the optimal strategy is given by the formulae*

$$B_t = \int_0^t g(V_s)\, dk_s, \tag{4.8.27}$$

$$c_t = \left(W_\xi(V_t)\right)^{1/(\gamma-1)}. \tag{4.8.28}$$

Proof. We follow the same line of arguments as in the proof of the Merton theorem. Applying the Itô formula and taking into account that $g_i W'(x) = 0$ for $x \in \partial K_0$, we obtain that

$$e^{-\beta t} W(V_t) + J_t^\pi = W(x) + \sigma \int_0^t V_t^2 W_\eta(V_t)\, dw_t. \tag{4.8.29}$$

The integrals with respect to dt and dk disappeared. To obtain the result, it remains to check that the stochastic integral above is a martingale (hence, its expectation is zero) and verify that, for a certain sequence of real numbers $t_n \uparrow \infty$,

$$\lim_{n\to\infty} e^{-\beta t_n} EW(V_{t_n}) = 0. \tag{4.8.30}$$

Due to the homotheticity property of the derivative of the Bellman function following from Lemma 4.7.5, we have the inequality $|W'(y)| \le \kappa |y|^{\gamma-1}$, where κ is the bound for the derivative of W on the intersection of the set K_0 with the line $\xi + \eta = 1$. Thus,

$$\big|\eta W_\eta(y)\big| \le \kappa |y|^\gamma \le \kappa\big(1 + |y|\big), \quad y \in K_0,$$

and the absolute value of the integrand is dominated by a linear function of the phase variable. Using the exponential bound of Proposition 4.2.1, we infer that the stochastic integral in (4.8.29) is a martingale.

To accomplish the proof, we need the inequality

$$W_\xi(y) \ge \kappa |y|^{\gamma-1}, \quad y \in K_0,$$

also implied by the homotheticity. Here the constant $\kappa > 0$ is the minimum of the partial derivative W_ξ on the intersection of the set K_0 with the line $\xi + \eta = 1$. It is strictly positive: the derivative of the Bellman function W in the direction g_1 is positive, the derivative in the radial direction is strictly positive, and the vector e_1 lies between these two directions.

Using these observations, we have the following chain of inequalities with a varying constant:

$$E\int_0^\infty e^{-\beta t} W(V_t)\, dt \le \kappa E \int_0^\infty e^{-\beta t} |V_t|^\gamma\, dt \le \kappa E \int_0^\infty e^{-\beta t} u(c_t)\, dt \le W(x).$$

Since W is finite, this obviously implies the existence of a sequence $t_n \uparrow \infty$ for which (4.8.30) holds. □

Remark. The case where $x \in K_i$ is easily reduced to the one treated in the theorem. It is sufficient to modify the process B by adding the initial jump

$$\Delta B_0 = \inf\{s \ge 0 : \ x - s g_i \in K_0\}, \quad x \in K_i,\ i = 1, 2. \tag{4.8.31}$$

The function W on the set K_i is constant along the direction g_i for $i = 1, 2$. Thus, such a modification gives a strategy resulting in the value $W(x + \Delta_0)$

coinciding with $W(x)$. Notice that in $\operatorname{int} K_1$ (resp., in $\operatorname{int} K_2$) the changes of the initial endowment means the buying of stock (resp. the selling of stock), while in $\operatorname{int} K_0$ there are no transactions. This explains the abbreviations BS, SS, and NT used in the literature for the corresponding regions.

Using the structure of optimal control, we improve a bit Proposition 4.3.5, which gives us an upper bound for the Bellman function in the case where the solution of the classical Merton problem is finite. It happens that, "usually," in the model with transaction costs the bound is strict, and this fact plays an important role to locate more precisely the boundaries of the no-transaction cone K_0 (see the next subsection). The precise statement is as follows.

Proposition 4.8.12 *Suppose that $\kappa_M > 0$. Let $p = (p_1, p_2) \in K^*$ and $p_1 = 1$. If $(1-\gamma)\sigma^2 \neq \mu$, then*

$$W(x) < \mathbf{m}u(px) = \frac{1}{\gamma}\kappa_M^{\gamma-1}(px)^\gamma \quad \forall x \in \operatorname{int} K. \tag{4.8.32}$$

If $(1-\gamma)\sigma^2 = \mu$, then $e_2 \in K_0$.

Proof. Proposition 4.3.5 says that the function $\varphi(x) = \mathbf{m}u(px)$ is a supersolution of the HJB equation, and, therefore,

$$\mathcal{L}_0\varphi + u^*(\varphi_\xi) = -\frac{\gamma}{2(1-\gamma)\sigma^2}\left[\frac{(1-\gamma)\sigma^2 p_2\eta}{\xi + p_2\eta} - \mu\right]^2 \phi(x) \le 0.$$

The equality holds if and only if

$$\mu\xi - \left[(1-\gamma)\sigma^2 - \mu\right]p_2\eta = 0. \tag{4.8.33}$$

Let us plug-in the optimal process $V = (V^1, V^2)$ (corresponding to the strategy given by (4.8.27) and (4.8.28), eventually with an initial transfer) into the function φ. Applying the Itô formula, we get

$$\begin{aligned}
e^{-\beta t}\varphi(V_t) &= \varphi(V_0) + \int_{]0,t]} e^{-\beta s}\varphi_\eta(V_s)\sigma V_s^2\, dw_s + \int_{]0,t]} e^{-\beta s}\varphi'(V_s)g(V_s)\, dk_s \\
&\quad + \int_{]0,t]} e^{-\beta s}\left[\mathcal{L}_0\varphi(V_s) - c_s\varphi_\xi(V_s) + u(c_s)\right] ds - \int_{]0,t]} e^{-\beta s}u(c_s)\, ds \\
&\le \varphi(x) + \int_{]0,t]} e^{-\beta s}\varphi_\eta(V_s)\sigma V_s^2\, dw_s \\
&\quad + \int_{]0,t]} e^{-\beta s}\left[\mathcal{L}_0\varphi(V_s) + u^*\left(\varphi_\xi(V_s)\right)\right] ds - \int_{]0,t]} e^{-\beta s}u(c_s)\, ds.
\end{aligned}$$

The above bound holds because $\varphi(V_0) \le \varphi(x)$ due to losses which occur at the initial transfer, $g_i\varphi'(x) \le 0$ for $x \in \partial K_0$, and

$$-c_s\varphi_\xi(V_s) + u(c_s) \le u^*\left(\varphi_\xi(V_s)\right).$$

By the same arguments as in the previous proof, we infer that the expectation of the stochastic integral is zero and $Ee^{-\beta t}\varphi(V_t) \to 0$ as $t \to \infty$. It follows that

$$W(x) \le \varphi(x) + E\int_{]0,t]} e^{-\beta s}\left[\mathcal{L}_0\varphi(V_s) + u^*\left(\varphi_\xi(V_s)\right)\right] ds \le \varphi(x).$$

In the case $(1-\gamma)\sigma^2 \ne \mu$, the second inequality is always strict: otherwise the integrant is a negligible process, i.e., according to (4.8.33), we would have

$$\mu V^1 - \left[(1-\gamma)\sigma^2 - \mu\right]p_2 V^2 = 0.$$

This identity is impossible because the left-side is a semimartingale with non-trivial diffusion component.

If $(1-\gamma)\sigma^2 = \mu$, we have, necessarily, that $V_t^1 = 0$ for all $t > 0$. Thus, the process V after the initial jump evolves along the axis of ordinates, and therefore $e_2 \in K_0$. □

4.8.6 Precisions on the No-Transaction Region

We established already that the no-transaction region $K_0 = \text{cone}\{\tilde g_1, \tilde g_2\}$ has a nonempty interior and lies strictly above the axis of abscises. Now we give some bounds on the position of the generator $\tilde g_2$.

The following simple lemma ensures us that, whatever is $\lambda > 0$, the interval $[z_1, z_2]$ (depending on λ) lies inside the fixed interval, namely, $[0, 2\theta+1]$, $\theta = \beta/(1-\gamma)$. We shall need this fact for the asymptotic analysis as $\lambda \to 0$.

Lemma 4.8.13 *We have*

$$0 < z_2 \le 1 + \frac{2\theta}{1+\lambda+2\theta\lambda}. \qquad (4.8.34)$$

Proof. The first inequality holds because $z_2 > z_1 > 0$. According to Proposition 4.8.2, we have the inclusion $K_2 \subseteq \text{cone}\{g_2, Ng_2 + e_2\}$, where the constant $N = 2\theta(1+\lambda)^{-1}$. The ray generated by $Ng_2 + e_2$ intersects the line $\{(\xi,\eta):\ \xi + \eta = 1\}$ at the point $(1-z, z)$, where

$$z = 1 + \frac{N}{1+N\lambda} = 1 + \frac{2\theta}{1+\lambda+2\theta\lambda}.$$

Since the point $(1-z_2, z_2)$ is the intersection of the boundary ray separating K_2 and K_0 with the aforementioned line, we have from obvious geometric considerations that $z_2 \le z$, which is exactly the second inequality. □

With minor efforts, we can get a more precise information about the positions of z_2 and z_1.

Recall that, according to Corollary 4.8.3, on the cones K_i, $i = 1, 2$, the Bellman function W has the form $W(x) = a_i(p_i x)^\gamma$, where

$$a_1 \geq \kappa_*^{\gamma-1}, \qquad a_2 \geq \kappa_*^{\gamma-1}\frac{1}{(1+\lambda)^\gamma},$$

with $\kappa_* = \beta/(1-\gamma)$. Now we are able to say a bit more: the inequality for a_2 is strict! Indeed, suppose that we have the equality for a_2. In virtue of (4.7.7), for such a value of a_2, the function $W(x)$ in the upper half-plane dominates the function $a_2(p_2 x)^\gamma$. But we know that the latter dominates $W(x)$ on the whole cone K and coincides with $W(x)$ exactly on the cone K_2. Combining these facts, we get that $K_2 \supseteq K \cap \{x = (\xi, \eta) : \eta \geq 0\}$, which is impossible.

Now we give sharper bounds for z_i.

Proposition 4.8.14 (a) *We always have the inequality*

$$z_2 < \frac{\mu(1+\lambda)}{\frac{1}{2}(1-\gamma)\sigma^2 + \mu\lambda}. \tag{4.8.35}$$

(b) *If $\kappa_M > 0$ and $\theta \neq 1$ (i.e., $(1-\gamma)\sigma^2 \neq \mu$), then*

$$z_2 > \frac{\mu(1+\lambda)}{(1-\gamma)\sigma^2 + \mu\lambda}. \tag{4.8.36}$$

(c) *If $\kappa_M > 0$ and $\theta = 1$, then $z_2 = 1$.*
(d) *If $\kappa_M > 0$ and $(1-\gamma)\sigma^2 > \lambda\mu$, then*

$$z_1 < \frac{\mu\lambda}{(1-\gamma)\sigma^2(1+\lambda) - \mu\lambda}. \tag{4.8.37}$$

Proof. (a) Let us consider the function $\psi(x) = a_2 u(p_2 x)$, which is the solution of the HJB equation in K_2. Since $a_2 > \kappa_*^{\gamma-1}(1+\lambda)^{-\gamma}$, we have, for every $x \in \operatorname{int} K$, the strict inequality $u^*(\psi_\xi(x)) - \beta\psi(x) < 0$ and, therefore, the bound

$$\mathcal{L}_0\psi(x) + u^*\big(\psi_\xi(x)\big) < a_2(p_2 x)^{\gamma-2}\eta\left[\frac{1}{2}\sigma^2(\gamma-1)\eta + \mu(p_2 x)\right].$$

The expression in the square bracket is less or equal to zero when the point $x = (1-z, z) \in K$ and z dominates the right-hand side of (4.8.35). This observation makes the bound (4.8.35) obvious.

(b) Let us examine the right-hand side of the identity

$$\begin{aligned}\mathcal{L}_0\psi(x) &+ u^*\big(\psi_\xi(x)\big)\\ &= -\frac{\gamma}{2(1-\gamma)\sigma^2}\left[\frac{(1-\gamma)\sigma^2\eta}{(1+\lambda)\xi+\eta} - \mu\right]^2\psi(x)\\ &\quad + \left(\frac{1}{2}\frac{\gamma}{1-\gamma}\frac{\mu^2}{\sigma^2} - \beta + (1-\gamma)a_2^{1/(\gamma-1)}(1+\lambda)^{\frac{\gamma}{\gamma-1}}\right)\psi(x).\end{aligned}$$

In virtue of Proposition 4.8.12, under the assumed hypotheses, the coefficient $(\dots)$ is strictly positive. Thus, the function given by the expression $[\dots]$ cannot vanish at points of the set $\mathrm{int}K_2$ where the function ψ is the solution of the HJB equation. It has the positive sign on this set (its values tend to $+\infty$ as $x=(\xi,\eta)$ approaches a point of the outer boundary of K_2 other than zero). Moreover, the continuity considerations imply that $[\dots]$ is strictly positive also on the inner boundary (of course, except the origin), in particular, at the point $(1-z_2,z_2)$. This last property is equivalent to inequality (4.8.36).

(c) In this case the Proposition 4.8.5 says that the coefficient $(\dots)=0$ and $z_2\le 1$. On the other hand, according to Proposition 4.8.12, $z_2\le 1$.

(d) If $(1-\gamma)\sigma^2=\mu$, inequality (4.8.37) is reduced to $z_1<1$. However, we already know that $z_1<z_2$ always and $z_2=1$ in the considered case as we just proved. Suppose that $(1-\gamma)\sigma^2\ne\mu$. For $\varphi(x)=a_1u(p_1(x))$, we have the identity

$$\mathcal{L}_0\psi(x)+u^*(\psi_\xi(x))=-\frac{\gamma}{2(1-\gamma)\sigma^2}\left[\frac{(1-\gamma)\sigma^2(1+\lambda)\eta}{\xi+(1+\lambda)\eta}-\mu\right]^2\varphi(x)$$
$$+\left(\frac{1}{2}\frac{\gamma}{1-\gamma}\frac{\mu^2}{\sigma^2}-\beta+(1-\gamma)a_1^{\frac{1}{\gamma-1}}\right)\varphi(x).$$

Proposition 4.8.12 provides us the information that the second term is strictly positive on $\mathrm{int}\,K_1$, and we derive the required inequality (4.8.37) by the same arguments as in (a). □

4.9 Liquidity Premium

4.9.1 Non-Robustness with Respect to Transaction Costs

According to Theorem 4.1.1, in the Merton two-asset model of frictionless financial market, the optimal expected utility of the unit wealth invested in a portfolio is given by the formula

$$\mathbf{m}=W_M(1)=\kappa_M^{\gamma-1}/\gamma,$$

where

$$\kappa_M:=\frac{1}{1-\gamma}\left(\beta-\frac{1}{2}\frac{\gamma}{1-\gamma}\frac{\mu^2}{\sigma^2}\right); \tag{4.9.1}$$

it is assumed that the model parameters are such that $\kappa_M>0$. The optimal strategy prescribes to rebalance continuously the portfolio to keep the proportion of stock to the total value at the constant level $\theta=(1-\gamma)^{-1}\mu/\sigma^2$, called in the literature the *Merton proportion*.

It is natural to expect that in the market with friction, even following an appropriate optimal strategy, the investor cannot achieve the above performance. It would be interesting to know to which extent the presence of transaction costs deteriorate the portfolio performance. Unfortunately, the solution of the Davis–Norman problem does not admit such an explicit expression, and the comparison between two results seems to be complicated. It can be done asymptotically for small transaction costs. Assume that the initial endowment of the investor is in θ units of stock and $1-\theta$ units on money, where θ is the Merton proportion. The next theorem due to Shreve asserts that the discrepancy between the two optimal values, in general, is of (exact) order $\lambda^{2/3}$ as the transaction cost coefficient λ tends to zero. Thus, the model is not robust in the sense that the discrepancy increases infinitely fast when the transaction costs appear. The only exception is the case $\theta = 1$, where the portfolio has zero position in money, and the stock is sold only to consume. This case will be considered separately at the end of the section.

In our presentation, in order to have simpler formulae, we assume that both operations, buying stock and selling stock, are charged equally.

Theorem 4.9.1 *Suppose that $\theta \neq 1$. Then there are constants κ_1, $\kappa_2 > 0$, independent of λ, such that*

$$\mathbf{m} - \kappa_2\lambda^{2/3} \le W(1-\theta,\theta) \le \mathbf{m} - \kappa_1\lambda^{2/3} \tag{4.9.2}$$

for all sufficiently small $\lambda > 0$.

Proof. Recall that the function $\psi(z) := W(1-z,z)$ is concave, continuously differentiable on the interval $]-1/\lambda, 1+1/\lambda[$, and its second derivative may have (jump) discontinuities only at two (distinct) points z_1, z_2. It satisfies, in the classical sense, everywhere except at these two points, the HJB equation

$$\max\{\ell_1\psi(z), \ell_0\psi(z), \ell_2\psi(z)\} = 0 \tag{4.9.3}$$

involving the first-order operators

$$\ell_1\psi(z) = -\lambda\gamma\psi(z) + (1+\lambda z)\psi'(z), \qquad \ell_2\psi(z) = -\lambda\gamma\psi - (1+\lambda-\lambda z)\psi'(z),$$

and the second-order operator

$$\ell_0\psi(z) = f_2(z)\psi''(z) + f_1(z)\psi'(z) + f_0(z)\psi(z) + \frac{1-\gamma}{\gamma}\left[\gamma\psi(z) - z\psi'(z)\right]^{\frac{\gamma}{\gamma-1}}$$

with the coefficients

$$\begin{aligned}
f_2(z) &= \frac{1}{2}\sigma^2 z^2(1-z)^2,\\
f_1(z) &= -\sigma^2(1-\gamma)z(1-z)(z-\theta),\\
f_0(z) &= -\frac{1}{2}\sigma^2\gamma(1-\gamma)(z-\theta)^2 - (1-\gamma)(\gamma\mathbf{m})^{\frac{1}{\gamma-1}}.
\end{aligned}$$

The structure of ψ is as follows. Outside of the interval $]z_1, z_2[$, the function $\psi^{1/\gamma}$ coincides with two linear functions: on $[-1/\lambda, z_1]$ with a function proportional to $(1+\lambda z)$, thus, vanishing at $-1/\lambda$; on $[z_2, 1+1/\lambda]$ with a function proportional to $(1+\lambda-\lambda z)$, thus, vanishing at $1+1/\lambda$. On the interval $]z_1, z_2[$ the function ψ is the classical solution of the second-order differential equation $\ell_0\psi(z) = 0$.

For a fixed λ, the graph of ψ is an arc-shaped curve, flattening and with the increasing base as λ tends to zero. We may expect that the "interpolation" interval degenerates into the single point θ and that the maximal value of ψ converges to $\mathbf{m}$.

Having in mind this behavior, we approximate ψ from above and below by C^1-functions $\tilde{\psi}$ having a similar shape. We interpolate their "linear in the power γ" parts by parabolas and choose, for these functions, the interpolation intervals $]\tilde{z}_1, \tilde{z}_2[$ containing θ and being of the length of order $\lambda^{1/3}$.

Namely, for $r > 0$, we put

$$Q(z) := \mathbf{m} - r\lambda^{2/3} - (z-\theta)^2\lambda^{2/3},$$

$\tilde{z}_1 := \theta - \delta_1\lambda^{1/3}$, and $\tilde{z}_2 := \theta + \delta_2\lambda^{1/3}$, where the bounded strictly positive coefficients $\delta_i = \delta_i(\lambda; r)$ will be chosen to guarantee the continuity of the first derivatives of the function $\tilde{\psi}(z) = \tilde{\psi}(z, \lambda; r)$. The latter is defined (for sufficiently small λ) by the formula

$$\begin{aligned} \tilde{\psi}(z) := {} & \frac{Q(\tilde{z}_1)}{(1+\lambda\tilde{z}_1)^\gamma}(1+\lambda z)^\gamma I_{[-1/\lambda, \tilde{z}_1[}(z) + Q(z) I_{[\tilde{z}_1, \tilde{z}_2]}(z) \\ & + \frac{Q(\tilde{z}_2)}{(1+\lambda-\lambda\tilde{z}_2)^\gamma}(1+\lambda-\lambda z)^\gamma I_{]\tilde{z}_2, 1+1/\lambda]}(z). \end{aligned} \qquad (4.9.4)$$

Its first derivative:

$$\tilde{\psi}'(z) = \begin{cases} \frac{\gamma\lambda}{1+\lambda z}\tilde{\psi}(z), & z \in]-1/\lambda, \tilde{z}_1[, \\ -2\lambda^{2/3}(z-\theta), & z \in]\tilde{z}_1, \tilde{z}_2[, \\ -\frac{\gamma\lambda}{1+\lambda-\lambda z}\tilde{\psi}(z), & z \in]\tilde{z}_2, 1+1/\lambda[. \end{cases}$$

Its second derivative:

$$\tilde{\psi}''(z) = \begin{cases} -\frac{\gamma(1-\gamma)\lambda^2}{(1+\lambda z)^2}\tilde{\psi}(z), & z \in]-1/\lambda, \tilde{z}_1[, \\ -2\lambda^{2/3}, & z \in]\tilde{z}_1, \tilde{z}_2[, \\ -\frac{\gamma(1-\gamma)\lambda^2}{(1+\lambda-\lambda z)^2}\tilde{\psi}(z), & z \in]\tilde{z}_2, 1+1/\lambda[. \end{cases}$$

The second derivative is constant over the central interval. It is continuous on the external intervals, and its limits at the points $\tilde{z}_i$ coincide with the corresponding one-sided derivatives.

The result will be established if we find, independently on λ, two values of the parameter $r > 0$ such that the corresponding $\delta_i(\lambda)$ belongs to a bounded interval, $\tilde{\psi}(z, \lambda; r)$ is a supersolution of the HJB equation for the smaller value (say, r_1) and a subsolution for the larger value (say, r_2) whatever are $\lambda \le \lambda_0$; the threshold λ_0 may depend on r_i. Indeed, in virtue of the comparison lemma given after the proof of the theorem, the function ψ will lie between these two functions, and the corresponding inequalities for the values calculated at the point θ yield (4.9.2). Note that the supersolution and the subsolution are understood in an "almost" classical sense: the corresponding inequalities for $\max_i \ell\tilde{\psi}(z)$ should hold everywhere except, maybe, at the points $\tilde{z}_i$.

We have

$$\begin{aligned} &\tilde{\psi}(\tilde{z}_i) = \mathbf{m} - r\lambda^{2/3} - \delta_i^2\lambda^{4/3}, \quad i = 1,2,\\ &D^+\tilde{\psi}(z_1) = 2\delta_1\lambda, \qquad D^-\tilde{\psi}(z_2) = -2\delta_2,\\ &D^-\tilde{\psi}(z_1) = \frac{\gamma\lambda}{1+\lambda\tilde{z}_1}Q(\tilde{z}_1), \qquad D^+\tilde{\psi}(z_2) = -\frac{\gamma\lambda}{1+\lambda-\lambda\tilde{z}_2}Q(\tilde{z}_2). \end{aligned}$$

The requirement that $\tilde{\psi}$ is continuously differentiable means that the right and left derivatives of $\tilde{\psi}$ at each point $\tilde{z}_i$ coincide. It is met when δ_i are the (positive) roots of the corresponding quadratic equations

$$A_i(\lambda)\delta^2 - B_i(\lambda)\delta + C_i(\lambda) = 0.$$

Their coefficients are as follows:

$$\begin{aligned} &A_1(\lambda) = (2-\gamma)\lambda^{4/3}, \qquad B_1(\lambda) = 2(1+\lambda\theta), \qquad C_1(\lambda) = \gamma\big[\mathbf{m} - r\lambda^{2/3}\big];\\ &A_2(\lambda) = (2+\gamma)\lambda^{4/3}, \qquad B_2(\lambda) = 2(1+\lambda-\lambda\theta), \qquad C_2(\lambda) = C_1(\lambda). \end{aligned}$$

Note that, for λ close to zero, the left-hand sides of these equations are positive for $\delta = B_i(\lambda)/C_i(\lambda)$ and negative for $\delta = 2B_i(\lambda)/C_i(\lambda)$. Thus, we can find roots $\delta_i(\lambda)$ in this interval. Asymptotically,

$$\delta_i(\lambda) \in \big[2/(\gamma\mathbf{m}), 4/(\gamma\mathbf{m}) + o(1)\big], \tag{4.9.5}$$

and, therefore, $\delta_i(\lambda) \in [2/(\gamma\mathbf{m}), 5/(\gamma\mathbf{m})]$ for λ less than a certain λ_0 depending on r.

Let us check that, for "r small" (respectively, "r large"), the function $\tilde{\psi}$ is a supersolution (respectively, subsolution) of the HJB equation (4.9.3).

In principle, we have to verify 18 inequalities. Luckily, most of them are trivial or easy. We organize our analysis in three parts.

1. The interval $]-1/\lambda, \tilde{z}_1[$.

We have here $\ell_1\tilde{\psi} = 0$, and, hence, the subsolution inequality ≥ 0 is obvious. Moreover, since $\tilde{\psi}$ and $\tilde{\psi}'$ are both positive, the inequality $\ell_2\tilde{\psi} \le 0$ always holds. To check the supersolution inequality ≤ 0, it remains to verify, on the considered interval, that $\ell_0\tilde{\psi} \le 0$.

Note that the function $\tilde{\psi}(z)$ on the considered interval is proportional to $(1+\lambda z)^{\gamma}$, and, therefore,

$$\gamma\tilde{\psi}(z) - z\tilde{\psi}'(z) = \gamma\frac{1}{1+\lambda z}\tilde{\psi}(z) = \text{const} \times (1+\lambda z)^{\gamma-1}.$$

Since $u^*(p) = (1-\gamma)/\gamma p^{\gamma/(\gamma-1)}$, the nonlinear term is proportional to $\tilde{\psi}$, i.e.,

$$u^*\left(\gamma\tilde{\psi}(z) - z\tilde{\psi}'(z)\right) = \kappa_\lambda \tilde{\psi}(z).$$

Since $\tilde{\psi}(\tilde{z}_1) = Q(\tilde{z}_1)$, we can determine the constant κ_λ. Namely,

$$\kappa_\lambda = (1+\lambda\tilde{z}_1)^{\frac{\gamma}{1-\gamma}}(1-\gamma)\gamma^{\frac{1}{\gamma-1}}\left[Q(\tilde{z}_1)\right]^{\frac{1}{\gamma-1}}.$$

Our definitions imply that $\tilde{z}_1 = \theta + o(1)$, $Q(\tilde{z}_1) = \mathbf{m} - r\lambda^{2/3} + o(\lambda^{2/3})$. The derivative $Q'(1) = 1$, and, hence, κ_λ has the following asymptotic expansion:

$$\kappa_\lambda = (1-\gamma)(\gamma\mathbf{m})^{\frac{1}{\gamma-1}}\left(1 + \frac{1}{1-\gamma}\frac{r}{\mathbf{m}}\lambda^{2/3}\right) + o(\lambda^{2/3}).$$

Inspecting the function $\ell_0\tilde{\psi}$, we see that the term $f_2(z)\tilde{\psi}''(z)$ is always negative. If $z \le 0$, the coefficient f_1 is negative, and so is $f_1(z)\tilde{\psi}'(z)$. Omitting these two terms and taking into account that, for $z \in]-1/\lambda, \tilde{z}_1[$,

$$\begin{aligned} f_0(z) &= -\frac{1}{2}\sigma^2\gamma(1-\gamma)(z-\theta)^2 - (1-\gamma)(\gamma\mathbf{m})^{\frac{1}{\gamma-1}} \\ &\le -\frac{1}{2}\sigma^2\gamma(1-\gamma)\delta_1^2\lambda^{2/3} - (1-\gamma)(\gamma\mathbf{m})^{\frac{1}{\gamma-1}}, \end{aligned}$$

we arrive, on the interval $]-1/\lambda, 0]$, at the bound

$$\frac{\ell_0\tilde{\psi}(z)}{\tilde{\psi}(z)} \le f_0(z) + \kappa_\lambda \le \frac{1}{2}\sigma^2\gamma(1-\gamma)\left(\kappa r - \frac{1}{(\gamma\mathbf{m})^2}\right)\lambda^{2/3} + o(\lambda^{2/3}),$$

which holds for all $\lambda \le \lambda_0$ (the threshold is chosen to insure that there exists a positive root $\delta_1 \ge 1/(\gamma\mathbf{m})$). Here the constant $\kappa > 0$, and, hence, for "small" r, the coefficient of the main term is strictly negative.

On the interval $]0, \tilde{z}_1[$ we cannot omit the term with the first derivative, but this does not affect the resulting asymptotic expansion: on this interval, $f_1(z)$ is bounded, and

$$f_1(z)\frac{\tilde{\psi}'(z)}{\tilde{\psi}(z)} = f_1(z)\frac{\gamma\lambda}{1+\lambda z} = o(\lambda)$$

uniformly in z.

2. The interval $]\tilde{z}_2, 1+1/\lambda[$.

The reasoning is completely analogous to the one given above.

3. The interval $]\tilde{z}_1, \tilde{z}_2[$.

Here the function

$$\ell_1\tilde{\psi}(z) = \ell_1 Q(z) = -\lambda\gamma Q(z) + (1+\lambda z)Q'(z)$$

is quadratic in z, and its coefficients at z^2 and z are, respectively, $\lambda^{5/3}(\gamma-2)$ and $2\lambda^{5/3}(-\gamma\theta+\theta-1/\lambda)$. The point

$$z_{\max}(\lambda) := \frac{\theta(1-\gamma)-1/\lambda}{2-\gamma},$$

where $\ell_1 Q(z)$ attains its maximum, tends to $-\infty$ as $\lambda \to 0$ and, hence, lies leftwards to the interval $[\tilde{z}_1, z_2]$ for sufficiently small λ. But this means that, on this interval, $\ell_1\tilde{\psi}$ decreases, i.e., $\ell_1\tilde{\psi}(z) \le \ell_1\tilde{\psi}(\tilde{z}_1) = 0$.

Exactly by the same arguments we check that $\ell_2\tilde{\psi}(z) \le 0$.

Summarizing: in the central interval, the super- or subsolution property for $\tilde{\psi}$ is equivalent, respectively, to the inequality $\ell_0\tilde{\psi} \le 0$ or $\ell_0\tilde{\psi} \ge 0$.

On the considered interval (degenerating in the limit to the point θ) we have that $Q(z) = \mathbf{m} - r\lambda^{2/3} + o(\lambda^{2/3})$, $Q'(z) = o(\lambda^{2/3})$ uniformly in z and $Q'' = -2\lambda^{2/3}$. With this, we get readily that

$$\ell_0\tilde{\psi}(z) = -\sigma^2\theta(1-\theta)\lambda^{2/3} + r\frac{1}{2}\sigma^2\gamma(1-\gamma)(z-\theta)^2 + r\kappa\lambda^{2/3} + o\big(\lambda^{2/3}\big)$$

with some constant $\kappa > 0$. Since $(z-\theta)^2 \le 5(\gamma\mathbf{m})\lambda^{2/3}$, this representation makes clear that in the case $\theta \ne 1$ we can choose a small value r such that $\ell_0\tilde{\psi} \le 0$ for sufficiently small λ. The opposite inequality for large values of r holds obviously for any θ. □

Remark. The condition $\theta \ne 1$ is needed only at the very end of the proof. Moreover, the subsolution property of the approximation is not violated even in the exceptional case, and, therefore, due to the comparison lemma below, the left inequality in (4.9.2) remains valid. In contrast to this, for $\theta = 1$, the right inequality fails because, as we show later, in this case the difference $\psi(\theta) - \mathbf{m}$ converges to zero with rate λ.

Now we are back to the comparison lemma. It is extremely simple for the C^1-functions which are twice continuously differentiable everywhere except a finite number of points where the limits of the second derivatives exist and coincide with the one-sided second derivatives. A subtlety is that one of the functions to be compared is ψ, which is simultaneously a super- and subsolution, but for which we cannot guarantee the mentioned behavior of the second derivatives at the point $z = 1$ (in the case $z_2 = 1$). However, $(z_n - 1)^2\psi''(z_n) \to 0$ for certain sequences $z_n \uparrow 0$ and $z_n \downarrow 0$. But this implies that the super- and subsolution inequalities hold also at the point $z = 1$ with the degenerate operator ℓ_0 defined by

$$\ell_0\psi(1) = f_0(1)\psi(1) + u^*\big(\gamma\psi(1) - \lambda\psi'(1)\big).$$

In the formulation below we suppose that ψ_1 and ψ_2 possess this property of the second derivative at $z = 1$.

Lemma 4.9.2 *Let ψ_1 be a supersolution, and ψ_2 be a subsolution with the same boundary condition. Then $\psi_1 \geq \psi_2$.*

Proof. Let us consider a point z_0 where the difference $\psi_2 - \psi_1$ attains its maximum. If the claim fails, then $\psi_2(z_0) > \psi_1(z_0)$. Since $\psi_2'(z_0) = \psi_1'(z_0)$ and ψ_1 is a supersolution, $l_i\psi_2(z_0) < l_i\psi_1(z_0) \leq 0$ for $i = 1, 2$. Suppose first that at z_0 the second derivatives are continuous. Then $\psi_2''(z_0) \leq \psi_1''(z_0)$. Taking into account the signs of the coefficients $f_2(z_0)$ and $f_0(z_0)$ and using the fact that u^* is decreasing, we infer that $l_0\psi_2(z_0) < l_0\psi_1(z_0) \leq 0$. Thus, all $l_i\psi_2(z_0)$ are strictly less than zero, in contradiction with the subsolution property. The general case $z_0 \neq 1$ is not much different: the one-sided second derivatives of $\psi_2 - \psi_1$ are negative at z_0, and we obtain, as before, that $l_0\psi_2(z_0+) < 0$. This means that the subsolution property of ψ_2 is violated in a neighborhood of z_0. At last, if $z_0 = 1$, we arrive at a violation of the subsolution property at this point due to the remark preceding the lemma. □

4.9.2 First-Order Asymptotic Expansion

A more elaborated analysis based on the same kind of approximation, but with the interpolating polynomials of the fourth order of the form

$$\begin{aligned} Q(z) := \mathbf{m} - r_2\lambda^{2/3} - r_3\lambda - r_4\lambda^{4/3} \\ - \rho_1(z-\theta)\lambda - \rho_2(z-\theta)^2\lambda^{2/3} + \rho_3(z-\theta)^2\lambda^{1/3} - \rho_4(z-\theta)^4, \end{aligned}$$

allows us to establish for the discrepancy the exact asymptotics of order $\lambda^{2/3}$ with an explicit expression for the constant.

Theorem 4.9.3 *Suppose that $\theta \neq 1$. Then*

$$W(1-\theta, \theta) - \mathbf{m} = r_2\lambda^{2/3} + o(\lambda^{2/3}), \tag{4.9.6}$$

where

$$r_2 = \left(\frac{9}{32}(1-\gamma)\theta^4(1-\theta)^4\right)^{1/3} (\gamma\mathbf{m})^{1+\frac{1}{1-\gamma}}\sigma^2. \tag{4.9.7}$$

Proof. The arguments require a bit patience, but for the reader accustomed already with the structure of coefficients and notations, it is quite a routine. The approximating functions are again given by formula (4.9.4) but with the interpolating polynomial $Q(z)$ of the fourth order with the maximum attained at θ, namely, with

$$Q(z) := \mathbf{m} - r_2\lambda^{2/3} - r_3\lambda - \rho_2(z-\theta)^2\lambda^{2/3} - \rho_4(z-\theta)^4. \tag{4.9.8}$$

Note that the first part is a series expansion in powers of $\lambda^{1/3}$ with the coefficients r_i, the second part, more involved, is an expansion in increasing powers of $z-\theta$ with the coefficients ρ_i multiplied by decreasing powers of $\lambda^{1/3}$. The coefficients r_1, r_4, and ρ_1, ρ_3 are already taken zero. We show that a good choice for other constants is the following:

$$\rho_2 := \frac{1}{\sigma^2\theta^2(1-\theta)^2}(\gamma\mathbf{m})^{\frac{1}{\gamma-1}}r_2, \tag{4.9.9}$$

$$\rho_4 := -\frac{1}{12}\frac{\gamma(1-\gamma)}{\theta^2(1-\theta)^2}\mathbf{m}. \tag{4.9.10}$$

The parameter r_3 is "free": it serves to produce sub- and supersolutions.

The extremities of the central interval will be

$$\tilde{z}_1 = \theta - \nu\lambda^{1/3} + O(\lambda^{2/3}), \qquad \tilde{z}_2 = \theta + \nu\lambda^{1/3} + O(\lambda^{2/3}), \tag{4.9.11}$$

with the positive constant ν determined by the relation

$$\frac{1}{2}\sigma^2\gamma(1-\gamma)\nu^2 = (\gamma\mathbf{m})^{\frac{1}{\gamma-1}}\mathbf{m}^{-1}r_2. \tag{4.9.12}$$

The parameters should be chosen to ensure that $\tilde{\psi} \in C^1$.

Our analysis will go in the inverse direction to that in the proof of the previous theorem. Though we have already listed the explicit values of the constants, we shall see in a clear and successive way how they appear to eliminate terms of lower orders.

The derivatives of $\tilde{\psi}$ are given by the formulae

$$\tilde{\psi}'(z) = \begin{cases} \frac{\gamma\lambda}{1+\lambda z}\tilde{\psi}(z), & z \in]-1/\lambda, \tilde{z}_1[, \\ -2[\rho_2(z-\theta)\lambda^{2/3} + 2\rho_4(z-\theta)^3], & z \in]\tilde{z}_1, \tilde{z}_2[, \\ -\frac{\gamma\lambda}{1+\lambda-\lambda z}\tilde{\psi}(z), & z \in]\tilde{z}_2, 1+1/\lambda[, \end{cases}$$

$$\tilde{\psi}''(z) = \begin{cases} -\frac{\gamma(1-\gamma)\lambda^2}{(1+\lambda z)^2}\tilde{\psi}(z), & z \in]-1/\lambda, \tilde{z}_1[, \\ -2[\rho_2\lambda^{2/3} + 6\rho_4(z-\theta)^2], & z \in]\tilde{z}_1, \tilde{z}_2[, \\ -\frac{\gamma(1-\gamma)\lambda^2}{(1+\lambda-\lambda z)^2}\psi(z), & z \in]\tilde{z}_2, 1+1/\lambda[. \end{cases}$$

First, we consider the approximations on the interpolation interval $[\tilde{z}_1, \tilde{z}_2]$ and relate constants in such a way that the sign of the inequality for $\ell_0\tilde{\psi}$ will be determined by the sign of the coefficient at λ.

Using the symbol $\approx$ to denote equalities which hold up to $O(\lambda^{4/3})$ (uniformly in z) and replacing by the symbol ... the coefficients for which explicit expressions are of no importance, we represent the asymptotic expansions for the linear terms of the operator in the following transparent form:

$$f_2(z)\tilde{\psi}''(z) \approx -\sigma^2\left[\theta^2(1-\theta)^2 + \ldots(z-\theta)\right]\left[\rho_2\lambda^{2/3} + 6\rho_4(z-\theta)^2\right],$$

$$f_1(z)\tilde{\psi}'(z) \approx 0,$$

$$f_0(z)\tilde{\psi}(z) \approx -\frac{1}{2}\sigma^2\gamma(1-\gamma)(z-\theta)^2\mathbf{m} - (1-\gamma)(\gamma\mathbf{m})^{\frac{1}{\gamma-1}}\left(\mathbf{m} - r_2\lambda^{2/3} - r_3\lambda\right).$$

Since

$$\gamma\tilde{\psi}(z) - z\tilde{\psi}'(z) \approx \gamma\big(\mathbf{m} - r_2\lambda^{2/3} - r_3\lambda\big) + \ldots(z-\theta)\lambda^{2/3},$$

the nonlinear term admits the expansion

$$\begin{aligned} &u^*\left(\gamma\tilde{\psi}(z) - z\tilde{\psi}'(z)\right) \\ &\approx \frac{1-\gamma}{\gamma}(\gamma\mathbf{m})^{\frac{\gamma}{\gamma-1}} + (\gamma\mathbf{m})^{\frac{1}{\gamma-1}}\big(\gamma r_2\lambda^{2/3} + \gamma r_3\lambda + \ldots(z-\theta)\lambda^{2/3}\big). \end{aligned}$$

Summing up, we obtain the approximation

$$\ell_0\tilde{\psi}(z) \approx \ldots(z-\theta)\lambda^{2/3} + \ldots(z-\theta)^3 + r_3\gamma(\gamma\mathbf{m})^{\frac{1}{\gamma-1}}\lambda;$$

the coefficients at $\lambda^{2/3}$ and $(z-\theta)^2$ are zero because of our choice of ρ_2 and ρ_4, see relations (4.9.9) and (4.9.10).

On the considered interval the width of which is controlled by the parameter ν, the absolute value of the first two terms of the right-hand side is dominated by $\kappa_\nu\lambda + o(\lambda)$, where κ_ν is a constant. This observation leads to the following important conclusion: whatever is ν, one can always find r_3 sufficiently large in absolute value such that $\ell_0\tilde{\psi} \ge 0$ or $\ell_0\tilde{\psi} \le 0$ in dependence of whether r_3 is positive or negative (of course, for $\lambda \le \lambda_0$). Automatically, r_3 takes the control over the sign of $\max_i \ell_i\tilde{\psi}$ because the functions $\ell_1\tilde{\psi}$ and $\ell_2\tilde{\psi}$ on the central interval are negative for small λ. The latter property follows easily from the asymptotic expansions. Indeed,

$$\ell_1\tilde{\psi}(\tilde{z}_1) = -\lambda\gamma Q(\tilde{z}_1) + (1+\lambda\tilde{z}_1)Q'(\tilde{z}_1) = -\gamma\mathbf{m} + \lambda + o(\lambda),$$

while, for the derivative, we have

$$\big[\ell_1\tilde{\psi}(z)\big]' = \lambda(1-\gamma)Q'(z) + (1+\lambda\tilde{z})Q''(z) = -2\rho_2\lambda^{2/3} + o\big(\lambda^{2/3}\big).$$

Thus, the function $\ell_1\tilde{\psi}$ is negative on $[\tilde{z}_1, \tilde{z}_2]$, being negative at the left extremity and decreasing on the whole interval. The arguments for $\ell_2\tilde{\psi}$ are exactly the same.

Let us examine now the situation on the interval $]-1/\lambda, \tilde{z}_1[$. As was explained in the previous proof, the only nontrivial part is to establish the supersolution inequality $\ell_0\tilde{\psi} \le 0$, which we expect to hold for large negative values of r_3.

We have

$$\frac{\ell_0\tilde{\psi}(z)}{\tilde{\psi}(z)} = -f_2(z)\frac{\gamma(1-\gamma)\lambda^2}{(1+\lambda z)^2} + f_1(z)\frac{\gamma\lambda}{1+\lambda z} + f_0(z) + \kappa_\lambda, \qquad (4.9.13)$$

where the coefficient

$$\kappa_\lambda = (1+\lambda\tilde{z}_1)^{\frac{\gamma}{1-\gamma}}(1-\gamma)\gamma^{\frac{1}{\gamma-1}}\big[Q(\tilde{z}_1)\big]^{\frac{1}{\gamma-1}}$$

$$\approx (1-\gamma)(\gamma\mathbf{m})^{\frac{1}{\gamma-1}}\left[1 + \frac{1}{1-\gamma}\mathbf{m}^{-1}r_2\lambda^{2/3} + \left(\frac{1}{1-\gamma}\mathbf{m}^{-1}r_3 - \frac{\gamma}{1-\gamma}\nu\right)\lambda\right].$$

Outside the interpolation interval we have the inequality

$$f_0(z) \le -\frac{1}{2}\sigma^2\gamma(1-\gamma)\big(\nu^2\lambda^{2/3} - 2\nu\eta\lambda\big) - (1-\gamma)(\gamma\mathbf{m})^{\frac{1}{\gamma-1}} + O\big(\lambda^{4/3}\big),$$

where $\eta = |\eta_1| \vee |\eta_2|$.

The first term in the left-hand side of (4.9.13) is negative. On the subinterval $]-1/\lambda, 0]$ the second term is also negative, and, hence,

$$\frac{\ell_0\tilde{\psi}(z)}{\tilde{\psi}(z)} \le f_0(z) + \kappa_\lambda \le \big((\gamma\mathbf{m})^{\frac{1}{\gamma-1}}\mathbf{m}^{-1}r_3 + \kappa\big)\lambda + o(\lambda)$$

because (4.9.12) is aimed to eliminate the coefficient at $\lambda^{2/3}$ in the right-hand side. The value of the constant κ is of no importance. On the subinterval $]0, \tilde{z}_1[$, where we can affirm only that

$$\frac{\ell_0\tilde{\psi}(z)}{\tilde{\psi}(z)} \le f_1(z)\frac{\gamma\lambda}{1+\lambda z} + f_0(z) + \kappa_\lambda;$$

the structure of the resulting estimate remains the same since here the coefficient $f_1(z)$ is bounded. With this, it is clear that $\ell_0\tilde{\psi} \le 0$ for large negative values of the parameter r_3 whatever is λ smaller than some threshold value.

The situation on the other external interval is exactly the same.

Until now we did not need any specific value of r_2. It appears from the conditions of the C^1-fit at the points $\tilde{z}_i$, which are as follows:

$$\frac{\gamma\lambda}{1+\lambda\tilde{z}_1}Q(\tilde{z}_1) + 2\rho_2(\tilde{z}_1-\theta)\lambda^{2/3} + 4\rho_4(\tilde{z}_1-\theta)^3 = 0,$$
$$\frac{\gamma\lambda}{1+\lambda-\lambda\tilde{z}_2}Q(\tilde{z}_2) - 2\rho_2(\tilde{z}_2-\theta)\lambda^{2/3} - 4\rho_4(\tilde{z}_2-\theta)^3 = 0.$$

Take formally $\tilde{z}_i = \nu\lambda^{1/3} + \eta\lambda^{2/3}$ and consider the asymptotic expansions of the left-hand sides of these relations (denoted by $F_i(\tilde{z}_i)$) in powers of $\lambda^{1/3}$. Equating the coefficients at λ, we obtain the same relation for both identities:

$$\gamma\mathbf{m} - 2\rho_2\nu - 4\rho_4\nu^3 = 0.$$

In virtue of (4.9.9) and (4.9.12), the coefficient ρ_2 is proportional to ν^2, and, therefore, this is a linear equation for ν^3. Its solution is

$$\nu^3 = \frac{3}{2}\frac{\theta^2(1-\theta)^2}{1-\gamma}.$$

Expressing r_2 from (4.9.12), we arrive at formula (4.9.7) for the coefficient r_2 in the formulation of the theorem.

Examine, e.g., the case $i = 1$. Clearly, the coefficient at $\lambda^{4/3}$ is a (non-degenerate) linear function of η_1, which vanishes at some value η_1^0. Due to our choice of ν, this coefficient determines, for sufficiently small λ, the signs of $F_1(\nu\lambda^{1/3} + (\eta_1^0+1)\lambda^{1/3})$ and $F_1(\nu\lambda^{1/3} + (\eta_1^0-1)\lambda^{1/3})$. Since they are opposite, the continuity implies that there is $\eta_1(\varepsilon) \in [\eta_1^0 - 1, \eta_1^0 + 1]$ such that $F_1(\nu\lambda^{1/3} + \eta_1(\varepsilon)\lambda^{1/3}) = 0$. Thus, we established the existence of the points $\tilde{z}_i$ satisfying (4.9.11) and ensuring the smooth fit of the interpolation.

The proof is completed. □

In the above reasoning we have proved a bit more than it was claimed in the formulation of the theorem. Namely, we have shown that ψ lies between two arch-shaped functions $\tilde{\psi}_1$ and $\tilde{\psi}_2$ depending on the parameter λ and converging to each other uniformly with rate λ. With this, we can easily get the asymptotics of the extremities z_i of the no-transaction interval. The following theorem asserts that, asymptotically, the length of the latter is proportional to $\lambda^{1/3}$. The no-transaction region opens wider very quickly with the introduction of transaction costs!

Theorem 4.9.4 *Suppose that $\theta \neq 1$. Then*

$$z_1 = \theta - \nu\lambda^{1/3} + O(\lambda^{2/3}), \qquad z_2 = \theta - \nu\lambda^{1/3} + O(\lambda^{2/3}). \tag{4.9.14}$$

Proof. Recall that $z_1 > 0$ and z_2 is bounded from above, i.e., $[z_1, z_2]$ lies in a certain fixed interval $[\zeta_1, \zeta_2]$ containing θ and not depending on λ. The derivative of the concave function ψ on this fixed interval decreases from its maximal (positive) value at ζ_1 to its minimal (negative) value at ζ_2 The interval $[\tilde{z}_1, \tilde{z}_2]$ shrinks to θ. For sufficiently small λ, the functions $\tilde{\psi}_1^{1/\gamma}$ and $\tilde{\psi}_2^{1/\gamma}$ on both intervals external to $[\zeta_1, \zeta_2]$ are linear, as well as the function $\psi^{1/\gamma}$ lying between them. Taking into account that the derivatives of $\tilde{\psi}_i^{1/\gamma}$ at ζ_1 and ζ_2 are of order λ, we conclude that

$$\sup_{z\in[\zeta_1,\zeta_2]} |\psi'(z)| = O(\lambda).$$

On the interval $[\zeta_1, \zeta_2]$,

$$u^*(\gamma\psi(z) - z\psi'(z)) = \frac{1-\gamma}{\gamma}(\gamma\mathbf{m})^{\frac{\gamma}{\gamma-1}} + (\gamma\mathbf{m})^{\frac{1}{\gamma-1}}\gamma r_2\lambda^{2/3} + O(\lambda).$$

Clearly,

$$f_0(z_i)\psi(z_i) = -\frac{1}{2}\sigma^2\gamma(1-\gamma)\mathbf{m}(z_i-\theta)^2 - (1-\gamma)(\gamma\mathbf{m})^{\frac{1}{\gamma-1}}(\mathbf{m} - r_2\lambda^{2/3}) + O(\lambda).$$

Recall that the equation $\ell_0\psi(z_i) = 0$ is always fulfilled under the convention that the term with the second derivative (may be not existing) is omitted at $z_i = 1$. In the case $z_i \neq 1$ the second derivative of ψ does exist. Since ψ is given by explicit formulae to the left of z_1 and to the right of z_2, we get easily that $\psi''(z_i) = O(\lambda^2)$. Thus, the terms with the second and first derivatives are negligible, and

$$\begin{aligned}\ell_0\psi(z_i) &= f_0(z_i)\psi(z_i) + u^*\big(\gamma\psi(z_i) - z\psi'(z_i)\big) + O(\lambda) \\ &= -\frac{1}{2}\sigma^2\gamma(1-\gamma)\mathbf{m}(z_i-\theta)^2 + (\gamma\mathbf{m})^{\frac{1}{\gamma-1}}r_2\lambda^{2/3} + O(\lambda).\end{aligned}$$

Expressing r_2 via ν from (4.9.12) and equating the left-hand side to zero, we obtain that, necessarily,

$$(z_i-\theta)^2 = \nu^2\lambda^{2/3} + O(\lambda) = \nu^2\lambda^{2/3}(1+O(\lambda^{1/3}))$$

and, therefore,

$$z_i = \theta \pm \nu\lambda^{1/3} + O(\lambda^{2/3})$$

with minus for z_1 and plus for z_2. □

4.9.3 Exceptional Case: $\theta = 1$

We consider now a very particular situation where $\theta = 1$ and the arguments of the previous subsection cannot be used.

However, according to Proposition 4.8.5, for this case we have in the cone $K \cap \{x = (\xi,\eta) : \ \xi \le 0\}$ the explicit formula

$$W(x) = \mathbf{m}\frac{1}{(1+\lambda)^\gamma}(p_2x)^\gamma,$$

and, therefore,

$$\psi(\theta) = \psi(1) = W(e_2) = \mathbf{m}\frac{1}{(1+\lambda)^\gamma} = \mathbf{m} - \mathbf{m}\gamma\lambda + o(\lambda).$$

Moreover, $z_2 = 1$. For z_1, the arguments of the proof of the above theorem can be repeated with $r_2 = 0$ until the last step. Its appropriate modification shows that $z_1 = 1 + O(\lambda^{1/2})$.

5

Appendix

5.1 Facts from Convex Analysis

By definition, a subset K in $\mathbf{R}^n$ (or in a linear space X) is a *cone* if it is convex and stable under multiplication by the nonnegative constants. It defines the partial ordering

$$x \geq_K y \quad \Longleftrightarrow \quad x - y \in K;$$

in particular, $x \geq_K 0$ means that $x \in K$.

A closed cone K is *proper* if the linear space $F := K \cap (-K) = \{0\}$, i.e., if the relations $x \geq_K 0$ and $x \leq_K 0$ imply that $x = 0$.

Let K be a closed cone, and let $\pi : \mathbf{R}^n \to \mathbf{R}^n/F$ be the canonical mapping onto the quotient space. Then πK is a proper closed cone.

For a set C, we denote by $\operatorname{cone} C$ the set of all conic combinations of elements of C. If C is convex, then $\operatorname{cone} C = \bigcup_{\lambda \geq 0} \lambda C$.

Let K be a cone. Its *dual positive cone*

$$K^* := \big\{ z \in \mathbf{R}^n : \ zx \geq 0 \ \forall x \in K \big\}$$

is closed. The *polar cone* K° is defined using the opposite inequality,[1] i.e., $K^\circ = -K^*$; K is closed if and only if $K = K^{**}$.

We use the notation $\operatorname{int} K$ for the interior of K and $\operatorname{ri} K$ for the relative interior (i.e., the interior in $K - K$, the linear subspace generated by K).

Recall that, in a finite-dimensional Euclidean space, the convex hull of a compact set is a compact.

A closed cone K in the Euclidean space $\mathbf{R}^n$ is proper if and only if there is a compact convex set C such that $0 \notin C$ and $K = \operatorname{cone} C$. One can take as C the convex hull of the intersection of K with the unit sphere $\{x \in \mathbf{R}^n : \ |x| = 1\}$.

A closed cone K is proper if and only if $\operatorname{int} K^* \neq \emptyset$.

[1] In the literature one can find examples where both K° and K^* are called dual cones.

Yu. Kabanov, M. Safarian, *Markets with Transaction Costs*,
Springer Finance, DOI 10.1007/978-3-540-68121-2_5,

We have

$$\operatorname{ri} K^* = \{w: \ wx > 0 \ \forall x \in K, \ x \neq F\};$$

in particular, if K is proper, then

$$\operatorname{int} K^* = \{w: \ wx > 0 \ \forall x \in K, \ x \neq 0\}.$$

By definition, the cone K is *polyhedral* if it is the intersection of a finite number of half-spaces $\{x: \ p_i x \geq 0\}$, $p_i \in \mathbf{R}^n$, $i = 1, \ldots, N$.

The Farkas–Minkowski–Weyl theorem:

A cone is polyhedral if and only if it is finitely generated.

Intuitively, this theorem, giving an alternative but equivalent definition of a polyhedral cone, is almost obvious, but its proof is not a just straightforward exercise (unlike other previously formulated statements) and requires certain efforts. Easy to remember, it provides a freedom to chose an appropriate definition to establish needed properties. For example, the closedness of a polyhedral cone is trivial from the initial definition. On the other hand, the property that *the (arithmetic) sum of polyhedral cones is again a polyhedral cone* is clear from the alternative definition: the union of generating sets for each cone is a generating set for the sum.

The following result is a direct generalization of the Stiemke lemma.

Lemma 5.1.1 *Let K and R be closed cones in $\mathbf{R}^n$. Assume that K is proper. Then*

$$R \cap K = \{0\} \quad \Longleftrightarrow \quad (-R^*) \cap \operatorname{int} K^* \neq \emptyset.$$

Proof. ($\Leftarrow$) The existence of w such that $wx \leq 0$ for all $x \in R$ and $wy > 0$ for all y in $K \setminus \{0\}$ obviously implies that R and $K \setminus \{0\}$ are disjoint.

($\Rightarrow$) Let C be a convex compact set such that $0 \notin C$ and $K = \operatorname{cone} C$. By the separation theorem (for the case where one set is closed and another is compact), there is a nonzero $z \in \mathbf{R}^n$ such that

$$\sup_{x \in R} zx < \inf_{y \in C} zy.$$

Since R is a cone, the left-hand side of this inequality is zero, and hence $z \in -R^*$ and, also, $zy > 0$ for all $y \in C$. The latter property implies that $zy > 0$ for $z \in K$, $z \neq 0$, and we have $z \in \operatorname{int} K$. □

In the classical Stiemke lemma, $R = \{y \in \mathbf{R}^n : \ y = Bx, \ x \in \mathbf{R}^d\}$, where B is a linear mapping, and $K = \mathbf{R}^n_+$. Usually, it is formulated as the alternative:

Either there is $x \in \mathbf{R}^d$ such that $Bx \geq_K 0$ and $Bx \neq 0$ or there is $y \in \mathbf{R}^n$ with strictly positive components such that $B^* y = 0$.

Let A be a convex set with nonempty interior, and let U be an open set. Then $A \cap U \neq \emptyset$ if and only if $\operatorname{int} A \cap U \neq \emptyset$. Combining this fact with the Stiemke lemma, we get the following assertion:

Lemma 5.1.2 *Let K and R be proper closed cones in $\mathbf{R}^n$. Then*

$$R \cap K = \{0\} \quad \Longleftrightarrow \quad (-\operatorname{int} R^*) \cap \operatorname{int} K^* \neq \emptyset.$$

Lemma 5.1.1 can be slightly generalized.

Let π be the natural projection of $\mathbf{R}^n$ onto the quotient space $\mathbf{R}^n/F$.

Theorem 5.1.3 *Let K and R be closed cones in $\mathbf{R}^n$. Assume that the cone πR is closed. Then*

$$R \cap K \subseteq F \quad \Longleftrightarrow \quad (-R^*) \cap \operatorname{ri} K^* \neq \emptyset.$$

Proof. It is easy to see that $\pi(R \cap K) = \pi R \cap \pi K$ and, hence,

$$R \cap K \subseteq F \quad \Longleftrightarrow \quad \pi R \cap \pi K = \{0\}.$$

By Lemma 5.1.1,

$$\pi R \cap \pi K = \{0\} \quad \Longleftrightarrow \quad (-\pi R)^* \cap \operatorname{int} (\pi K)^* \neq \emptyset.$$

Since $(\pi R)^* = \pi^{*-1} R^*$ and $\operatorname{int} (\pi K)^* = \pi^{*-1}(\operatorname{ri} K^*)$, the condition in the right-hand side can be written as

$$\pi^{*-1}\big((-R^*) \cap \operatorname{ri} K^*\big) \neq \emptyset$$

or, equivalently,

$$(-R^*) \cap \operatorname{ri} K^* \cap \operatorname{Im} \pi^* \neq \emptyset.$$

But $\operatorname{Im} \pi^* = (K \cap (-K))^* = K^* - K^* \supseteq \operatorname{ri} K^*$, and we get the result. □

Notice that if R is polyhedral, then πR is also polyhedral and, hence, closed.

Lemma 5.1.4 *Let K_1 and K_2 be closed cones in $\mathbf{R}^n$ such that the cone $K_1^* + K_2^*$ is closed. Then $(K_1 \cap K_2)^* = K_1^* + K_2^*$.*

Proof. The inclusion $\supseteq$ follows immediately from the definition of the dual cone. To prove the converse, suppose that $y \in (K_1 \cap K_2)^*$ but does not belong to the convex closed cone $K_1^* + K_2^*$. By the separation theorem there exists x such that $xy < 0$ and $xy_i \geq 0$ for all $y_i \in K_i^*$. But the latter property means that $x \in K_i^{**} = K_i$ for $i = 1, 2$, i.e., $x \in K_1 \cap K_2$. Thus, $xy \geq 0$, a contradiction. □

Remark. If K_1 and K_2 are polyhedral cones, then $K_1^* + K_2^*$ is polyhedral, and the identity of the above lemma holds. Since in the left-hand side we always have a closed set, the identity fails when $K_1^* + K_2^*$ is not closed.

Let X be a Hilbert space, and let $f : X \to \mathbf{R} \cup \{\infty\}$. The *effective domain* $\operatorname{dom} f := \{x : \ f(x) < \infty\}$. The *dual* or *conjugate* to f is the convex function $f^*(y) := \sup_x [yx - f(x)]$. The *Fenchel inequality* is an immediate corollary of this definition: $f(x) + f^*(y) \geq yx$.

5.2 Césaro Convergence

5.2.1 Komlós Theorem

The following Komlós theorem, asserting that a sequence of random variables bounded in L^1 contains a subsequence converging in Césaro sense a.s., is very useful in various applications, especially, in proofs of the existence of optimal controls.

Theorem 5.2.1 *Let (ξ_n) be a sequence of random variables on $(\Omega, \mathcal{F}, P)$ bounded in L^1, i.e., with $\sup_n E|\xi_n| < \infty$. Then there exist a random variable $\xi \in L^1$ and a subsequence (ξ_{n_k}) Césaro convergent to ξ a.s., that is, $k^{-1}\sum_{i=1}^k \xi_{n_i} \to \xi$ a.s. Moreover, the subsequence (ξ_n) can be chosen in such a way that any its further subsequence is also Césaro convergent to ξ a.s.*

5.2.2 Application to Convex Minimization in L^1

Let $\underline{J} := \inf_{\xi \in \mathcal{Z}} Ef(\xi)$ be the value of the minimization problem

$$Ef(\xi) \to \min \quad \text{on } \mathcal{Z}, \tag{5.2.1}$$

where $\mathcal{Z}$ is a nonempty convex set of probability densities, and $f : \mathbf{R}_+ \to \mathbf{R}$ is a convex (hence, continuous) function such that $f \geq -c$ and $f(x)/x \to \infty$ as $x \to \infty$.

Proposition 5.2.2 *If $\mathcal{Z}$ is closed in L^1 and $\underline{J} < \infty$, then there exists $\underline{\xi} \in \mathcal{Z}$ such that $\underline{J} = Ef(\underline{\xi})$.*

Proof. Take a sequence $\xi_j \in \mathcal{Z}$ such that $Ef(\xi_j) \to \underline{J}$. Since $\xi_j \geq 0$ and $E\xi_j = 1$, in virtue of the Komlós theorem, there is a subsequence j_k such that $\tilde{\xi}_n := n^{-1}\sum_{k=1}^n \xi_{j_k}$ converge a.s. to a certain $\underline{\xi} \in L^1$. Due to the Fatou lemma and convexity of f,

$$Ef(\underline{\xi}) = E \lim f(\tilde{\xi}_n) \leq \liminf Ef(\tilde{\xi}_n) \leq \lim \frac{1}{n}\sum_{k=1}^n Ef(\xi_{j_k}) = \underline{J}.$$

The de la Vallée-Poussin criterion ensures that the sequence $(\tilde{\xi}_n)$ is uniformly integrable and, hence, converges also in L^1. Thus, $\underline{\xi} \in \mathcal{Z}$. □

If f is strictly convex, then, obviously, the minimizer is unique.

5.2.3 Von Weizsäcker Theorem

The next result due to von Weizsäcker is a beautiful extension of the Komlós theorem to sequences of random variables from L^0_+. It is less known, and we give its proof here.

Theorem 5.2.3 *Let (ξ_n) be a sequence of positive random variables. Then there exist a random variable ξ taking values in $[0,\infty]$ and a subsequence (ξ_{n_k}) such that all subsequences of the latter are Césaro converging a.s. to ξ.*

Proof. It consists of several steps.

Lemma 5.2.4 *Let (ξ_n) be a sequence of positive random variables unbounded in probability. Then there exist a set B with $P(B) > 0$ and a subsequence (ξ_{n_k}) such that all its further subsequences are Césaro converging a.s. on B to infinity.*

Proof. Recall that a set G of random variable is bounded in probability if

$$\sup_{\eta\in G} P\big(|\eta| \geq N\big) \to 0, \quad N \to \infty.$$

Since in our case this property does not hold, we may assume, passing, if necessary, to a subsequence, that there is $\varepsilon > 0$ such that $\sup_n P(B_n) > \varepsilon$ where $B_n := \{\xi_n > n\}$. Applying the Komlós theorem to the sequence (I_{B_n}), we may assume that there is a random variable η such that, for all subsequences,

$$\frac{1}{m}\sum_{k=1}^{m} I_{B_{n_k}} \to \eta \quad \text{a.s.}$$

Clearly, $0 \leq \eta \leq 1$ and, by dominated convergence, $E\eta \geq \varepsilon$. Thus, $P(B) \geq \varepsilon$ where $B := \{\eta > 0\}$. Then, for every subsequence (ξ_{n_k}) and every $N > 0$,

$$\liminf_m \frac{1}{m}\sum_{k=1}^{m} \xi_{n_k} \geq \liminf_m \frac{1}{m}\sum_{k=1}^{m} N I_{B_{n_k}} \geq N\eta \quad \text{a.s.},$$

and, therefore, (ξ_{n_k}) is Césaro converging a.s. on B to infinity. □

The key step is to find a "maximal" set B satisfying, with a certain subsequence of the original sequence, the property declared above. The formal framework can be describe as follows. Let us consider the set $\mathcal{R}$ of pairs $(B,(\xi_{n_k}))$, where $B \in \mathcal{F}$ and all subsequences of (ξ_{n_k}) are Césaro converging a.s. on B to infinity. We introduce on $\mathcal{R}$ a partial ordering by letting $(\tilde B,(\xi_{\tilde n_k})) \succeq (B,(\xi_{n_k}))$ if $\tilde B \supseteq B$ and $(\xi_{\tilde n_k})$ is a subsequence of (ξ_{n_k}).

Lemma 5.2.5 *There is a pair $(B,(\xi_{n_k})) \in \mathcal{R}$ such that $P(\tilde B) = P(B)$ for any pair $(\tilde B,(\xi_{\tilde n_k})) \in \mathcal{R}$ dominating $(B,(\xi_{n_k}))$.*

Proof. Let $B_0 = \emptyset$. Trivially, the pair $(B_0,(\xi_n))$ is in $\mathcal{R}$. We construct in $\mathcal{R}$ recursively a sequence $(B_l,(\xi_{n_{l,k}}))$, $l = 0,1,\dots$, increasing with respect to the partial ordering. Suppose that it is already defined up to a number l. Let

$$\mathcal{R}_l := \big\{(\tilde B,(\xi_{\tilde n_k})) \in \mathcal{R}:\ (\tilde B,(\xi_{\tilde n_k})) \succeq (B_l,(\xi_{n_{l,k}}))\big\},$$

and let $a_l := \sup\{P(\tilde{B}) : (\tilde{B}, (\xi_{\tilde{n}_k})) \in \mathcal{R}_l\}$. We take as $(B_{l+1}, (\xi_{n_{l+1,k}}))$ any element of $\mathcal{R}_l$ for which $P(B_{l+1}) \geq a_l - 1/l$. The pair $(B, \xi_{n_{k,k}})$, where $B = \bigcup B_l$, has the desired property. Indeed, suppose, on the contrary, that there is a pair $(\tilde{B}, (\xi_{\tilde{n}_k}))$ such that $(\tilde{B}, (\xi_{\tilde{n}_k})) \succeq (B, (\xi_{n_k}))$ and $P(\tilde{B}) > P(B) + 1/l$ for some $l \in \mathbf{N}$. Since $(\tilde{B}, (\xi_{\tilde{n}_k})) \in \mathcal{R}_l$, we have the inequality $a_l \geq P(\tilde{B})$. Thus, $P(B) < a_l - 1/l \leq P(B_{l+1})$, which is a contradiction. □

Now we are able to complete the proof. Choose the "maximal" pair $(B, (\xi_{n_k}))$ as in the above lemma. Without loss of generality we may assume that all the subsequences of (ξ_n) are Césaro converging a.s. on B to infinity and we cannot enlarge B to keep this property even passing to a subsequence. Let $A := B^c$. By Lemma 5.2.4 the sequence $(\xi_n I_A)$ is bounded in probability. Combining the Komlós theorem with a diagonal procedure, we may also assume that, for each $N \in \mathbf{N}$, there is a random variables $\zeta_N \leq N$ such that the sequence $(\xi_n \wedge N)$ and all its subsequences are Césaro converging a.s. to ζ_N. Obviously, ζ_N are increasing to a certain limit ζ. Taking into account that

$$\liminf_m \frac{1}{m} \sum_{k=1}^m \xi_{n_k} \geq \sup_N \zeta_N = \zeta,$$

we conclude that $\zeta < \infty$ a.s. on A (otherwise we could enlarge B by adding the set $A \cap \{\zeta = \infty\}$).

Put $A_m := A \cap \{\zeta \leq m\}$. Using the dominated convergence and obvious inequalities, we have

$$\liminf_n EI_{A_m}(\xi_n \wedge N) \leq \lim_n EI_{A_m} \sum_{k=1}^n (\xi_k \wedge N) = EI_{A_m}\zeta_N \leq EI_{A_m}\zeta \leq m.$$

Passing again to a suitable subsequence, we may assume that

$$\liminf_n EI_{A_m}(\xi_n \wedge N) \leq m$$

for all integer N and m. Since the sequence $(\xi_n I_A)$ is bounded in probability, there are integers N_k such that $P(\xi_n I_A > N_k) \leq 2^{-k}$ for all n. Let n_k be such that $EI_{A_m}(\xi_n \wedge N_k) \leq 2m$ for all $n \geq n_k$. Let us consider the decomposition

$$I_{A_m}\xi_{n_k} = \eta'_{k,m} + \eta''_{k,m},$$

where $\eta'_{k,m} := I_{A_m}(\xi_{n_k} \wedge N_k)$ and $\eta''_{k,m} := I_{A_m} I_{\{\xi_{n_k} > N_k\}}(\xi_{n_k} - N_k)$. Notice that $P(\eta''_{k,m} > 0) \leq 2^{-k}$. By the Borel–Cantelli lemma the sequences $(\eta''_{k,m}(\omega))_{k\in\mathbf{N}}$ have only a finite number of nonzero terms for almost all ω. By construction, $E\eta'_{k,m} \leq 2m$. The Komlós theorem applied to $(\eta'_{k,m})_{k\in\mathbf{N}}$ shows that we may assume that (ξ_{n_k}) and all its subsequences are Césaro converging a.s. on A_m (to a finite limit). Using the diagonal procedure, we can easily construct a subsequence for which the same property holds on the union of A_m, i.e., on the set A. This proves the theorem. □

Of course, the result remains true if the sequence (ξ_n) is bounded from below (by a constant or a finite random variable).

5.2.4 Application to Convex Minimization in L^0

Again let $\underline{J} := \inf_{\xi \in \mathcal{Z}} Ef(\xi)$ be the value of the minimization problem

$$Ef(\xi) \to \min \quad \text{on } \mathcal{Z}, \tag{5.2.2}$$

where now $\mathcal{Z}$ is a nonempty convex set of positive random variables, and $f : \mathbf{R}_+ \to \mathbf{R}$ is a convex function bounded from below with $f(\infty) = \infty$.

Proposition 5.2.6 *If $\mathcal{Z}$ is closed in L^0 and $\underline{J} < \infty$, then there exists $\underline{\xi} \in \mathcal{Z}$ such that $\underline{J} = Ef(\underline{\xi})$.*

Proof. Take a sequence $\xi_j \in \mathcal{Z}$ such that $Ef(\xi_j) \to \underline{J}$. In virtue of the von Weizsäcker theorem, there is a subsequence j_k such that $\tilde{\xi}_n := n^{-1} \sum_{k=1}^n \xi_{j_k}$ converge a.s. to a certain random variable $\underline{\xi}$ which may take infinite values. Due to the Fatou lemma and convexity of f, we obtain, as in the proof of Proposition 5.2.2, that $Ef(\underline{\xi}) \le \underline{J} < \infty$. It follows that ξ is finite and hence an element of $\mathcal{Z}$ minimizing the functional. □

5.2.5 Delbaen–Schachermayer Lemma

The above theorems are easy to memorize and apply in various situations. However, their proofs are rather lengthy. The following assertion due to Delbaen and Schachermayer can be used for the same purposes as the much more delicate von Weizsäcker theorem. It has an important advantage that its proof is elementary and short.

For a sequence (ξ_n), we define the set $\mathcal{T}_n := \mathcal{T}_n(\xi) := \operatorname{conv}\{\xi_k,\ k \ge n\}$, the convex hull of its tail.

Lemma 5.2.7 *Let (ξ_n) be a sequence of positive random variables. Then there exist a sequence $\eta_n \in \mathcal{T}_n$ and a random variable η with values in $[0, \infty]$ such that $\eta_n \to \eta$ a.s.*

Proof. The sequence $J_n := \inf_{\eta \subset \mathcal{T}_n} Ee^{-\eta}$ increases to some $J \le 1$. Let us take $\eta_n \in \mathcal{T}_n$ with $Ee^{-\eta_n} \le J_n + 1/n$. It is easy to see that, for any $\varepsilon > 0$. there is $\delta = \delta(\varepsilon) > 0$ such that

$$e^{-(x+y)/2} \le (e^{-x} + e^{-y})/2 - \delta I_{B_\varepsilon}(x, y),$$

where $B_\varepsilon := \{(x, y) \in \mathbf{R}_+^2 :\ |x - y| \ge \varepsilon,\ x \wedge y \le 1/\varepsilon\}$. Therefore,

$$J_{n\wedge m} \le Ee^{-(\eta_n + \eta_m/2} \le \big(Ee^{-\eta_n} + Ee^{-\eta_m}\big)/2 - \delta P\big((\eta_n, \eta_m) \in B_\varepsilon\big).$$

It follows that $\lim_{m,n\to\infty} P((\eta_n, \eta_m) \in B_\varepsilon) = 0$. We infer from the inequality

$$E\big|e^{-\eta_n} - e^{-\eta_m}\big| \le \varepsilon + 2e^{-1/\varepsilon} + P\big((\eta_n, \eta_m) \in B_\varepsilon\big)$$

that $e^{-\eta_n}$ is a Cauchy sequence in L^1. It remains to recall that a sequence convergent in L^1 (hence, in L^0) contains a subsequence convergent a.s. □

Remark. The limit points of a set in L^0 bounded in probability are finite random variables. Thus, if a certain set $\mathcal{T}_n$ is bounded in probability, then $\eta < \infty$.

We give a sufficient condition ensuring that the limit η is strictly positive.

Lemma 5.2.8 *Let $P(\xi_n \geq \alpha) \geq \alpha > 0$ for all $n \geq n_0$. Then $P(\eta > 0) > 0$.*

Proof. If $P(\xi \geq \alpha) \geq \alpha$, then

$$Ee^{-\xi} \leq 1 - \alpha + \alpha e^{-\alpha}.$$

It follows from the Jensen inequality that the same bound (strictly less than the unity) holds for any ξ which is a convex combination of ξ_n, $n \geq n_0$. By the dominated convergence it holds also for the limit points of $\mathcal{T}_{n_0}$. But the inequality $Ee^{-\xi} < 1$ implies that $P(\xi > 0) > 0$. □

5.3 Facts from Probability

5.3.1 Essential Supremum

For any family $\{\xi_\alpha\}_{\alpha \in J}$ of scalar random variables (which may take also infinite values), there exists a random variable η with the following properties:

(1) $\eta \geq \zeta_\alpha$ for all α;
(2) if $\eta' \geq \zeta_\alpha$ for all α, then $\eta' \geq \eta$.

Obviously, this random variable (more precisely, a class of equivalence) is unique and denoted by $\operatorname{ess\,sup}_{\alpha \in J} \zeta_\alpha$.

The proof of existence is easy. It suffices to consider the case where all ζ_α take values in a bounded interval (indeed, if $\tilde{\eta}$ is the essential supremum for the family $\{\tilde{\zeta}_\alpha\}$ with $\tilde{\zeta}_\alpha := \arctan \zeta_\alpha$, then $\tan \tilde{\eta}$ is the essential supremum for $\{\zeta_\alpha\}$). Let $a := \sup_I E\zeta_I$, where I runs through the set of finite subsets of J, and ζ_I stands for $\sup_{\alpha \in I} \zeta_\alpha$. Take a sequence of I_n such that $E\zeta_{I_n} \to a$. Replacing, if necessary, I_n by $\bigcup_{k \leq n} I_k$, we may assume without loss of generality that $I_n \uparrow I_\infty$. Then $\zeta_{I_n} \uparrow \eta := \zeta_{I_\infty}$. Taking into account that $E\eta = a$ and using the monotone convergence, it is easy to verify that η is the essential supremum. Notice that $\eta = \sup_{\alpha \in I_\infty} \zeta_\alpha$, where I_∞ is a *countable* subset of I.

The above arguments show clearly that if the family $\{\zeta_\alpha\}$ is directed upward (i.e., for all α_1, α_2, there is α such that $\zeta_\alpha \geq \zeta_{\alpha_1} \vee \zeta_{\alpha_2}$), then one can find an *increasing* sequence ζ_{α_n} such that $\lim_n \zeta_{\alpha_n} = \operatorname{ess\,sup}_\alpha \zeta_\alpha$. An immediate consequence of this observation is the following:

Proposition 5.3.1 *Assume that the family $\{\zeta_\alpha\}$ is directed upwards and $\zeta_\alpha \geq \zeta$, where $E(|\zeta||\mathcal{G}) < \infty$. Then*

$$E\Big(\operatorname*{ess\,sup}_\alpha \zeta_\alpha | \mathcal{G}\Big) = \operatorname*{ess\,sup}_\alpha E(\zeta_\alpha|\mathcal{G}).$$

5.3.2 Generalized Martingales

The classical definition of a martingale consists of two parts: the integrability property and the property involving conditional expectations. The former can be replaced by a weaker one, namely, by the existence of the conditional expectations needed for the latter. This leads to a notion of generalized martingale, which, in a discrete-time setting, coincides with that of local martingale.

Let $(\Omega, \mathcal{F}, P)$ be a probability space equipped with a discrete-time filtration $\mathbf{F} = (\mathcal{F}_t)$, $t = 0, 1, \ldots$. An adapted process $X = (X_t)$ is a *generalized martingale* if $E(|X_{t+1}||\mathcal{F}_t) < \infty$ and $E(X_{t+1}|\mathcal{F}_t) = X_t$ for $t = 0, 1, \ldots$.

Proposition 5.3.2 *Let X be an adapted process with $X_0 \in L^1$. Then X is a generalized martingale if and only if X is a local martingale.*

Proof. Suppose that X is a local martingale, i.e., there is a sequence of stopping times $\tau_n \uparrow \infty$ (a.s.) such that the stopped processes $X^{\tau_n} = (X_{\tau_n \wedge t})$ are martingales. The set $\{\tau_n \geq t+1\}$ is in $\mathcal{F}_t$, and

$$EI_{\{\tau_n \geq t+1\}} E\big(|X_{t+1}||\mathcal{F}_t\big) = EI_{\{\tau_n \geq t+1\}} |X_{t+1}| = EI_{\{\tau_n \geq t+1\}} \big|X_{t+1}^{\tau_n}\big| < \infty.$$

Thus, $E(|X_{t+1}||\mathcal{F}_t)$ is finite (a.s.) on each set $\tau_n \geq t+1$ and, therefore, on their union, which is of full measure. Moreover, for any $\Gamma \in \mathcal{F}_t$, we have, in virtue of the martingale property of X^{τ_n},

$$\begin{aligned} EI_{\{\tau_n \geq t+1\}} I_\Gamma E(X_{t+1}|\mathcal{F}_t) &= EI_{\{\tau_n \geq t+1\}} I_\Gamma X_{t+1} \\ &= EI_{\{\tau_n \geq t+1\}} I_\Gamma X_{t+1}^{\tau_n} = EI_{\{\tau_n \geq t+1\}} I_\Gamma X_t^{\tau_n}. \end{aligned}$$

This means that $E(X_{t+1}|\mathcal{F}_t) = X_t$ on each set $\tau_n \geq t+1$, i.e., almost everywhere. Hence, X is a generalized martingale.

The proof of the converse is based on the following observation: if M is a martingale and H is a predictable process, then the process $H \cdot M$ (the increments of which are $H_t \Delta M_t$) is a local martingale. Indeed, the random variables $\tau_n := \inf\{t : \ |H_{t+1}| \geq n\}$ are stopping times increasing to infinity, the increment of the stopped process $H \cdot M^{\tau_n}$, being of the form $I_{\{\tau_n \leq t\}} H_t \Delta M_t$, is bounded by the integrable random variable $n|\Delta M_t|$, and its conditional expectation with respect to $\mathcal{F}_{t-1}$ is zero.

Represent an arbitrary generalized martingale X as $X = X_0 + H \cdot M$ with $H_t = 1 + E(|\Delta X_t||\mathcal{F}_{t-1})$ and $M = H^{-1} \cdot X$. The increment ΔM_t takes values in the interval $[-1, 1]$ and its conditional expectation with respect to $\mathcal{F}_{t-1}$ is zero. Thus, M is a martingale, $H \cdot M$ is a local martingale and so is X.

Remark. We could conclude because $X_0 \in L^1$. In general, the class of generalized martingales is larger than the class of local martingales. The reason is that the definition of a generalized martingale does not require the integrability of its initial value. Thus, X shifted by any $\mathcal{F}_0$-measurable random variable remains a generalized martingale, while the local martingale may not. Of course, both classes coincide under the assumption (frequent in the literature) that $\mathcal{F}_0$ is trivial.

Proposition 5.3.3 *Let X be a local martingale with $X_0 \in L^1$. Suppose that $EX_T^- < \infty$. Then X is a martingale.*

Proof. Adding to X the martingale $E(X_T^-|\mathcal{F}_t)$, we may assume without loss of generality that $X_T \geq 0$. By the above, X is a generalized martingale, and, therefore, $X_{T-1} = E(X_{T-1}|\mathcal{F}_T) \geq 0$. It follows that the whole process $X \geq 0$. The Fatou lemma applied to the equality $EX_{\tau_n \wedge t} = EX_0$ implies that $EX_t \leq EX_0 < \infty$, i.e., $X_t \in L^1$. □

5.3.3 Equivalent Probabilities

Let $(\Omega, \mathcal{F}, P)$ be a probability space with filtration $(\mathcal{F}_t)_{t \leq T}$, and let Q be a probability measure equivalent to P. Let $\rho = (\rho_t)$ be the corresponding density process, i.e., the martingale with $\rho_t = E(dQ/dP|\mathcal{F}_t)$.

Lemma 5.3.4 *A process M is a Q-martingale (resp. local Q-martingale) if and only if ρM is a P-martingale (resp. local P-martingale).*

Proof. Using the definition of the density and the martingale property of ρ, we get that

$$E_Q|M_t| := E\rho_T|M_t| = E\rho_t|M_t| = E|\rho_t M_t|,$$

implying the simultaneous finiteness of the first and last terms in this chain of equalities. If this is the case, then, for any $\Gamma \in \mathcal{F}_s$, $s \leq t$,

$$E_Q M_t I_\Gamma = E\rho_T M_t I_\Gamma = E\rho_t M_t I_\Gamma.$$

If $M \in \mathcal{M}(Q)$, the left-hand side above is $E_Q M_s I_\Gamma = E\rho_T M_s I_\Gamma = E\rho_s M_s I_\Gamma$, and we obtain the martingale property of ρM. Conversely, if $\rho M \in \mathcal{M}$, the left-hand side is equal to $E\rho_T M_s I_\Gamma = E_Q M_s I_\Gamma$, and, hence, $M \in \mathcal{M}(Q)$. The extension to the local martingales is obvious. □

Lemma 5.3.5 *Let $\{\xi_n\}$ be a sequence in L^0_+. Then there exists a sequence of strictly positive reals such that $\sum_n a_n \xi_n < \infty$.*

Proof. Take $b_n > 0$ such that $P(\xi_n > b_n) \leq 1/n^2$. Then the assertion holds with $a_n := 1/(b_n n^2)$ because, in virtue of the Borel–Cantelli lemma, for almost all ω, only a finite number of $\xi_n(\omega)$ are larger than b_n. □

Proposition 5.3.6 *Let $\{\xi_n\}$ be an at most countable set in L^0. Then there exists a probability measure $Q \sim P$ such that the density $\rho = dQ/dP$ is bounded and all $\xi_n \in L^1(Q)$. Moreover, if ξ_n converges a.s. to a finite random variable ξ, then one can choose Q so that $E_Q|\xi_n - \xi| \to 0$.*

Proof. The assertion is obvious for a single random variable ξ: one can take $Q = \rho P$ with $\rho = c(1 + |\xi|)e^{-(1+|\xi|)}$, where c is a normalizing constant. So it holds also for a finite set. The countable case follows from the lemma above. In the case of convergent sequence, one can take Q such that the finite random variable $\sup_n |\xi_n - \xi|$ is also integrable, ensuring the required convergence in $L^1(Q)$. □

5.3.4 Snell Envelopes of Q-Martingales

In the usual setting of discrete-time model with finite horizon, consider the set $\mathcal{Q}$ of probability measures equivalent to P, denoting by $\mathcal{Z}$ the set of corresponding density processes. We say that $\mathcal{Q}$ is *stable under concatenation* if, for all elements $\rho^1, \rho^2 \in \mathcal{Z}$ and all $\Gamma \in \mathcal{F}_s$, $s \le T$, it contains also the process $\rho = (\rho_t)_{t\le T}$ with

$$\rho_t := \rho_t^1 I_{\{t\le s\}} + \rho_t^1 I_\Gamma I_{\{t>s\}} + \rho_t^2 \frac{\rho_s^1}{\rho_s^2} I_{\Gamma^c} I_{\{t>s\}}.$$

Clearly, ρ, being constructed in this way from strictly positive martingales, is also a martingale; the concatenation property does not depend on the choice of the reference measure.

Note that the sets Q^e (of equivalent martingale measures for some process S) and Z_l^e possess the concatenation property.

Proposition 5.3.7 *Suppose that $\mathcal{Q}$ is stable under concatenation. Let $\xi \ge 0$ be a random variable such that $\sup_{Q\in\mathcal{Q}} E_Q\xi < \infty$. Then the process*

$$X_t = \operatorname*{ess\,sup}_{Q\in\mathcal{Q}} E_Q(\xi|\mathcal{F}_t) \tag{5.3.1}$$

is a Q-supermartingale whatever is $Q \in \mathcal{Q}$.

Proof. Since the assertion does not depend on the choice of the reference measure, we assume that P is an element of $\mathcal{Q}$ and check the supermartingale property $E(X_s|\mathcal{F}_{s-1}) \le X_{s-1}$ with respect to this measure.

Let us consider the set $\mathcal{Z}_s$ of local densities from $\mathcal{Z}$ which are equal to the unity for $t \le s$. The process identically equal to the unity is in $\mathcal{Z}$. By the concatenation property the process $I_{\{t\le s\}} + (\rho_t/\rho_s)I_{\{t>s\}}$ belongs to $\mathcal{Z}_t$ whatever is $\rho \in \mathcal{Z}$. Expressing the conditional expectation in terms of the reference measure and the corresponding density, using this observation, we obtain that

$$X_s = \operatorname*{ess\,sup}_{\rho\in\mathcal{Z}_s} E(\rho_T\xi|\mathcal{F}_s).$$

Put $\zeta_\rho := E(\rho_T\xi|\mathcal{F}_s)$. By the concatenation property $\{\zeta_\rho\}_{\rho\in\mathcal{Z}_s}$ is directed upward: $\zeta_\rho = \zeta_{\rho^1} \vee \zeta_{\rho^2}$ for

$$\rho_t = I_{\{t\le s\}} + \big(I_\Gamma \rho_t^1 + I_{\Gamma^c}\rho_t^2\big)I_{\{t>s\}}$$

with $\Gamma := \{\zeta_{\rho^1} \ge \zeta_{\rho^2}\}$. Using Proposition 5.3.1 and the inclusion $\mathcal{Z}_s \subseteq \mathcal{Z}_{s-1}$, we get that

$$E(X_s|\mathcal{F}_{s-1}) = \operatorname*{ess\,sup}_{\rho\in\mathcal{Z}_s} E(\rho_T\xi|\mathcal{F}_{s-1}) \le \operatorname*{ess\,sup}_{\rho\in\mathcal{Z}_{s-1}} E(\rho_T\xi|\mathcal{F}_{s-1}),$$

and we conclude. □

The above result can be easily extended to include the classical formulation of the Snell envelope, which was introduced in the theory of optimal stopping (for the case where $\mathcal{Q}$ is a singleton $\{P\}$).

Let $\mathcal{T}_t$ denote the set of stopping times with values in the set $\{t, t+1, \dots, T\}$.

Proposition 5.3.8 *Suppose that $\mathcal{Q}$ is stable under concatenation. Let $Y_t \geq 0$ be an adapted process such that $\sup_{Q\in\mathcal{Q}} E_Q Y_t < \infty$ for $t = 0, 1, \dots, T$. Then the process*

$$X_t = \operatorname*{ess\,sup}_{Q\in\mathcal{Q},\, \tau\in\mathcal{T}_t} E_Q(Y_\tau|\mathcal{F}_t) \tag{5.3.2}$$

is a Q-supermartingale whatever is $Q \in \mathcal{Q}$. Moreover, X is the smallest process which is a Q-supermartingale for all Q and which dominates Y.

Proof. The supermartingale properties is verified by the same argument as in the previous proposition. One should only check that, for a fixed s, the family of random variables $\zeta_\rho := E(\rho_T Y \mathcal{T}|\mathcal{F}_s)$, $\rho \in \mathcal{Z}_s$, $\tau \in \mathcal{T}_s$, is directed upward. This is an easy exercise. Finally, if $\tilde{Y}$ is a Q-supermartingale for all Q, then the corresponding process $\tilde{X}$ coincides with Y. The operator defined by the right-hand side of (5.3.2) is monotone, that is, a larger input process results in a larger output. Combining these two properties, we obtain the concluding assertion. □

5.4 Measurable Selection

Measurable spaces $(E, \mathcal{E})$ and $(E', \mathcal{E}')$ are *Borel isomorphic* if there exists a bijection $f : E \to E'$ such that $f(A) \in \mathcal{E}'$ and $f^{-1}(A') \in \mathcal{E}$ whatever are $A \in \mathcal{E}$ and $A' \in \mathcal{E}'$.

A measurable space (Borel) isomorphic to a Borel subset of a Polish space (i.e., complete separable metric space) is called Borel (or Lusin, or standard measurable) space. The basic fact: any infinite Borel space is isomorphic either to $\mathbf{N}$ or $\mathbf{R}_+$. Exactly this result allows one to reduce the proof of the existence of regular conditional distributions to the scalar case where one can work comfortably with distribution functions and use the linear ordering of the real line. Similarly, it works in the proof of the following measurable selection theorem.

Theorem 5.4.1 *Let $(\Omega, \mathcal{F}, P)$ be a complete probability space, let $(E, \mathcal{E})$ be a Borel space, and let $\Gamma \subset \Omega \times E$ be an element of the σ-algebra $\mathcal{F} \otimes \mathcal{E}$. Then the projection $Pr_\Omega \Gamma$ of Γ onto Ω is an element of $\mathcal{F}$, and there exists an E-valued random variable ξ such that $\xi(\omega) \in \Gamma_\omega$ for all non-empty ω-sections Γ_ω of Γ.*

In applications, the y-axis is usually a Polish space, while the x-axis can be a σ-finite complete measurable space (this extension is obvious).

Let $(E, \mathcal{E})$ be a Polish space. A set-valued mapping $\omega \mapsto \Gamma(\omega)$, $\Gamma(\omega) \subseteq E$ is *measurable* if its graph $\{(\omega, x) : x \in \Gamma(\omega)\}$ is measurable; a measurable selector ξ of the graph is called *measurable selector* (or simply *selector*) of the set-valued mapping; the terminology *measurable selection*, *selection* is frequent in the modern literature. The distance between two measurable set-valued mappings, as functions of ω, is a random variable. Measurability properties of a set-valued mapping the values of which are sets in $(\mathbf{R}^d, \mathcal{B}^d)$ are preserved by major operations of convex analysis. In particular, if Γ is a measurable set-valued mapping, the mapping$\omega \mapsto \operatorname{conv} \Gamma(\omega)$ is also measurable; for the convex-valued Γ, so is the dual mapping $\omega \mapsto \Gamma^\circ(\omega)$, the mapping $\omega \mapsto \operatorname{ri} \Gamma(\omega)$, etc.

We denote by $L^p(\Gamma)$ or, in the need, $L^p(\Gamma, \mathcal{F})$, $L^p(\Gamma, \mathcal{F}, P)$, the set of all L^p-selectors of a measurable set-valued mapping Γ. If the values of Γ are convex sets (respectively, cones) then, $L^p(\Gamma)$ is a convex subset (respectively, a cone) in L^p.

Let Γ be a measurable mapping the values of which are closed non-empty subsets of $\mathbf{R}^d$. Knowing that such a mapping admits a measurable selector, it is easy to infer that Γ admits a countable family of measurable selectors $\{\xi_n\}$ such that the values $\xi_n(\omega)$ are dense in the section $\Gamma(\omega)$ for each $\omega \in \Omega$, i.e., $\Gamma(\omega) = \overline{\{\xi_n(\omega)\}}$. This family is called *Castaing representation* of Γ. Note that if $L^0(\Gamma)$ is a closed cone, and $\{\xi_n\}$ is a Castaing representation of Γ, then the set of conic combinations of elements of $\{\xi_n\}$ with rational coefficients is also a Castaing representation of Γ. It follows that $L^0(\Gamma)$ is a closed cone in L^0 if and only if almost all values of Γ are closed cones. A similar assertion relates the convexity of values of Γ and of the set of its selectors.

The following useful lemma shows that such a Castaing representation being fixed, all other selectors can be approximated by members of this family.

Lemma 5.4.2 *Let $\xi_n \in L^p$, $p \in [0, \infty[$, and $\Gamma(\omega) = \overline{\{\xi_n(\omega)\}}$. Then $L^p(\Gamma)$ is a closure in L^p of the set of random variables of the form $\sum I_{A_i} \xi_i$, where $\{A_i\}$ is a finite measurable partition of Ω.*

Proof. Let $p \in]0, \infty[$, let $\xi \in L^p$ be a measurable selector of Γ and let $\varepsilon > 0$. Starting from the sets $\{|\xi - \xi_i| < \varepsilon/2\}$, we construct a measurable countable partition $\{B_i\}$ of Ω such that $|\xi - \xi_i| < \varepsilon/2$ on B_i. Choose n such that

$$\sum_{i \geq n+1} E I_{B_i} |\xi - \xi_1|^p \leq \varepsilon/2$$

and put $A_1 := B_1 \cup (\bigcup_{i \geq n+1} B_i)$, $A_i := B_i$, $i = 2, \ldots, n$. Then

$$E \left| \sum_{i=1}^n I_{A_i} (\xi - \xi_i) \right|^p = \sum_{i=1}^n E I_{B_i} |\xi - \xi_i|^p + \sum_{i=n+1}^\infty E I_{B_i} |\xi - \xi_1|^p < \varepsilon,$$

and we get the result.

The case $p = 0$ is a corollary of the previous one. Indeed, there is $\tilde{P} \sim P$ such that ξ and all ξ_n a $\tilde{P}$-integrable. Thus, we can approximate ξ in $L^1(\tilde{P})$ hence, in $L^0(\tilde{P})$. But the convergence in $\tilde{P}$-probability coincides with the convergence in P-probability. □

The next assertion gives a characterization of closed subsets in L^p which are formed by the selectors of a measurable set-valued mapping. To formulate it we use the following definition. A subset $\Xi \in L^p$ is called *decomposable* if with two its elements ξ_1, ξ_2 it contains also $\xi_1 I_A + \xi_2 I_{A^c}$ whatever is $A \in \mathcal{F}$. It is easily seen that in this case $\sum \xi_i I_{A_i}$ belongs to Ξ for every finite measurable partition of Ω and $\xi_i \in \Xi$, $i = 1, \dots, n$.

Proposition 5.4.3 *Let Ξ be a closed subset of $L^p(\mathbf{R}^d)$, $p \in [0, \infty[$. Then $\Xi = L^p(\Gamma)$ for some Γ the values of which are closed sets if and only if Ξ is decomposable.*

Proof. Only the "if" part needs arguments. Let consider the case $p \in]0, \infty[$. Let $\{x_i\}$ be a countable dense subset of $\mathbf{R}^d$, and let $a_i := \inf_{\eta \in \Xi} E|\eta - x_i|^p$. By definition there exists $\eta_{ij} \in \Xi$ such that $E|\eta_{ij} - x_i|^p \le a_i + 1/j$. We consider the mapping Γ the values of which are closures of the sets $\{\eta_{ij}(\omega)\}$. The inclusion $L^p(\Gamma) \subseteq \Xi$ follows from the lemma above and the fact that Ξ is decomposable. To prove the reverse inclusion, we suppose that there is $\xi \in \Xi$ which is not an element of $L^p(\Gamma)$, and, hence for some $\delta \in]0, 1]$, the set $A := \{|\xi - \eta_{ij}| > \delta \ \forall i, j\}$ is of strictly positive probability. Fix i such that the set $B := A \cap \{|\xi - x_i| < \delta/3\}$ also is of strictly positive probability and put $\eta'_j := \xi I_B + \eta_{ij} I_{B^c}$. Then $\eta'_j \in \Xi$. On the set B we have

$$|x_i - \eta_{ij}| \ge |\xi - \eta_{ij}| - |\xi - x_i| \ge 2\delta/3.$$

Thus,

$$\begin{aligned} E|x_i - \eta_{ij}|^p - a_i &\ge E\big(|x_i - \eta_{ij}|^p - |x_i - \eta'_j|^p\big) \\ &= E\big(|x_i - \eta_{ij}|^p - |x_i - \xi|^p\big) I_B \\ &\ge \big((2\delta/3)^p - (\delta/3)^p\big) P(B) \not\to 0, \quad j \to \infty, \end{aligned}$$

and this is a contradiction with the choice of η_{ij}.

For the case $p = 0$ we get the claim using the same arguments but replacing the function $|x|^p$ by the function $|x| \wedge 1$. □

Proposition 5.4.4 *Let $\mathcal{G}$ be a (complete) sub-σ-algebra of $\mathcal{F}$. Let Γ be a measurable mapping the values of which are nonempty closed convex subsets of the unit ball in $\mathbf{R}^d$. Then there exists a $\mathcal{G}$-measurable mapping $E(\Gamma|\mathcal{G})$ the values of which are nonempty convex compact subsets of the unit ball in $\mathbf{R}^d$ and the set of its $\mathcal{G}$-measurable a.s. selectors coincides with the set of $\mathcal{G}$-conditional expectations of a.s. selectors of Γ.*

Proof. For $d = 1$, the result is almost obvious. In this case $\Gamma(\omega) = [\alpha(\omega), \beta(\omega)]$, where α and β are random variables. Then $E(\Gamma|\mathcal{G}) = [E(\alpha|\mathcal{G}), E(\beta|\mathcal{G})]$ (the conditional expectations here can be chosen as the integrals with respect to regular conditional distributions). Indeed, the inequalities $\alpha \le \xi \le \beta$ imply the corresponding inequalities for the $\mathcal{G}$-conditional expectations. On the other hand, a $\mathcal{G}$-measurable selector $\tilde{\xi}$ of the newly defined mapping is of the form $\tilde{\xi} = \tilde{\lambda} E(\alpha|\mathcal{G}) + (1-\tilde{\lambda})E(\beta|\mathcal{G})$, where λ is a $\mathcal{G}$-measurable random variable taking values in the interval $[0, 1]$. Therefore, $\tilde{\xi} = E(\tilde{\lambda}\alpha + (1 - \tilde{\lambda})\beta|\mathcal{G})$.

In the general case we use the previous proposition. Indeed, let Ξ be the set of random variables $E(\eta|\mathcal{G})$ where η runs over the set of measurable selectors of Γ. The boundedness and convexity of Γ ensure that Ξ is closed in L^1. Indeed, let $\xi_n := E(\eta_n|\mathcal{G})$ converge in L^1 to ξ. Since $|\eta_n| \le 1$, we can apply the Komlós theorem and find a measurable selector η of Γ such that $\xi = E(\eta|\mathcal{G})$. Note that Ξ is decomposable on the space $(\Omega, \mathcal{G}, P)$. Thus, $\Xi = L^1(\tilde{\Gamma}) = L^\infty(\tilde{\Gamma})$ for some $\mathcal{G}$-measurable mapping $\tilde{\Gamma}$ the values of which are nonempty convex compact subsets of the unit ball in $\mathbf{R}^d$, and we conclude. □

5.5 Fatou-Convergence and Bipolar Theorem in L^0

Usually, bipolar theorems giving dual descriptions of convex sets are formulated for locally convex spaces. Unfortunately, the infinite-dimensional space L^0 does not belong to the latter class. Nevertheless, for some specific types of convex sets in L^0, one can give a dual description. We need a relatively simple theorem, which can be easily obtained from the usual bipolar theorem in L^∞.

Let K be a measurable multifunction on Ω whose values are convex closed cones in $\mathbf{R}^d$ containing $\mathbf{R}^d_+$. We denote by L^0_b the cone in $L^0(\mathbf{R}^d)$ formed by random variables ξ such that $\xi + \kappa_\xi \mathbf{1} \in L^0(K)$ for some constant κ_ξ, i.e., elements of L^0_b are bounded from below in the sense of the partial ordering induced by K. We shall use the terminology "Fatou-convergence" in relation with this partial ordering.

Let $A \ne \emptyset$ be a convex subset in L^0_b, and $A^\infty := A \cap L^\infty$. For $\eta \in L^1(K^*)$ and $\xi \in L^0_b$, the expectation $E\xi\eta$ is well defined ($\xi\eta \ge -\kappa_\xi|\eta|$). We put

$$f(\eta) := \sup_{\xi \in A} E\eta\xi.$$

Now we recall the following fact.

Proposition 5.5.1 *Let $F \subseteq L^\infty$ be a convex set. Then*

$$F \text{ is weak* closed} \Longleftrightarrow F \cap \{\xi : \ \|\xi\|_\infty \le \kappa\} \text{ is closed in probability for every } \kappa.$$

Proof. The classical Krein–Šmulian theorem (see, e.g., [166]) says that a convex set in the dual to a Banach space is weak* closed (i.e., closed in $\sigma\{L^\infty, L^1\}$)

if and only if its intersection with every ball around the origin is weak* closed. Thus, we may assume that F is bounded. Now the dominated convergence works, and the limit of a sequence convergent in probability is a limit of a weak* convergent sequence. Thus, the implication $\Rightarrow$ holds. To prove the reverse, we consider F as a subset of L^2. Being closed in probability, it is closed also in L^2. But a closed convex set in a Hilbert space is weakly closed. Thus, F is closed in $\sigma(L^\infty, L^2)$ and, hence, in $\sigma(L^\infty, L^1)$. $\square$

As corollary we have:

Lemma 5.5.2 *If A is Fatou-closed, then the set A^∞ is weak* closed.*

Proof. A sequence convergent in probability contains a subsequence convergent almost surely. But $\{\xi : \ \|\xi\|_\infty \le \kappa\} + \kappa\mathbf{1} \subseteq L^0(\mathbf{R}^d_+) \subseteq L^0(K)$. Thus, a bounded sequence convergent a.s. is Fatou-convergent. Therefore, an intersection of A^∞ with balls is closed in probability, and we conclude by the above proposition. $\square$

Theorem 5.5.3 *Assume that a set A is Fatou-closed, A^∞ is dense in A with respect to the Fatou-convergence, and there exists $\xi_0 \in A^\infty$ such that $\xi_0 - L^\infty(K) \subseteq A^\infty$. Then*

$$A = \{\xi \in L^0_b : \ E\xi\eta \le f(\eta) \ \forall \eta \in L^1(K^*)\}. \tag{5.5.1}$$

Proof. It is sufficient to verify that

$$A^\infty = \{\xi \in L^\infty : \ E\xi\eta \le f(\eta) \ \forall \eta \in L^1(K^*)\}. \tag{5.5.2}$$

To check the only nontrivial inclusion "$\supseteq$", take $\zeta \in L^\infty \setminus A^\infty$. Since A^∞ is weak* closed, by the Hahn–Banach theorem there is $\eta \in L^1$ such that

$$\sup_{\xi \in A^\infty} E\xi\eta < E\zeta\eta. \tag{5.5.3}$$

Considering the r.v.'s of the form $\xi_0 - \xi$ with $\xi \in L^\infty(K)$, we deduce from here that

$$\inf_{\xi \in L^\infty(K)} E\xi\eta > -\infty.$$

As K is a cone, the infimum is equal to zero. By the usual measurable selection argument we deduce from this that $\eta \in L^1(K^*)$. The set A^∞ being Fatou-dense in A, the supremum in (5.5.3) can be taken over A, and the relation (5.5.3) means that ζ does not belong to the set given by the right-hand side of (5.5.2). $\square$

5.6 Skorokhod Problem and SDE with Reflections

We give here a short introduction to the Skorokhod problem, which is needed to ensure the existence and uniqueness of the optimal process in the Davis–Norman consumption–investment problem with transaction costs. Having in mind this application, we restrict ourselves by considering the simplest case of a convex domain with smooth boundary.

5.6.1 Deterministic Skorokhod Problem

The deterministic Skorokhod problem for a convex domain $G \subseteq \mathbf{R}^d$ can be formulated as follows.

We are given a closed convex subset G with nonempty interior and a set-valued mapping $x \mapsto \Gamma(x)$ defined on the boundary ∂G. The values of Γ are subsets of the unit sphere $\partial \mathcal{O}_1(0) := \{y \in \mathbf{R}^d : \ |y| = 1\}$. The elements of $\Gamma(x)$ are interpreted as directions of reflections at point x of the boundary. It is convenient to extend Γ to the whole space by putting $\Gamma(x) = \{0\}$ for $x \notin \partial G$. The "input" is a d-dimensional continuous function $f = (f_t)$ such that $f_0 \in G$.

The solution of the Skorokhod problem is a pair of $\mathbf{R}^d$-valued continuous functions (ξ, ϕ) satisfying the following conditions:

(1) $\xi = f + \phi$,
(2) ξ takes values in G, and $\xi_0 = f_0$;
(3) ϕ is of bounded variation, and $\dot{\phi}_t \in \Gamma(\xi_t)\ d|\phi|_t$-a.e., where $\dot{\phi}_t = d\phi_t/d|\phi|_t$.

For functions f admitting a unique pair with such properties, one can define the mapping $f \mapsto (\xi, \phi)$ called the *Skorokhod mapping*.

Note that if the Skorokhod mapping is defined for the set of continuous functions on each interval $[0, T]$, then it is also defined for the continuous functions on $\mathbf{R}_+$. One can consider the Skorokhod problem on the interval $[T, T_1]$. If the Skorokhod mappings are defined for all continuous functions on $[0, T]$ and $[T, T_1]$, then the Skorokhod mapping is defined for all continuous functions on $[0, T_1]$.

Example. The simplest but rather instructive case is the Skorokhod problem in a half-space with normal reflection. Let $G = \{x \in \mathbf{R}^d : \ xe_1 \geq 0\}$, and let $\Gamma(x) = e_1$ when $x \in \partial G$. This problem admits an explicit solution. To write it, we introduce some notation. Let E be the space of d-dimensional continuous functions (on $[0, T]$ or $\mathbf{R}_+$) with the uniform norm $\|.\|$. Define the mapping $\Sigma : E \to E \times E$ by associating with $f \in E$ the function $\Sigma(f) = (\Sigma^1(f), \Sigma^2(f))$ with

$$\Sigma^1_t(f) := \left(f^1_t + \left(\inf_{s \leq t} f^1_t\right)^-, f^2_t, \dots, f^d_t\right),$$

$$\Sigma^2_t(f) := \left(\left(\inf_{s \leq t} f^1_t\right)^-, 0, \dots, 0\right).$$

Let f be an "input" function of the Skorokhod problem, i.e., a function with $f^1_0 \geq 0$. It is easily seen that the pair $(\Sigma^1(f), \Sigma^2(f))$ is the solution (note that the first component of $\Sigma^2(f)$ is an increasing function).

To check the uniqueness, let us consider two solutions (ξ, ϕ) and $(\bar{\xi}, \bar{\phi})$. We have the identity $\xi - \bar{\xi} = \phi - \bar{\phi}$. If $\bar{\xi}^1_t = 0$, then $\xi^1_t \geq \bar{\xi}^1_t$ trivially. If $\bar{\xi}^1_t > 0$, then $\bar{\phi}^1_t = 0$ and $\xi^1_t - \bar{\xi}^1_t = \phi^1_t \geq 0$ because the function ϕ^1 is increasing and starts from zero. Thus, $\xi^1 \geq \bar{\xi}^1$. By symmetry we also have the opposite inequality, i.e., $\xi^1 = \bar{\xi}^1$ and, automatically, $\phi^1 = \bar{\phi}^1$. Since other components of ϕ and $\bar{\phi}$ are all equal to zero, the considered Skorokhod problem has a unique solution.

Thus, for the considered problem, the Skorokhod mapping is defined for all inputs. Its following properties are obvious:

$$\left\|\Sigma^2(f) - \Sigma^2(g)\right\| \le \|f - g\|,$$
$$\left\|\Sigma^1(f) - \Sigma^1(g)\right\| \le 2\|f - g\|.$$

This means that Σ satisfies the functional Lipschitz condition. Also,

$$\sup_{u,v\in[s,t]} \left|\Sigma^2_u(f) - \Sigma^2_v(f)\right| \le \sup_{u,v\in[s,t]} |f_u - f_v|,$$
$$\sup_{u,v\in[s,t]} \left|\Sigma^1_u(f) - \Sigma^1_v(f)\right| \le 2 \sup_{u,v\in[s,t]} |f_u - f_v|,$$

i.e., the oscillation of the reflected process can be only twice as large as the oscillation of the input. For any input functions f and g and constants $a, b \ge 0$, we have the identity

$$\Sigma(af + bg) = a\Sigma(f) + b\Sigma(g).$$

Remark. It is not difficult to prove the existence and uniqueness theorem for the Skorokhod problem for an arbitrary convex domain with normal reflection on the boundary. In this case, $\Gamma(x)$ is the set of the inward normal vectors to supporting hyperplanes and it is not a singleton in the "corners." The result can be extended with appropriately modified formulations to the case where the "input" f is a càdlàg function (in particular, for the half-space, the above arguments need no changes). In contrast to this, the Skorokhod problem with oblique reflection when the domain has "corners" may have no solution.

5.6.2 Skorokhod Mapping

Using the method of local maps (or local coordinates), we show that the Skorokhod mapping is defined for all continuous functions in the case where G is a bounded domain with smooth boundary. We denote by $n(x)$ the inward normal vector at point $x \in \partial G$.

Theorem 5.6.1 *Let G be a bounded domain with smooth boundary. Suppose that, for $x \in \partial G$, the sets $\Gamma(x)$ are singletons $\{\gamma(x)\}$, the unit vectors $\gamma(x)$ depend continuously on x, and $\gamma(x)n(x) \ge c$ for some constant $c > 0$. Then the Skorokhod mapping Σ is well defined for all continuous inputs and satisfies the Lipschitz condition.*

Proof. The smoothness assumptions on boundary and reflection imply that there exists a final covering of the boundary by balls $U_i = O_r(x_i)$, $i = 1, \dots, N$, admitting bijections $u_i : U_i \to \mathbf{R}^d$ with the following properties:

(i) u_i and u_i^{-1} are C^2 functions satisfying the Lipschitz condition;
(ii) $u_i(U_i) \cap \operatorname{int} G = \{u : \ ue_1 \geq 0\}$ and $u_i(U_i) \cap \partial G = \{u : \ ue_1 = 0\}$;
(iii) $\nabla u_i^1(x)\gamma(x) = 1$, $\nabla u_i^k(x)\gamma(x) = 0$, $k = 2, \dots, d$, for $x \in U_i \cap \partial G$.

The set $G \setminus \cap_i U_i$ is also compact and, hence, can be covered by a finite number of open balls contained in G. We denote by U_0 the union of these balls and associate with U_0 the identity mapping denoted by u_0. Let $d_i(x)$ be equal to the distance of x to the boundary of U_i if $x \in U_i$ and zero otherwise. Since the function $d(x) := \min_i d_i(x)$ is continuous on G, it attains its minimum $\delta > 0$. Let $k(x)$ denote the first index i for which $O_\delta(x) \subseteq U_i$.

Now the solution of the Skorokhod problem in G can be constructed by the following procedure.

On the first step we construct the solution on the interval $[0, t_1]$ where t_1 is the first instant when ξ leaves $U_{k(f_0)}$. On the second step we repeat the construction on the interval $[t_1, t_2]$ with the input function $f_t - f_{t_1} + \xi_{t_1}$ where t_2 is the first instant after t_1 when the solution ξ leaves $U_{k(\xi_{t_1})}$, and so on. If $k(f_0)$, then $\xi = f$ and $\phi = 0$. If $k(f_0) \geq 1$, we use the mapping $u_{k(f_0)}$, solve the Skorokhod problem in the half-space with normal reflection with the input $u_{k(f_0)}(f)$, and define the functions ξ and ϕ by applying the inverse mapping $u_{k(f_0)}^{-1}$ to the constructed solution. Note that the oscillations of ξ are controlled by the oscillations of inputs. Therefore, $t_n \to \infty$, and this procedure allows us to define the solution of the problem on any interval $[0, T]$. The uniqueness follows from the uniqueness of the solution for the half-space. Since all mappings involved in the construction satisfy the Lipschitz condition, so does the Skorokhod mapping. □

5.6.3 Stochastic Skorokhod Problem

The stochastic setting for the Skorokhod problem (often referred to as a stochastic differential equation with reflection) in the domain G with the reflection mapping Γ includes also a stochastic basis with a Wiener process w and the drift and diffusion coefficients $a(x)$ and $\sigma(x)$, $x \in G$. By definition, the (strong) solution is a pair (X, Z), where $X = (X_t)$ is an adapted continuous process evolving in $\bar{G}$, and $Z = (Z_t)$ is an adapted $\mathbf{R}^d$-valued continuous process of bounded variations such that:

(a) the following equality between process holds:

$$X_t = X_0 + \int_{[0,t]} a(X_t)\, dt + \int_{[0,t]} \sigma(X_t)\, dw_t + Z_t;$$

(b) $\dot{Z}_t \in \Gamma(X_t)\ d|Z_t|$-a.s.

First, we investigate the stochastic Skorokhod problem in the half-space with normal reflection. The corresponding result is simple and gives the idea.
□

Proposition 5.6.2 *Let $G = \{x \in \mathbf{R}^d : \; xe_1 \geq 0\}$, and let $\Gamma(x) = e_1$ when $x \in \partial G$. Suppose that the coefficients $a(x)$ and $\sigma(x)$ satisfy the Lipschitz condition. Then the stochastic Skorokhod problem admits a unique solution.*

Proof. Let us consider the stochastic differential equation

$$dY_t = \tilde{a}_t(Y)\, dt + \tilde{\sigma}_t(Y)\, dw_t, \quad Y_0 \in G, \tag{5.6.1}$$

where $\tilde{a}_t(Y) := a(\Sigma_t^1(Y))$ and $\tilde{\sigma}(Y) := \sigma(\Sigma_t^1(Y))$. Since a and σ satisfy the Lipschitz condition and so does the mapping Σ^1, the coefficients $\tilde{a}$ and $\tilde{\sigma}$ also satisfy the (functional) Lipschitz condition, and, therefore, the stochastic equation has a unique solution $Y = (Y_t)$. Recalling the identity

$$Y = \Sigma^1(Y) - \Sigma^2(Y),$$

we get that

$$d\Sigma_t^1(Y) = a\big(\Sigma_t^1(Y)\big)\, dt + \sigma\big(\Sigma_t^1(Y)\big)\, dw_t + d\Sigma_t^2(Y).$$

That is, the pair (X, Z) with $X = \Sigma_t^1(Y)$ and $Z = \Sigma_t^2(Y)$ is the solution of the Skorokhod problem in the half-space.

To prove the uniqueness, we consider two solutions (X, Z) and (X', Z'). Put $\tilde{X} = X - Z$, $\tilde{X}' = X' - Z'$ and note that $\Sigma_t^1(\tilde{X}) = X$, $\Sigma_t^1(\tilde{X}') = X'$. It follows that $\tilde{X}$ and $\tilde{X}'$ are solutions of the stochastic differential equation (5.6.1), for which the uniqueness theorem holds. Thus, $X - X' = Z - Z'$, and we derive from here that $X = X'$ and $Z = Z'$ by the same argument as that used above for the deterministic problem. □

Theorem 5.6.3 *Let G be a bounded domain with smooth boundary. Suppose that, for $x \in \partial G$, the sets $\Gamma(x)$ are singletons $\{\gamma(x)\}$, the unit vectors $\gamma(x)$ depend continuously on x, and $\gamma(x)n(x) \geq c$ for some constant $c > 0$. Suppose that the coefficients $a(x)$ and $\sigma(x)$ satisfy the Lipschitz condition. Then the stochastic Skorokhod problem admits a unique solution.*

Proof. The arguments are the same as in the proposition above and use the existence of the Skorokhod mapping for the considered case established in Theorem 5.6.1. The uniqueness follows from the uniqueness of the solution in the half-space. □

Bibliographical Comments

Chapter 1

The Black–Scholes formulae for pricing call-options were derived in the seminal paper [19] of 1973 using economic considerations combined with PDE arguments which resemble the derivation of the heat equation from the first physical principles. We uses a "modern" probabilistic approach, more natural for probability students having the first course in random processes, where the predictable representation theorem and the Girsanov theorem are (presumably!) basic facts. It is worth to recall that the needed predictable representation theorem for the Wiener process (together with others, more profound results) was formulated by J.M.C. Clark only in 1971, [38], though it can be easily derived from what is called now the "chaotic representation theorem" established in the famous paper [99] by K. Itô on multiple Wiener integrals of 1951 (by the way, this was indicated by Itô himself, who was the reviewer of [58]; Itô also pointed out his less known paper [100]). By the way, this theorem and its version for the Poisson process were the topic of the graduation project of the first author of this book in 1971, see [113, 114]. Unfortunately, the needed predictable representation property is very rare, and in the class of (stochastically continuous) processes with independent increments it holds only for the Wiener and Poisson processes. This fact was discovered by Dellacherie [58], who related it with the uniqueness of a martingale measure. For further development, see the fundamental book by Jacod [101].

As for the Girsanov theorem, we need only its rudimentary version, corresponding to a deterministic and even constant drift, which was known already to Cameron and Martin and whose result was generalized by Girsanov for the case of random drift; now Girsanov's name is used for all theorems on a change of characteristics under an absolute continuous change of probability measure, see, e.g., the paper [124] and monographs [101, 161, 102, 159] with further references therein.

Yu. Kabanov, M. Safarian, *Markets with Transaction Costs*,
Springer Finance, DOI 10.1007/978-3-540-68121-2,

We have no need to discuss here numerous economical and mathematical aspects of Black–Scholes formulae: for this, we send the reader to the books [18, 57, 144, 212, 173, 181], ...

Portfolio dynamics models with proportional transaction cost were discussed already in the 1970s. Their origin can be traced back to the consumption–investment model suggested by Magill and Constantinides in 1976 as a natural generalization of the Merton model of 1973, [169]. However, Leland's paper [156] of 1985 in *Journal of Finance* happened to be the most important for the industry because of its easy practical implementation: there was no need to change existing codes. Moreover, for typical parameters of the real-world stock markets, it gives quite a satisfactory precision, see, e.g., [12]. On the other hand, mathematicians were rather perplexed by Leland's heuristic arguments. To be honest, his claims should be considered only as conjectures. A correct one was the assertion that the terminal value of the value process approaches the pay-off in the case of "rescaling," where the transaction costs decrease as $n^{-1/2}$, where n is the number of portfolio revisions. Unfortunately, this was just mentioned in a footnote remark in [156], and no arguments were provided. K. Lott in his thesis [164] gave a rigorous proof of the convergence result in this case, which is of particular interest because the adjusted volatility does not depend on n. But the situation with nonrescaled case remained confusing. In our note [130] (circulating as a preprint from 1995) we disproved the main conjecture by Leland. It was shown that there is a discrepancy if the transaction costs do not depend on the number of revisions. We confirmed that if the transaction costs decrease as $n^{-\alpha}$ for any $\alpha \in]0, 1/2$, the terminal value of the portfolio approximates the pay-off of the call option. Initially, the note was submitted to *Journal of Finance* as a correction to [156], but it was refused with a comment that the journal is flooded by submissions, and the editor has a right to reject even good papers without explanations!

There is a lot of activity to extend the Leland approach to various situations: more general contingent claims and more general price processes, nonuniform grids as well as other schemes of portfolio revisions intended to improve the performance, see [164, 209, 179, 60–62]. The asymptotic analysis of these extensions is quite delicate and requires a lot of patience. One should take care that is some papers, e.g., [86], arguments are not rigorous, and extra hypotheses might be needed. The size of discrepancy for realistic values of transaction costs is discussed in [235, 236, 157].

Granditz and Schachinger [85] noted that in [130] the function f describing the random discrepancy $f(S_1)$ contains a removable discontinuity at the point K. Since S_1 takes the value K with zero probability, this does not matter for the answer but gives a hint that the approximation including the correction can be rather bad when at the maturity date the price of the stock approaches the strike. They started a study of the rate of convergence of this approximation and discovered that it should be $n^{-1/4}$. Pergamenshchikov in [179] confirmed this using more elaborate analysis: his limit theorem is a nontrivial and technically difficult result.

Lott also calculated the first-order asymptotic term of the quadratic deviation of the terminal value of a Leland-type portfolio from the pay-off. For the case of nonuniform grids, the corresponding expression for the first-order coefficient (at n^{-1}) was obtained using heuristic arguments in [86]. The main idea of the latter paper is to improve the portfolio performance by choosing the grid minimizing this coefficient, but, to our understanding, it remains an open problem. In Chap. 1 we give a rigorous derivation of the asymptotics of the mean-square error for convex pay-off functions and nonuniform grids following the paper [63] (see [81] for the case of call option) and also obtain a limiting distribution of the error by proving a functional limit theorem as in [64, 63]. Other ideas on approximate hedging can be found in [27, 153, 93, 95, 2, 143, 202, 168, 48, 229, 154, 26, 98].

We include a result that the super-replication problem with transaction costs in the classical model, where the price process is a geometric Brownian motion, has a trivial solution. It happened that the super-replication price of the options depending on the terminal value of the stock price is so high that the investor can just buy the stock at time zero and keep it until the exercise date, without any further trading. This is in a striking difference with the Black–Scholes frictionless market model, where the super-replication price coincides with a replication price requiring continuous trading. The proofs of this curious fact (conjectured by J.M.C. Clark and M. Davis in [46]) were given almost simultaneously by Soner, Shreve, and Cvitanić in [216] (using rather involved techniques based on convex analysis) and by Levental and Skorohod in [158] (using a direct probabilistic approach and covering much more general models for price processes). We give here a succinct presentation of the second proof. It is necessary to recall that even in models of frictionless market, when the price process is Lévy-driven, the super-replication problem has only the trivial solution. For the precise formulation, see Eberlein and Jacod [67] and [104, 105] for further development.

Chapter 2

The aim of this chapter is to provide a comprehensive account of what is sometimes called the Fundamental Theorem of Asset Pricing (FTAP) for frictionless markets and a development around. We collect here the results scattered in numerous publications. Our presentation is designed as to the first, preliminary, step to markets with friction. Mathematicians usually consider Harrison, Kreps, and Pliska (see [91, 92, 152]) as the pioneers of the theory, though people working in the financial economics have their own reason for saying that the studies on arbitrage theory started much earlier, having in mind, e.g., the Arbitrage Pricing Theory (APT) of Ross–Huberman [200, 97], see a short account in [115]. To our opinion, the credit given to the former is completely fair because, revealing the relation between the economically important notion of arbitrage and such a basic concept of the theory of stochastic processes as martingale, Harrison, Kreps, and Pliska placed the theory in the mainstream of modern mathematics and gave a huge impulse to subsequent development.

Essentially, the Harrison–Pliska theorem for the case of finite Ω is the Stiemke lemma of 1915, [220], on solutions of linear inequalities accompanied by the observation that the strictly positive normalized solution can be interpreted as the density of a martingale measure.

The theorem of Dalang–Morton–Willinger [45] is a much deeper result rather than just a previous theorem with a removed hypothesis. This is clear from the formulation which consists in a number of nontrivially related conditions. The beauty and importance of the DMW-theorem makes challenging to find a short proof suitable for lecture courses, and this is the reason of a number of publications where various aspects of this theorem were thoroughly inspected. The original proof was based on the reduction to a one-period model and used a measurable selection theorem with a boring verification of measurability. Stricker was the first who noticed the importance of the closedness of the set A_T of hedgeable claims, [217, 6], and this observation was fully exploited by Schachermayer in [203]. A kind of linear algebra with random coefficients was used in the proof by Kabanov and Kramkov combined with a simple argument to avoid a reference to a measurable selection theorem [120]. Jacod and Shiryaev [103] completed the list of equivalent conditions, adding, in particular, condition (h). The paper [103] should be mentioned together with the earlier study [223] in relations with the uniqueness of martingale measure and the market completeness. The corresponding result sometimes is referred to as the Second FTAP (though it is rather deceptive from the practical point of view: it happens that the uniqueness is an exceptional property). Rogers [193] introduced a brilliant idea to find the martingale density as a solution of a simple optimization problem (interpreted as the maximization of expected exponential utility in [57]); unfortunately, this idea was somehow spoiled in his original proof by more complicated measurable selection arguments needed to manipulate with conditional expectations. All these proofs are based on a reduction to the one-period case and cannot be extended to the model with friction.

In a search for NA-criteria for the latter, Kabanov and Stricker developed a method which allows one to avoid the intermediation of one-period NA-conditions and which works well also in the classical case. This was published in "Teachers' note" [131], and we follow here the lines of the latter. The important point in these studies was the problem how to avoid measurability issues. It was done here via Lemma 2.1.2, which should be credited to Engelbert and von Weizsäcker. In 2000 the first author in his lecture on arbitrage theory at Winter School on Stochastic Processes in Siegmundsburg complained that in many cases the arguments arrived at a need to find a measurable selector in the set of limit points of a random sequence. Since a general theorem is not considered to be in standard syllabus, one can try to construct the needed selector using some home-made tools. Engelbert and von Weizsäcker commented immediately that it is much better to construct in a measurable way a convergent subsequence using a standard recipe, which is of frequent use in the theory of random processes. This "magic" lemma on measurable subsequences allows us to simplify and standardize several proofs, in particular, to get a "fast" proof of the DMW-theorem using the idea of Rogers mentioned above.

Theorem 2.1.4 was established independently and almost simultaneously by Kreps [152], in financial context and by Yan [230] in a general mathematical context. Its key idea is using the so-called "exhaustion argument" (to find "the most positive" separating functional). Its origin can be traced back to the Halmos–Savage theorem [90] on a countable equivalent subfamily of dominated measures. The latter theorem, very simple, may serve as a reference accelerating the proof, but we prefer to give an arrangement based on the elegant Lemma 2.1.3 borrowed from the paper [87]. On generalizations of the Kreps–Yan and Halmos–Savage theorems, see [195, 145, 111]. Note the preferential use in mathematical finance of the spaces L^0, which is not locally convex, and L^∞, which is not reflexive. But both are invariant under equivalent change of probability measure. The strong dual of the latter is a space of rather complicated structure (described by Yosida and Hewitt, [231]). The advantage of its weak* dual, L^1, is that its elements can be interpreted as prices.

The closedness of the set R_T, which is obtained here as a by-product of our arguments for the DMW-theorem, for the first time, was noted by Stricker, who also proved the closedness of the space of discrete-time stochastic integrals (this follows from a much more general and complicated theorem due to Mémin [167]). The "comment" on absolute continuous martingale measures seems to be an outcome of discussions with G. Last. Theorem 2.1.10 is taken from [137]. The paper [41] deals with the Law of One Price for a multiperiod model with arbitrary Ω.

In this book we do not touch neither generalization of NA-criteria related with portfolio constrains, see, e.g., [71, 33, 109], nor more "exotic" cases like analogs for random fields [126].

The idea to use the optional decomposition theorem and (generalized) Snell envelopes for calculating super-replication price goes back to El Karoui and Quenez [70]. It was generalized to the case of locally bounded semimartingales by Kramkov [150] and, afterwards, to the case of arbitrary semimartingales by Kabanov and Föllmer [73] using an approach based on Lagrange multipliers. Optional decomposition is discussed also in [74–76, 219, 57]. It is worth to note that the Kramkov theorem does not cover the discrete-time optional decomposition theorem (Theorem 2.1.12), an easy result which we prove here following [73]. On the other hand, the local boundedness assumption can be removed using the general theorem on martingale measures with bounded densities from [133]. Unfortunately, the last assertion, so innocently looking, is far from being trivial, and the only available proof is of the same difficulty (and based on the same ideas) as the proof of the optional decomposition in [73]. We give here only its much simpler discrete-time version (Theorem 2.1.17) following [186].

Subsection on the duality between maximization of the expected exponential utility and minimization of the entropy over the set of equivalent martingale measures is based on the paper [134]. One can find necessary preliminaries in the textbooks and monographs on convex analysis, see, e.g., [191, 11, 192]. There are many results on "interesting" martingale measures and duality which we do not discuss here, see, e.g., [190, 84, 82, 78, 205, 218, 52].

The developed apparatus allows us to present in a rather succinct manner a synthesis of results on no-arbitrage conditions for discrete-time models with infinite horizon. Though the starting point is the paper by Schachermayer [204], we use also ideas from works [54, 56, 116, 133], and others. For the first time, Theorem 2.2.1 was explicitly formulated in [118], but it can be easily obtained as a byproduct of the no-arbitrage criterion given in [204]. For the study of total variation distance on the filtered space see [125, 102]. Theorem 2.2.2 is taken from [182]. Proposition 2.2.14 came from the note [138] inspired by the work [88].

We end the chapter by a brief account of the arbitrage theory in continuous time, which is beyond the scope of this book. We send the interested readers to the classical papers of Delbaen and Schachermayer [55, 56] and others collected in a revised form with extended comments in their book [57], cf. also with a more recent development in [116, 121, 136, 37].

Chapter 3

The first serious work on arbitrage theory in presence of transaction costs was done in 1995 by Jouini and Kallal [108], who considered a continuous-time model based on selling and buying prices. Under certain rather restrictive assumptions, their main theorem asserts that there is no-arbitrage if and only if there are a probability measure $Q \sim P$ and a process M evolving in the price spread such that M is a martingale with respect to Q. However, it was far from being clear how to extend this result to the case of the currency market where any asset can be exchanged directly to each other. The difficulty was a psychological one, rather than mathematical. Due to enormous success of arbitrage pricing, i.e., via calculating the expectation of the contingent claim with respect to a certain martingale measure (in a "risk-neutral" world), there was an impression that a corresponding object for models with transaction costs also should be based on the same concept. A more adequate approach was found in the study [117] (inspired by the paper [43], where a two-asset diffusion model was considered): one should think in terms of the martingale density rather than of the measure, and the natural analog of the latter (later baptized in [206] the *consistent price system*) is a martingale related to the dual of the solvency cone. The geometric approach to the modeling of financial markets with proportional transaction costs aimed hedging theorems and was developed further in [53] (discrete-time case) and [122] and [123] (continuous-time case); see also [210, 146, 147]. This experience allowed one to get criteria of absence of arbitrage. First, the model with finite number of states of the nature was analyzed in [132], where two kinds of *NA*-properties, namely, NA^w and NA^s were introduced. In a survey [115] it was indicated that not only the arguments become more transparent when the assets are accounted in physical units and not in units of numéraire, as was a custom in the waste literature, but, moreover, the modeling itself can be done without using the numéraire. The market with transaction costs can be viewed as a commodity market. The first NA^w-criteria for the general Ω were obtained in [128] under the assumption of efficient friction. It was understood that the consistent price system is just a martingale evolving in the duals to solvency cones (in physical units), and a rather natural condition is that this martingale does not hit the boundaries. It was shown in [206] that the existence of such a martingale, a *strictly consistent price system*, is equivalent to the NA^r-property of the model: there is the NA^w-property even for smaller transaction costs. Moreover, Schachermayer constructed a counterexample in which NA^s holds, but a consistent price system does not exist, and, therefore, a seemingly plausible extension of the DMW-theorem based on NA^w, the most natural no-arbitrage property, fails to be true. Schachermayer's approach was extended in [129] to the abstract framework of cone-valued processes developed in [128]; it was related with an interesting result of Penner [178], implying that in the model with constant transaction costs the NA^s is equivalent to the existence of a strictly consistent price system.

In all mentioned studies the arguments involve separation theorems, and, therefore, the closedness of the set of hedgeable claims appears to be important and even indispensable to ensure the no-arbitrage property as it is the case for frictionless markets. Surprisingly, as was shown by Rásonyi in [185, 188], it may happen that the NA^s-condition holds, but the mentioned set is not closed. Nevertheless, in the two-asset models NA^s is equivalent to the existence of a consistent price system. This unexpected result was established by Grigoriev [87]. New proofs of *NA*-criteria, in an abstract model, allowing one to circumvent a study of the closedness of the set of hedgeable claims were suggested by Rásonyi in [187] (under efficient friction) and, independently, by Rokhlin in [199] (in the general case). For ramifications related with abstract set-valued (not necessarily cone-valued) processes, see [8, 142] and recent papers by Rokhlin [196–198] and Pennanen and Penner [177]. Criteria for the *NA*2-property were obtained by [189]. The first paper on hedging of American-type contingent claims under transaction costs seems to be that of Chalasani and Jha [35]. These authors considered a two-asset model with finite probability space and used randomized stopping times. Our presentation follows the article by Bouchard and Temam [24], where the "standard" general discrete-time model was investigated, and an alternative description of the dual variable was suggested.

The model with transaction costs and incomplete information was considered by Bouchard in [21]. In our presentation we follow [50], where a bit different type of order coding was used, and some assumptions were relaxed.

Until now there are no studies on no-arbitrage conditions which are adequate to the situation in classical theory. Contrarily, the problem of super-replication was investigated by many authors for different models and using various methods. The first paper in this direction is due to Karatzas and Cvitanić [43] (two-asset diffusion model). The problem of minimal endowment in money to super-replicate a contingent claim depending on the terminal value of the diffusion price process was solved in [44] using the theory of viscosity solutions; see also [224] and [25] for further development in this direction. The duality approach for a description of the set of hedging endowments was initiated in the papers [122, 123] in the general framework of continuous semimartingales. The difference between these two papers is in the definition of admissibility. By an analogy with the frictionless case, it seems that the natural definition is the boundedness from below of the portfolio process measured in the units of the numéraire in the sense of the partial ordering induced by the solvency cone. It happened that this definition is too restrictive, and one can work only assuming the boundedness from below in terms of physical units. Note that this definition of the admissibility is similar to that used by Sin in his thesis [215] on frictionless markets. The arguments of the mentioned paper and those in [135] were based on the assumption that the price process is continuous. It was shown by Rásonyi [185] that one cannot preserve the existing formulation of the hedging theorem without modifying the definition of the portfolio process. Such a modification was suggested by Campi and Schachermayer [32]. In their setting the portfolio process is no more càdlàg but làdlàg and so is beyond the scope of the traditional stochastic calculus. Note that a stochastic calculus for processes of this kind was developed by Galtchouk [79, 80]. It seems that, for the market with transaction costs, the làdlàg-modeling is a correct one. As an indirect confirmation of this, the hedging theorem for American options (Theorem 3.6.20) may serve. In our presentation of hedging theorems for American options, we follow the paper [49]. For another approach, see [22]. The continuous-time requires some facts from the general theory of stochastic processes which can be found in the books [59, 101, 102, 183].

A sufficient condition for the existence of a consistent price system is taken from [138], though the paper [88] should be credited for the main idea. Section 3.7 of Chap. 3 is based on the paper [65].

We did not touch many other directions of the theory of transaction costs, like problems of equilibrium (see, e.g., [110]) or liquidity.

Chapter 4

In 1975, the first author of this book, just having finished his postgraduate studies and started his first research at CEMI on the stochastic maximum principle, noticed in the paper by Bismut [17][1] the reference to Merton [169]. Having read the latter and been rather impressed, he said to his senior colleague N. Krylov: "Look, this guy solved a nice problem—an explicit solution is found!" The reaction of Krylov (known for his sharp mind and direct language) was immediate: "It is trivial."

Indeed, from the mathematical point of view the Merton problem for power utility function is easy: it is clear that the Bellman function inherits the homogeneity of the same degree. Therefore, if it is finite, then it is the original utility function multiplied by a constant. Thus, the Bellman equation can be reduced to an algebraic one to determine this constant if possible (i.e., in the case of finite solution). On the other hand, this simple problem had enormous influence to the whole development of mathematical finance. We refer the reader to the books [140, 141, 72] for detailed discussions, generalizations, and further references. However, we find convenient for our purposes to present the Merton problem in its simplest form.

The extension of the Merton model to portfolio optimization under transaction costs is due to Magill and Constantinides [165]. It has appeared soon, in 1976. Rather remarkably, these authors, using heuristic arguments, described correctly the structure of optimal control for the two-asset model. The problem was solved, in the mathematical sense, by Davis and Norman [47]. The further important step was done by Soner and Shreve [214], who used the theory of viscosity solution. Their approach was developed further and explained in papers [3, 4, 232–234, 14, 119], and others. The classical reference on viscosity solutions is "user's guide" by Crandall, Ishii, and Lyons [42]; see also books [72, 13]. The attractive feature is that a solution of the HJB equation in viscosity sense always exists, of course, under certain rather mild conditions. Moreover, one can start working with such solutions having only a basic knowledge on the whole concept. This is the reason why we prefer to give a brief "pragmatic" account of the theory. The proof of the uniqueness of the viscosity solution involving a Lyapunov function is taken from [119]. It seems that the uniqueness for the power utility function γc^γ with $\gamma < 0$ and logarithmic utility remains an open question.

Traditionally, reasoning used to check that the Bellman function is the viscosity solution of the HJB equation is based on the Dynamic Programming Principle (DPP) in a suitable form. Our analysis revealed that in the present setting it is difficult to find a reliable reference for the latter. Some authors, to circumvent difficulties, replace the initial model of strong solutions by a model of weak solutions, where the probability space and the driving Wiener process are also elements of control. As was shown in [127], even for regular control problems, such a replacement is not innocent and change structural properties of the model. Fortunately, for the considered model, the DPP can be easily proven under the assumption that the underlying space is a canonical one

[1] The first paper where BSDE (backward stochastic differential equation) was used in the context of financial modeling.

admitting translation and concatenation operators. Note also that the passage from DPP to the HJB equation requires some arguments related with the local character of differential operators. It is essential that a continuous process, being stopped when it leaves a closed set, remains in it. This is not the case with discontinuous processes, and one should be very attentive with extensions to Lévy-driven models, see, e.g., [15, 16, 77]. Our presentation of the Davis–Norman solution for the two-asset model is based on the paper [214], as well as the first part of the asymptotic analysis with vanishing transaction costs. The results on the exact asymptotic behavior is taken from the paper [106]; see also [213] and, for related questions, [10, 170, 194, 1, 107]. There exists a vast literature on various aspects of optimal control of portfolios under transaction costs: [221, 232–234, 155, 39, 40, 51, 112, 4, 9, 29–31, 174, 175, 149, 162, 163, 172, 139], ...

Appendix

One can observe that the development of theory of markets with transaction costs involve more and more results from convex geometry. Trying to limit necessary prerequisites, we choose the most relevant facts from the latter. For proofs, we recommend to consult monographs [191, 192, 184, 207, 11, 68, 176]. The reader may find interesting to look at the original papers on "alternatives" by Gordan [83] of 1983 and Stiemke [220] of 1915. Theorem 5.1.3 is taken from [115].

The theorem established by Komlós [148] gave rise to further development in various directions. Its proofs can be found in [36, 89, 225, 208, 127]; see also a discussion in [57]. The generalization to the sequences bounded from below was proven by von Weizsäcker in [226]. The "vulgar version" of the latter given by Lemma 5.2.7 is due to Delbaen and Schachermayer; it is taken from [54].

The coincidence of the concepts of local and generalized martingales with deterministic (or integrable) initial value was established by P.-A. Meyer; see the textbook by Shiryaev [211].

The usefulness of Proposition 5.3.3 was indicated by Shiryaev. Proposition 5.3.6 is due to Dellacherie. The extension of the Snell envelopes for families of probability measures stable under concatenation was suggested by El Karoui. For the use of Snell envelopes for optimal stopping problems we send the reader to the recent book [180].

The book [34] is the classical treatise on the measurable selection. Many useful information can be found in [227] and [228]. The construction of conditional expectations of set-valued mappings is taken from the book [171]; see also the original paper by Hiai and Umegaki [94].

Theorem 5.5.3 was established in [123]. For more delicate duality results in this direction, see [28, 237, 20], and [23]. Krein-Šmulian theorem is a classical result, [151], see [96] for a modern version. The books [201, 166] can be consulted for questions in functional analysis.

In our presentation of the Skorokhod problem (i.e., stochastic differential equations) we follow the method of local coordinates used by Anderson and Orey [5] (we are indebted to P. Dupuis for this reference). For further reading, see [222, 160, 66, 7, 69].

References

1. Albanese, C. and Tompaidis, S., Small transaction cost asymptotics and dynamic hedging, *Eur. J. Oper. Res.* **185**:3 (2008), 1404–1414.
2. Ahn, H., Dayal, M., Grannan, E. and Swindle, G., Option replication with transaction costs: General diffusion limits, *Ann. Appl. Probab.* **8**:3 (1998), 676–707.
3. Akian, M., Menaldi, J. L. and Sulem, A., On an investment-consumption model with transaction costs, *SIAM J. Control Optim.* **34**:1 (1996), 329–364.
4. Akian, M., Sulem, A. and Taksar, M. I., Dynamic optimization of long-term growth rate for a portfolio with transaction costs and logarithmic utility, *Math. Finance* **11**:2 (2001), 153–188.
5. Anderson, R. F. and Orey, S., Small random perturbations of dynamical systems with reflecting boundary, *Nagoya Math. J.* **60** (1976), 189–216.
6. Ansel, J. P. and Stricker, C., Couverture des actifs contingents, *Ann. Inst. Henri Poincaré. Sér. Probab. Stat.* **30**:2 (1994), 303–315.
7. Anulova, S. V. and Liptser, R. S., Diffusion approximation for processes with normal reflection, *Teor. Veroyatn. Primen.* **35**:3 (1990), 417–430. English translation: *Theory Probab. Appl.* **35**:3 (1991), 411–423.
8. Astic, F. and Touzi, N., No arbitrage conditions and liquidity, *J. Math. Econ.* **43**:6 (2007), 692–708.
9. Atkinson, C., Pliska, S. and Wilmott, P., Portfolio management with transaction costs, *Proc. R. Soc. Lond., Ser. A* **453**:1958 (1997), 551–562.
10. Atkinson, C. and Wilmott, P., Portfolio management with transaction costs: an asymptotic analysis, *Math. Finance* **5** (1995), 357–367.
11. Aubin, J.-P., *Applied Functional Analysis*, Wiley-Interscience, New York, 1977.
12. Avellaneda, M. and Paras, A., Optimal hedging portfolios for derivative securities in the presence of large transaction costs, *Appl. Math. Finance* **2** (1994), 165–193.

Yu. Kabanov, M. Safarian, *Markets with Transaction Costs*,
Springer Finance, DOI 10.1007/978-3-540-68121-2,

13. Bardi, M. and Dolcetta, C., *Optimal Control and Viscosity Solutions of Hamilton–Jacobi–Bellman Equations*, Birkhäuser, Boston, 1997.
14. Barles, G. and Soner, H. M., Option pricing with transaction costs and a non-linear Black–Scholes equation, *Finance Stoch.* **2**:4 (1998), 369–397.
15. Benth, F. E., Karlsen, K. H. and Reikvam, K., Portfolio optimization in a Lévy market with intertemporal substitution and transaction costs, *Stoch. Stoch. Rep.* **74**:3–4 (2002), 517–569.
16. Benth, F. E., Karlsen, K. H. and Reikvam, K., Optimal portfolio management rules in a non-Gaussian market with durability and intertemporal substitution, *Finance Stoch.* **5**:4 (2001), 447–467.
17. Bismut, J.-M., Growth and optimal intertemporal allocation of risks, *J. Econ. Theory* **10**:2 (1975), 239–257.
18. Björk, T., *Arbitrage Theory in Continuous Time*, Oxford University Press, Oxford, 1998.
19. Black, F. and Scholes, M., The pricing of options and corporate liabilities, *J. Polit. Econ.* **81** (1973), 637–659.
20. Bouchard, B., A version of the $\mathcal{G}$-conditional bipolar theorem in $L^0(\mathbf{R}^d_+; \Omega, \mathcal{F}, P)$, *J. Theor. Probab.* **18**:2 (2005), 439–467.
21. Bouchard, B., No-arbitrage in discrete-time markets with proportional transaction costs and general information structure, *Finance Stoch.* **10**:2 (2006), 276–297.
22. Bouchard, B. and Chassagneux, J.-F., Representation of continuous linear forms on the set of làdlàg processes and the pricing of American claims under proportional costs, *Electron. J. Probab.* **14** (2009), 612–632.
23. Bouchard, B. and Mazliak, L., A multidimensional bipolar theorem in $L^0(\mathbf{R}^d, \Omega, \mathcal{F}, P)$, *Stoch. Process. Their Appl.* **107**:2 (2003), 213–231.
24. Bouchard, B. and Temam, E., On the hedging of American options in discrete time markets with proportional transaction costs, *Electron. J. Probab.* **10**:22 (2005), 746–760 (electronic).
25. Bouchard-Denize, B. and Touzi, N., Explicit solution of the multivariate super-replication problem under transaction costs, *Ann. Appl. Probab.* **10** (2000), 685–708.
26. Bouchard, B., Kabanov, Yu. M. and Touzi, N., Option pricing by large risk aversion utility under transaction costs, *Decis. Econ. Finance* **24**:2 (2002), 127–136.
27. Boyle, F. and Vorst, T., Option replication in discrete time with transaction costs, *J. Finance* **47** (1992), 271–293.
28. Brannath, W. and Schachermayer, W., A bipolar theorem for $L^0_+(\Omega, \mathcal{F}, \mathbf{P})$. *Sémin. Probab. XXXIII, Lect. Notes Math.* **1709** (1999), 349–354.
29. Cadenillas, A., Consumption–investment problems with transaction costs: survey and open problems, *Math. Meth. Oper. Res.* **51**:1 (2000), 43–68.

30. Cadenillas, A. and Haussmann, U., The stochastic maximum principle for a singular control problem, *Stoch. Stoch. Rep.* **49** (1994), 211–237.
31. Cadenillas, A. and Pliska, S., Optimal trading of a security when there are taxes and transaction costs, *Finance Stoch.* **3**:3 (1999), 137–165.
32. Campi, L. and Schachermayer, W., A super-replication theorem in Kabanov's model of transaction costs, *Finance Stoch.* **10**:4 (2006), 579–596.
33. Carassus, L., Pham, H. and Touzi, N., No arbitrage in discrete time under portfolio constraints, *Math. Finance* **11** (2001), 315–329.
34. Castaing, C. and Valadier, M., *Convex Analysis and Measurable Multifunctions*, Lecture Notes Math. **580**, Springer, Berlin–Heidelberg–New York, 1977.
35. Chalasani, P. and Jha, S., Randomized stopping times and American option pricing with transaction costs, *Math. Finance* **11**:1 (2001), 33–77.
36. Chatterji, S. D., Un principe de sous-suites dans la théorie des probabilités. *Sémin. Probab. VI., Lect. Notes Math.* **258** (1972), 72–89.
37. Cherny, A. S., General arbitrage pricing model: probability and possibility approaches, *Sém. Probab. XL, Lect. Notes Math.* **1899** (2007), 415–481.
38. Clark, J. M. C., The representation of functionals of Brownian motion by stochastic integrals, *Ann. Math. Stat.* **41** (1970), 1282–1295.
39. Constantinides, G. M., Capital market equilibrium with transaction costs, *J. Polit. Econ.* **94** (1996), 842–862.
40. Constantinides, G. M. and Zariphopoulou, T., Bounds on prices of contingent claims in intertemporal economy with transaction costs, *Finance Stoch.* **3**:3 (1999), 345–369.
41. Courtault, J.-M., Kabanov, Yu. M., Delbaen, F. and Stricker, Ch., On the law of one price, *Finance Stoch.* **8**:4 (2004), 525–530.
42. Crandall, M., Ishii, H. and Lions, P.-L., User's guide to viscosity solutions of second order partial differential equations, *Bull. Am. Math. Soc.* **277** (1983), 1–42.
43. Cvitanić, J. and Karatzas, I., Hedging and portfolio optimization under transaction costs: a martingale approach, *Math. Finance* **6**:2 (1996), 133–165.
44. Cvitanić, J., Pham, H. and Touzi, N., A closed-form solution to the problem of super-replication under transaction costs, *Finance Stoch.* **3**:1 (1999), 35–54.
45. Dalang, R. C., Morton, A. and Willinger, W., Equivalent martingale measures and no-arbitrage in stochastic securities market models, *Stoch. Stoch. Rep.* **29** (1990), 185–201.
46. Davis, M. H. A. and Clark, J. M. C., A note on super-replicating strategies, *Philos. Trans. R. Soc. Lond. A* **347** (1994), 485–494.
47. Davis, M. and Norman, A., Portfolio selection with transaction costs, *Math. Oper. Res.* **15** (1990), 676–713.
48. Davis, M., Panas, V. and Zariphopoulou, T., European option pricing under transaction costs, *SIAM J. Control Optim.* **31** (1993), 470–493.

49. De Vallière, D., Denis, E. and Kabanov, Y. M., Hedging of American options under transaction costs, *Finance Stoch.* **13**:1 (2009), 105–119.
50. De Vallière, D., Kabanov, Yu. M. and Stricker, Ch., No-arbitrage properties for financial markets with transaction costs and incomplete information, *Finance Stoch.* **11**:2 (2007), 237–251.
51. Deelstra, G., Pham, H. and Touzi, N., Dual formulation of the utility maximization problem under transaction costs, *Ann. Appl. Probab.* **11**:4 (2001), 1353–1383.
52. Delbaen, F., Grandits, P., Rheinländer, Th., Samperi, D., Schweizer, M. and Stricker, Ch., Exponential hedging and entropic penalties, *Math. Finance* **12**:2 (2002), 99–123.
53. Delbaen, F., Kabanov, Yu. M. and Valkeila, E., Hedging under transaction costs in currency markets: a discrete-time model, *Math. Finance* **12**:1 (2002), 45–61.
54. Delbaen, F. and Schachermayer, W., A general version of the fundamental theorem of asset pricing, *Math. Ann.* **300** (1994), 463–520.
55. Delbaen, F. and Schachermayer, W., A compactness principle for bounded sequence of martingales with applications, in *Proceedings of the Seminar of Stochastic Analysis, Random Fields and Applications*, 1999.
56. Delbaen, F. and Schachermayer, W., The Fundamental Theorem of Asset Pricing for unbounded stochastic processes, *Math. Ann.* **312** (1998), 215–250.
57. Delbaen, F. and Schachermayer, W., *The Mathematics of Arbitrage*, Springer, Berlin–Heidelberg–New York, 2006.
58. Dellacherie, C., Intégrales stochastiques par rapport aux processus de Wiener ou de Poisson, *Sémin. Probab. VIII, Lect. Notes Math.* **381** (1974), 25–26. Correction in: *Sémin. Probab. XI, Lect. Notes Math.* **465** (1975), 494.
59. Dellacherie, C. and Meyer, P.-A., *Probabilités et Potentiel*, Hermann, Paris, 1980.
60. Denis, E., *Marchés avec coûts de transaction: approximation de Leland et arbitrage*, Thèse, Université de Franche-Comté, Besançon, 2008.
61. Denis, E., Leland's approximations for concave pay-off functions, in Kijima, M., Egami, M., Tanaka, K. and Muromachi, Y. (eds.), *Recent Advances in Financial Engineering*, pp. 107–117, World Scientific, Singapore, 2009.
62. Denis, E., *Modified Leland's strategy for constant transaction costs rate*, Preprint, Besançon, 2009/19.
63. Denis, E. and Kabanov, Yu. M., Mean square error for the Leland–Lott hedging strategy: convex pay-offs, *Finance Stoch.* **14** (2009).
64. Denis, E. and Kabanov, Yu. M., *Functional limit theorem for Leland–Lott hedging strategy*, Working paper, 2009.

65. Denis, E. and Kabanov, Yu. M., *Consistent price systems and arbitrage opportunities of the second kind in models with transaction costs*, Working paper, 2009.
66. Dupuis, P. and Ishii, H., SDEs with oblique reflection on nonsmooth domains, *Ann. Probab.* **21** (1993), 554–580.
67. Eberlein, E. and Jacod, J., On the range of options prices, *Finance Stoch.* **1**:2 (1997), 131–140.
68. Ekeland, I. and Temam, R., *Convex Analysis and Variational Problems*, SIAM, Philadelphia, 1999.
69. El Karoui, Processus de réflexion dans $\mathbf{R}^n$, *Sémin. Probab. IX, Lect. Notes Math.* **465** (1975), 534–554.
70. El Karoui, N. and Quenez, M.-C., Dynamic programming and pricing of contingent claims in an incomplete market, *SIAM J. Control Optim.* **33**:1 (1995), 27–66.
71. Evstigneev, I. V., Schürger, K. and Taksar, M., On the fundamental theorem of asset pricing: random constraints and bang-bang no-arbitrage criteria, *Math. Finance*, **14**:2 (2004), 201–221.
72. Fleming, W. and Soner, H. M., *Controlled Markov Processes and Viscosity Solutions*, Springer, Berlin–Heidelberg–New York, 1993.
73. Föllmer, H. and Kabanov, Yu. M., Optional decomposition and Lagrange multipliers, *Finance Stoch.* **2**:1 (1998), 69–81.
74. Föllmer, H. and Kabanov, Yu. M., Optional decomposition theorems in discrete time, in *Atti del convegno in onore di Oliviero Lessi, Padova, 25–26 marzo 1996*, pp. 47–68.
75. Föllmer, H. and Kramkov, D. O., Optional decomposition theorem under constraints, *Probab. Theory Relat. Fields* **109**:1 (1997), 1–25.
76. Föllmer, H. and Schied, A., *Stochastic Finance*, De Gruyter, Berlin, 2002.
77. Framstad, N. C., Øksendal, B. and Sulem, A., Optimal consumption and portfolio in a jump diffusion market with proportional transaction costs, *J. Math. Econ.* **35** (2001), 233–257.
78. Frittelli, M., On minimal entropy martingale measure and the valuation problem in incomplete markets, *Math. Finance* **10**:1 (2000), 39–52.
79. Galtchouk, L. I., Optional martingales, *Mat. Sb.* **112**:4 (1980), 483–512 (in Russian). English translation: **40** (1981), 435–468.
80. Galtchouk, L. I., Stochastic integrals with respect to optional semimartingales and random measures, *Teor. Veroyatn. Primen.* **29**:1 (1984), 93–107 (in Russian). English translation: *Theory Probab. Appl.* **29** (1985), 93–108.
81. Gamys, M. and Kabanov, Yu. M., Mean square error for the Leland–Lott hedging strategy, in Kijima, M., Egami, M., Tanaka, K. and Muromachi, Y. (eds.), *Recent Advances in Financial Engineering*, pp. 1–25, World Scientific, Singapore, 2009.

82. Goll, Th. and Rüschendorf, L., Minimax and minimal distance martingale measures and their relationship to portfolio minimization, *Finance Stoch.* **5**:4 (2001), 557–581.
83. Gordan, P., Über die Auflösung linearer Gleichungen mit reelen Koeffizienten, *Math. Ann.* **6** (1873), 23–28.
84. Granditz, P. and Rheinländer, T., On the minimal entropy martingale measure, *Ann. Probab.* **30**:3 (2002), 1003–1038.
85. Granditz, P. and Schachinger, W., Leland's approach to option pricing: The evolution of discontinuity, *Math. Finance* **11**:3 (2001), 347–355.
86. Grannan, E. R. and Swindle, G. H., Minimizing transaction costs of option hedging strategies, *Math. Finance* **6**:4 (1996), 341–364.
87. Grigoriev, P., On low dimensional case in the fundamental asset pricing theorem under transaction costs, *Stat. Decis.* **23**:1 (2005), 33–48.
88. Guasoni, P., Rásonyi, M. and Schachermayer, W., Consistent price systems and face-lifting pricing under transaction costs, *Ann. Appl. Probab.* **18**:2 (2008), 491–520.
89. Hall, P. and Heyde, C. C., *Martingale Limit Theory and Its Applications*, Academic Press, San Diego, 1980.
90. Halmos, P. and Savage, L., Application of the Radon–Nikodym theorem to the theory of sufficient statistics, *Ann. Math. Stat.* **20**:2 (1949), 225–241.
91. Harrison, J. M. and Kreps, D., Martingales and arbitrage in multiperiod securities markets, *J. Econ. Theory* **20** (1979), 381–408.
92. Harrison, M. and Pliska, S., Martingales and stochastic integrals in the theory of continuous trading, *Stoch. Process. Their Appl.* **11** (1981), 215–260.
93. Henrotte, P., *Transaction costs and duplication strategies*, Stanford University, Working paper, 1993.
94. Hiai, F. and Umegaki, H., Integrals, conditional expectations, and martingales of multivalued functions, *J. Multivar. Anal.* **7** (1977), 149–182.
95. Hodges, S. and Neuberger, A., Optimal replication of contingent claims under transaction costs, *Rev. Futures Mark.* **8** (1989), 222–239.
96. Horváth, J., *Topological Vector Spaces and Distributions. V.1*, Addison-Wesley, Reading, 1966.
97. Huberman, G., A simple approach to Arbitrage Pricing Theory, *J. Econ. Theory* **28** (1982), 183–191.
98. Irle, A. and Sass, J., Good portfolio strategies under transaction costs: a renewal theoretic approach, in Grossinho, M. R. et al. (eds.), *Stochastic Finance*, Springer, Berlin, 2006.
99. Ito, K., Multiple Wiener integral, *J. Math. Soc. Jpn.* **3** (1951), 157–169.
100. Ito, K., Spectral type of the shift transformation of differential processes with stationary increments, *Trans. Am. Math. Soc.* **81** (1956), 253–263.
101. Jacod, J., *Calcul Stochastiques et Problèmes de Martingales*, Lecture Notes in Mathematics, Springer, Berlin–Heidelberg–New York, 1979.
102. Jacod, J. and Shiryaev, A. N., *Limit Theorems for Stochastic Processes*, Springer, Berlin–Heidelberg–New York, 1987.

103. Jacod, J. and Shiryaev, A. N., Local martingales and the fundamental asset pricing theorem in the discrete-time case, *Finance Stoch.* **2**:3 (1998), 259–273.
104. Jakubenas, P., On option pricing in certain incomplete markets, *Tr. Mat. Inst. Steklova* **237** (2002), 123–142. Stokhast. Finans. Mat. English translation: *Proc. Steklov Inst. Math.* **237** (2002), 114–133.
105. Jakubenas, P., Levental, S. and Ryznar, M., The super-replication problem via probabilistic methods, *Ann. Appl. Probab.* **13** (2003), 742–773.
106. Janeček, K. and Shreve, S., Asymptotic analysis for optimal investment and consumption with transaction costs, *Finance Stoch.* **8**:2 (2004), 181–206.
107. Jang, B.-G., Koo, K. H., Liu, H. and Loewenstein, M., *Transaction cost can have a first-order effect on liqudity premium*, Preprint.
108. Jouini, E. and Kallal, H., Martingales and arbitrage in securities markets with transaction costs, *J. Econ. Theory* **66** (1995), 178–197.
109. Jouini, E. and Kallal, H., Arbitrage in securities markets with short sale constraints, *Math. Finance* **5**:3 (1995), 197–232.
110. Jouini, E. and Kallal, H., Viability and equilibrium in securities markets with frictions, *Math. Finance* **9**:3 (1999), 275–292.
111. Jouini, E., Napp, C. and Schachermayer, W., Arbitrage and state price deflators in a general intertemporal framework, *J. Math. Econ.* **41**:6 (2005), 722–734.
112. Jouini, E. and Porte, V., *Efficient trading strategies with transaction costs*, Preprint.
113. Kabanov, Yu. M., Representation of functionals of Wiener and Poisson processes as stochastic integrals, *Teor. Veroyatn. Primen.* **18**:2 (1973), 376–380. English translation: *Theory Probab. Appl.* **18**:2 (1974), 362–365.
114. Kabanov, Yu. M., Integral representation for functionals of the processes with independent increments, *Teor. Veroyatn. Primen.* **19**:4 (1974), 889–893. English translation: *Theory Probab. Appl.* **19**:4 (1975), 853–857.
115. Kabanov, Yu. M., Arbitrage theory, in *Handbooks in Mathematical Finance. Option Pricing: Theory and Practice*, pp. 3–42, Cambridge University Press, Cambridge, 2001.
116. Kabanov, Yu. M., On the FTAP of Kreps–Delbaen–Schachermayer, in *Statistics and Control of Random Processes. The Liptser Festschrift. Proceedings of Steklov Mathematical Institute Seminar*, pp. 191–203, World Scientific, Singapore, 1997.
117. Kabanov, Yu. M., Hedging and liquidation under transaction costs in currency markets, *Finance Stoch.* **3**:2 (1999), 237–248.
118. Kabanov, Yu. M., In discrete time a local martingale is a martingale under an equivalent probability measure, *Finance Stoch.* **12**:3 (2008), 293–297.

119. Kabanov, Yu. M. and Klüppelberg, C., A geometric approach to portfolio optimization in models with transaction costs, *Finance Stoch.* **8**:2 (2004), 207–227.
120. Kabanov, Yu. M. and Kramkov, D. O., No-arbitrage and equivalent martingale measure: an elementary proof of the Harrison–Pliska theorem, *Teor. Veroyatn. Primen.* **39**:3 (1994), 635–640. English translation: *Theory Probab. Appl.* **39**:3 (1995), 523–527.
121. Kabanov, Yu. M. and Kramkov, D. O., Asymptotic arbitrage in large financial markets, *Finance Stoch.* **2**:2 (1998), 143–172.
122. Kabanov, Yu. M. and Last, G., Hedging in a model with transaction costs, *Proc. Steklov Inst. Math.* **237** (2002), 208–214.
123. Kabanov, Yu. M. and Last, G., Hedging under transaction costs in currency markets: a continuous-time model, *Math. Finance* **12**:1 (2002), 63–70.
124. Kabanov, Yu. M., Liptser, R. Sh. and Shiryaev, A. N., Absolute continuity and singularity of locally absolute continuous probability distributions. Part I, II, *Mat. Sb.* **107**:3 (1978), 364–415; **108**:1 (1979), 32–61. English translation: *Math. USSR Sb.* **35**:5 (1979), 631–680; **36**:1 (1980), 31–58.
125. Kabanov, Yu. M., Liptser, R. Sh. and Shiryaev, A. N., On the variation distance for probability measures defined on a filtered space, *Probab. Theory Relat. Fields* **71**:1 (1986), 19–36.
126. Kabanov, Yu. M., Mishura, Yu. and Sakhno, L., Multiparameter generalizations of the Dalang–Morton–Willinger theorem, in Kabanov, Yu. M., Stoyanov, J. and Liptser, R. (eds.), *From Stochastic Calculus to Mathematical Finance. The Shiryaev Festschrift*, pp. 333–341, Springer, Berlin–Heidelberg–New York, 2006.
127. Kabanov, Yu. M. and Pergamenshchikov, S., *Two-scale Stochastic Systems. Asymptotic Analysis and Control*, Springer, Berlin–Heidelberg–New York, 2003.
128. Kabanov, Yu. M., Rásonyi, M. and Stricker, Ch., No-arbitrage criteria for financial markets with efficient friction, *Finance Stoch.* **6**:3 (2002), 371–382.
129. Kabanov, Yu. M., Rásonyi, M. and Stricker, Ch., On the closedness of sums of convex cones in L^0 and the robust no-arbitrage property, *Finance Stoch.* **7**:3 (2003), 403–411.
130. Kabanov, Yu. M. and Safarian, M., On Leland's strategy of option pricing with transaction costs, *Finance Stoch.* **1**:3 (1997), 239–250.
131. Kabanov, Yu. M. and Stricker, Ch., A teachers' note on no-arbitrage criteria, *Sémin. Probab. XXXV, Lect. Notes Math.* **1755** (2001), 149–152.
132. Kabanov, Yu. M. and Stricker, Ch., The Harrison–Pliska arbitrage pricing theorem under transaction costs, *J. Math. Econ.* **35**:2 (2001), 185–196.

133. Kabanov, Yu. M. and Stricker, Ch., On equivalent martingale measures with bounded densities, *Sém. Probab. XXXV, Lect. Notes Math.* **1755** (2001), 139–148.
134. Kabanov, Yu. M. and Stricker, Ch., On the optimal portfolio for the exponential utility maximization: remarks to the six-author paper, *Math. Finance* **11**:2 (2002), 99–123.
135. Kabanov, Yu. M. and Stricker, Ch., Hedging of contingent claims under transaction costs, in Sandmann, K. and Schönbucher, Ph. (eds.), *Advances in Finance and Stochastics*, pp. 125–136, Springer, Berlin, 2002.
136. Kabanov, Yu. M. and Stricker, Ch., Remarks on the true no-arbitrage property, *Sémin. Probab. XXXVIII, Lect. Notes Math.* **1857** (2005), 186–194.
137. Kabanov, Yu. M. and Stricker, Ch., The Dalang–Morton–Willinger theorem under delayed and restricted information, *Sémin. Probab. XXXIX, Memoriam Paul-André Meyer, Lect. Notes Math.* **1874** (2006), 209–213.
138. Kabanov, Yu. M. and Stricker, Ch., On martingale selectors of cone-valued processes, *Sémin. Probab. XLI, Lect. Notes Math.* **1934** (2008), 439–442.
139. Kallsen, J. and Muhle-Karbe, J., *On using shadow prices in portfolio optimization with transaction costs*, Preprint, 2008.
140. Karatzas, I. and Shreve, S. E., *Brownian Motion and Stochastic Calculus*, Springer, Berlin–Heidelberg–New York, 1988.
141. Karatzas, I. and Shreve, S. E., *Methods of Mathematical Finance*, Springer, Berlin–Heidelberg–New York, 1998.
142. Kaval, K. and Molchanov, I., Link-save trading, *J. Math. Econ.* **42** (2006), 710–728.
143. Keller, Th., *Zur Bewertungstheorie von Angebotsoptionen im Kreditgeschäft: Berücksichtigung der Transaktionskosten*, Universität Dortmund, Dissertation, 2001.
144. Kijima, M., *Stochastic Processes with Applications to Finance*, Chapman and Hall, Boca Raton, 2003.
145. Klein, I. and Schachermayer, W., A quantitative and a dual version of the Halmos–Savage theorem with applications to mathematical finance, *Ann. Probab.* **24**:2 (1996), 867–881.
146. Koehl, P., Pham, H. and Touzi, N., Hedging in discrete time under transaction costs and continuous-time limit, *J. Appl. Probab.* **36**:1 (1999), 163–178.
147. Koehl, P., Pham, H. and Touzi, N., On super-replication in discrete time under transaction costs, *Teor. Veroyatn. Primen.* **45**:4 (2000), 783–788. English translation in *Theory Probab. Appl.* **45**:4 (2002), 667–673.
148. Komlós, J., A generalization of a problem of Steinhaus, *Acta Math. Sci. Hung.* **18** (1967), 217–229.
149. Korn, R., Portfolio optimization under strictly positive transaction costs, *Finance Stoch.* **2**:2 (1998), 85–114.

150. Kramkov, D. O., Optional decomposition of supermartingales and hedging in incomplete security markets, *Probab. Theory Relat. Fields* **105**:4 (1996), 459–479.
151. Krein, M. and Šmulian, V., On regularly convex sets in the space conjugate to a Banach space, *Ann. Math.* **41**:2 (1940), 556–583.
152. Kreps, D. M., Arbitrage and equilibrium in economies with infinitely many commodities, *J. Math. Econ.* **8** (1981), 15–35.
153. Kusuoka, S., Limit theorems on option replications with transaction costs, *Ann. Appl. Probab.* **5** (1995), 198–221.
154. Lamberton, D., Pham, H. and Schweizer, M., Local risk-minimization under transaction costs, *Math. Oper. Res.* **23**:3 (1998), 585–612.
155. Lehoczky, Shreve S., Absolutely continuous and singular stochastic control, *Stoch. Stoch. Rep.* **17** (1986), 91–109.
156. Leland, H., Option pricing and replication with transactions costs, *J. Finance* **40**:5 (1985), 1283–1301.
157. Leland, H., Comments on "Hedging errors with Leland's option model in the presence of transactions costs", *Finance Res. Lett.* **4**:3 (2007), 200–202.
158. Levental, S. and Skorohod, A. V., On the possibility of hedging options in the presence of transaction costs, *Ann. Appl. Probab.* **7** (1997), 410–443.
159. Liese, F. and Vajda, I., *Convex Statistical Distances*, Teubner, Leipzig, 1987.
160. Lions, P.-L. and Sznitman, A.-S., Stochastic differential equations with reflecting boundary conditions, *Commun. Pure Appl. Math.* **37**:4 (1984), 511–537.
161. Liptser, R. Sh. and Shiryaev, A. N., *Theory of Martingales*, Kluwer Academic, Dordrecht, 1989.
162. Liu, H., Optimal consumption and investment with transaction costs and multiple risky assets, *J. Finance* **59**:1 (2004), 289–338.
163. Liu, H. and Loewenstein, M., Optimal portfolio selection with transaction costs and finite horizons, *Rev. Financ. Stud.* **15** (2002), 805–835.
164. Lott, K., *Ein Verfahren zur Replikation von Optionen unter Transaktionkosten in stetiger Zeit*, Dissertation, Universität der Bundeswehr München, Institut für Mathematik und Datenverarbeitung, 1993.
165. Magill, M. J. P. and Constantinides, G. M., Portfolio selection with transaction costs, *J. Econ. Theory* **13** (1976), 245–263.
166. Megginson, R. E., *An Introduction to Banach Space Theory*, Springer, Berlin–Heidelberg–New York, 1998.
167. Mémin, J., Espaces de semimartingales et changement de probabilité, *Z. Wahrsch. Verw. Geb.* **52**:1 (1980), 9–39.
168. Mercurio, F. and Vorst, T., Option pricing and hedging in discrete time with transaction costs and incomplete market, in Dempster, M. A. H. and Pliska, S. (eds.), *Mathematics of Derivative Securities*, pp. 190–215, Cambridge University Press, Cambridge, 1997.

169. Merton, R. C., Optimum consumption and portfolio rules in a continuous time model, *J. Econ. Theory* **3** (1971), 373–413.
170. Mokkhavesa, S. and Atkinson, C., Perturbation solution of optimal portfolio theory with transaction costs for any utility function, *IMA J. Manag. Math.* **18**:1 (2008), 115–134.
171. Molchanov, I., *Theory of Random Sets*, Springer, Berlin–Heidelberg–New York, 2005.
172. Morton, A. and Pliska, S., Optimal portfolio management with fixed transaction costs, *Math. Finance* **5**:4 (1995), 337–356.
173. Musiela, M. and Rutkowski, M., *Martingale Methods in Financial Modelling*, Springer, Berlin–Heidelberg–New York, 1997.
174. Muthuraman, K. and Kumar, S., Multidimensional portfolio optimization with proportional transaction costs, *Math. Finance* **16**:4 (2006), 301–335.
175. Muthuraman, K. and Zha, H., Simulation-based portfolio optimization for large portfolios with transaction costs, *Math. Finance* **18**:1 (2008), 115–134.
176. Nikaido, H., *Convex Structures and Economic Theory*, Academic Press, San Diego, 1968.
177. Pennanen, T. and Penner, I., *Hedging of claims with physical delivery under convex transaction costs*, Preprint, arXiv:0810.2016.
178. Penner, I., *Arbitragefreiheit in Finanzmärkten mit Transaktionskosten*, Diplomarbeit, Humboldt–Universität zu Berlin, 2001.
179. Pergamenshchikov, S., Limit theorem for Leland's strategy, *Ann. Appl. Probab.* **13** (2003), 1099–1118.
180. Peskir, G. and Shiryaev, A. N., *Optimal Stopping and Free-Boundary Problems*, Birkhäuser, Basel–Boston–Berlin, 2006.
181. Pliska, S., *Introduction to Mathematical Finance*, Blackwell Publishers, Oxford, 1997.
182. Prokaj, V. and Rásonyi, M., Local and true martingales in discrete time, *Teor. Veroyatn. Primen.* (2009).
183. Protter, P., *Stochastic Integration and Differential Equations*, Springer, Berlin–Heidelberg–New York, 1990.
184. Pshenychnyi, B. N., *Convex Analysis and Extremal Problems*, Nauka, Moscow, 1980 (in Russian).
185. Rásonyi, M., *On certain problems of arbitrage theory in discrete time financial market models*, Thesis, Université de Franche-Comté, Besançon, 2002.
186. Rásonyi, M., A note on martingale measures with bounded densities, *Tr. Mat. Inst. Steklova* **237** (2002), 212–216. Stokh. Finans. Mat. English translation: *Proc. Steklov Inst. Math.* **237** (2002), 203–207.
187. Rásonyi, M., New methods in the arbitrage theory of financial markets with transaction costs, *Sémin. Probab. XXXVII, Lect. Notes Math.* **1934** (2008), 455–462.

188. Rásonyi, M., A remark on the superhedging theorem under transaction costs, *Sémin. Probab. XXXVII, Lect. Notes Math.* **1832** (2003), 394–398.
189. Rásonyi, M., Arbitrage with transaction costs revisited, in Delbaen, F., Rśonyi, M. and Stricker, Ch. (eds.), *Optimality and Risk—Modern Trends in Mathematical Finance*, Springer, Berlin–Heidelberg–New York, 2009.
190. Rheinländer, Th., *Optimal martingale measures and their applications in mathematical finance*, Thesis, Berlin Technical University, 1998.
191. Rockafellar, R. T., *Convex Analysis*, Princeton University Press, Princeton, 1970.
192. Rockafellar, R. T. and Wets, R., *Variational Analysis*, Springer, Berlin–Heidelberg–New York, 1998.
193. Rogers, L. C. G., Equivalent martingale measures and no-arbitrage, *Stoch. Stoch. Rep.* **51** (1994), 41–51.
194. Rogers, L. C. G., Why is the effect of proportional transaction costs $O(\delta^{2/3})$? in Yin, G. and Zhang, Q. (eds.), *Mathematics of Finance*, pp. 303–308, AMS Contemporary Mathematics Series **351**, AMS, Providence, 2004.
195. Rokhlin, D. B., The Kreps–Yan theorem for L^∞, *Int. J. Math. Math. Sci.* **17** (2005), 2749–2756.
196. Rokhlin, D. B., Martingale selection problem in the finite discrete time case, *Teor. Veroyatn. Primen.* **50**:3 (2005), 480–500. English translation: *Theory Probab. Appl.* **50**:3 (2006), 420–435.
197. Rokhlin, D. B., Constructive no-arbitrage criterion under transaction costs in the case of finite discrete time, *Teor. Veroyatn. Primen.* **52**:1 (2007), 41–59. English translation: *Theory Probab. Appl.* **52**:2 (2008), 93–107.
198. Rokhlin, D. B., A theorem on martingale selection for relatively open convex set-valued random sequences, *Math. Notes* **81** (2007), 543–548.
199. Rokhlin, D. B., Martingale selection problem and asset pricing in finite discrete time, *Electron. Commun. Probab.* **12** (2007), 1–8.
200. Ross, S. A., The arbitrage theory of asset pricing, *J. Econ. Theory* **13**:1 (1976), 341–360.
201. Rudin, W., *Functional Analysis*, McGraw-Hill, New York, 1973.
202. Safarian, M., *Optionspreisbildung und Absicherung von Optionen unter Berücksichtigung von Transaktionskosten*, Dissertation, Humboldt Universität zu Berlin, 1996.
203. Schachermayer, W., A Hilbert space proof of the fundamental theorem of asset pricing in finite discrete time, *Insurance: Math. Econ.* **11** (1992), 249–257.
204. Schachermayer, W., Martingale measures for discrete-time processes with infinite horizon, *Math. Finance* **4**:1 (1994), 25–55.
205. Schachermayer, W., Optimal investment in incomplete market when wealth may become negative, *Ann. Appl. Probab.* **11**:3 (2001), 694–734.

206. Schachermayer, W., The Fundamental Theorem of Asset Pricing under proportional transaction costs in finite discrete time, *Math. Finance* **14**:1 (2004), 19–48.
207. Schrijver, A., *Theory of Linear and Integer Programming*, Wiley, New York, 1986.
208. Schwartz, M., A new proof of a theorem of Komlós, *Acta Math. Hung.* **47** (1986), 181–185.
209. Sekine, J. and Yano, J., *Hedging errors of Leland's strategies with time-inhomogeneous rebalancing*, Preprint.
210. Shirakawa, H. and Konno, H., *Pricing of options under the proportional transaction costs*, Preprint, 1995.
211. Shiryaev, A. N., *Probability*, Springer, Berlin–Heidelberg–New York, 1984.
212. Shiryaev, A. N., *Essentials of Stochastic Finance*, World Scientific, Singapore, 1999.
213. Shreve, S., Liquidity premium for capital asset pricing with transaction costs, in Davis, M. H. A. et al. (eds.), *Mathematical Finance IMA* **65**, pp. 117–133, Springer, New York, 1995.
214. Shreve, S. and Soner, M., Optimal investment and consumption with transaction costs, *Ann. Appl. Probab.* **4**:3 (1994), 609–692.
215. Sin, C. A., *Strictly local martingales and hedge ratios on stochastic volatility models*, Ph.D. dissertation, Cornell University, 1996.
216. Soner, H. M., Shreve, S. E. and Cvitanić, J., There is no nontrivial hedging portfolio for option pricing with transaction costs, *Ann. Appl. Probab.* **5** (1995), 327–355.
217. Stricker, Ch., Arbitrage et lois de martingale, *Ann. Inst. Henri Poincaré. Probab. Stat.* **26**:3 (1990), 451–460.
218. Stricker, Ch., Simple strategies in the exponential utility maximization, *Sémin. Probab. XXXVI, Lect. Notes Math.* **1801** (2002), 415–418.
219. Stricker, Ch. and Yan, J. A., Some remarks on the optional decomposition theorem, *Sémin. Probab. XXXII, Lect. Notes Math.* **1686** (1998), 56–66.
220. Stiemke, E., Über positive Lösungen homogener linearer Gleichungen, *Math. Ann.* **76** (1915), 340–342.
221. Taksar, M., Klass, M. J. and Assaf, D., A diffusion model for optimal portfolio selection in the presence of brokerage fees, *Math. Oper. Res.* **13** (1988), 277–294.
222. Tanaka, H., Stochastic differential equations with reflecting boundary condition in convex regions, *Hiroshima Math. J.* **9** (1979), 163–177.
223. Taqqu, M. and Willinger, W., The analysis of finite security markets using martingales, *Adv. Appl. Probab.* **19**:1 (1987), 1–25.
224. Touzi, N., Super-replication under proportional transaction costs: from discrete to continuous-time models, *Math. Meth. Oper. Res.* **50** (1999), 297–320.

225. Trautner, R., A new proof of the Komlós–Révész theorem, *Probab. Theory Relat. Fields* **84** (1990), 281–287.
226. von Weizsäcker, H., Can one drop L^1-boundedness in Komlós subsequence theorem? *Am. Math. Mon.* **111**:10 (2004), 900–903.
227. Wagner, D., Survey of measurable selection theorems, *SIAM J. Control Optim.* **15**:5 (1977), 859–903.
228. Wagner, D., Survey of measurable selection theorems: An update. Measure theory, in *Proc. Conf., Oberwolfach 1979*, Lecture Notes Math. **794**, pp. 176–219 (1980).
229. Whalley, A. E. and Wilmott, P., An asymptotic analysis of an optimal hedging model for option pricing under transaction costs, *Math. Finance* **7** (1997), 307–324.
230. Yan, J. A., Caractérisation d'une classe d'ensembles convexes de L^1 et H^1, *Sémin. Probab. XIV, Lect. Notes Math.* **784** (1980), 220–222.
231. Yosida, K. and Hewitt, E., Finitely additive measures, *Trans. Am. Math. Soc.* **72** (1952), 46–66.
232. Zariphopoulou, T., Investment–consumption models with transaction fees and Markov-chain parameters, *SIAM J. Control Optim.* **30**:3 (1992), 613–636.
233. Zariphopoulou, T., Transaction cost in portfolio management and derivative pricing, in *Introduction to Mathematical Finance*, Proc. Sympos. Appl. Math. **57**, pp. 101–163, AMS, Providence, 1999.
234. Zariphopoulou, T., Stochastic control methods in asset pricing, in Kannan, D. and Lakshmikantham, V. (eds.), *Handbook of Stochastic Analysis and Applications*, Dekker, New York–Basel, 2002.
235. Zhao, Y. and Ziemba, W. T., Hedging errors with Leland's option model in the presence of transaction costs, *Finance Res. Lett.* **4**:1 (2007), 49–58.
236. Zhao, Y. and Ziemba, W. T., Comments on and corrigendum to "Hedging errors with Leland's option model in the presence of transaction costs", *Finance Res. Lett.* **4**:3 (2007), 196–199.
237. Žitković, G., A filtered version of the bipolar theorem of Brannath and Schachermayer, *J. Theor. Probab.* **15** (2002), 41–61.

Index

Yu. Kabanov, M. Safarian, *Markets with Transaction Costs*,
Springer Finance, DOI 10.1007/978-3-540-68121-2,